W0091913

Institute for Nonlinear Science

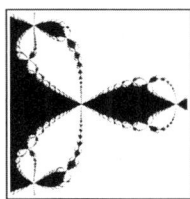

Institute for Nonlinear Science

Henry D.I. Abarbanel *Analysis of Chaotic Series* (1996)

Jordi García-Ojalvo, Jose M. Sancho *Noise in Spatially Extended Systems* (1999)

Leon Glass, Peter Hunter, Andrew McCullogh (Eds.) *Theory of Heart: Biomechanics, Biophysics, and Nonlinear Dynamic of Cardiac Function* (1991)

Mark Millonas (Ed.) *Fluctuations and Order: The New Synthesis* (1996)

Linda E. Reichl *The Transition to Chaos in Conservative Classical Systems: Quantum Manifestations* (1992)

Bruce West, Mauro Bologna, Paolo Grigolini *Physics of Fractal Operators* (2003)

Linda E. Reichl *The Transition to Chaos in Conservative Classical Systems: Quantum Manifestations,* Second Edition (2004)

Lawrence E. Larson, Jia-Ming Liu, Lev S. Tsimring (Eds.) *Digital Communications Using Chaos and Nonlinear Dynamics* (2006)

Lawrence E. Larson
Jia-Ming Liu
Lev S. Tsimring

Editors

Digital Communications Using Chaos and Nonlinear Dynamics

With 194 Illustrations

 Springer

Editors

Lawrence E. Larson
Department of Electrical and
Computer Engineering
and
Lev S. Tsimring
Institute for Nonlinear Science
University of California, San Diego
La Jolla, CA 92093-0402
USA
larson@ece.ucsd.edu
ltsimring@ucsd.edu

Jia-Ming Liu
Department of Electrical Engineering
University of California
Los Angeles, CA 90095-1594
USA
liu@ee.ucla.edu

Editorial Board

Institute for Nonlinear Science, University of California—San Diego
    Henry D.I. Abarbanel, Physics (Scripps Institution of Oceanography)
    Mikhail I. Rabinovich, Research
    Al Selverston, Biology
    Lev S. Tsimring, Research
    Lawrence E. Larson

Library of Congress Control Number: 2005934890

ISBN-10: 0-387-29787-1   e-ISBN: 0-387-29788-X
ISBN-13: 978-0387-29787-3

Printed on acid-free paper.

© 2006 Springer Science+Business Media, LLC
All rights reserved. This work may not be translated or copied in whole or in part without the
written permission of the publisher (Springer Science+Business Media, LLC, 233 Spring Street,
New York, NY 10013, USA), except for brief excerpts in connection with reviews or scholarly
analysis. Use in connection with any form of information storage and retrieval, electronic adap-
tation, computer software, or by similar or dissimilar methodology now known or hereafter
developed is forbidden.
The use in this publication of trade names, trademarks, service marks, and similar terms, even if
they are not identified as such, is not to be taken as an expression of opinion as to whether or
not they are subject to proprietary rights.

Printed in the United States of America.   (SBI)

9 8 7 6 5 4 3 2 1

springer.com

Preface

This book provides a summary of the research conducted at UCLA, Stanford University, and UCSD over the last five years in the area of nonlinear dynamics and chaos as applied to digital communications. At first blush, the term "chaotic communications" seems like an oxymoron; how could something as precise and deterministic as digital communications be chaotic?

But as this book will demonstrate, the application of chaos and nonlinear dynamics to communications provides many promising new directions in areas of coding, nonlinear optical communications, and ultra-wideband communications. The eleven chapters of the book summarize many of the promising new approaches that have been developed, and point the way to new research directions in this field.

Digital communications techniques have been continuously developed and refined for the past fifty years to the point where today they form the heart of a multi-hundred billion dollar per year industry employing hundreds of thousands of people on a worldwide basis. There is a continuing need for transmission and reception of digital signals at higher and higher data rates. There are a variety of physical limits that place an upper limit on these data rates, and so the question naturally arises: are there alternative communication techniques that can overcome some of these limitations?

Most digital communications today is carried out using electronic devices that are essentially "linear," and linear system theory has been used to continually refine their performance. In many cases, inherently nonlinear devices are linearized in order to achieve a certain level of linear system performance. However, as device technology reaches its fundamental limits, the natural question arises: can the intrinsic *nonlinearity* of electronic devices be exploited in some fundamental way to improve communications system performance?

One example of the type of improvement that can potentially be achieved with the judicious application of the intrinsic nonlinearity of an electronic device is the well-known use of *solitons* in fiber-optic transmission systems. The inherent nonlinearity of an optical fiber ensures that a digital pulse — a soliton — retains a constant shape over a large distance, and that the pulses do not diffuse or disperse during transmission. In this case, the nonlinearity of the fiber compensates for its dispersion, and vice versa. Solitons have many interesting and useful properties, including essentially distortionless propagation over enormous distances and "damage-free" soliton-soliton collision.

The potential advantages of operation of nonlinear devices for generation of digital communications signals include improved efficiency, lower dc power, lower probability of intercept, and lower probability of detection. These potential advantages have to be balanced against the need for efficient spectrum management, and the cost of this new technology.

Figure 0.1 shows a block diagram of a typical digital communications system. Each block in the transmission chain performs a unique function. The source encoding block takes the data provided by the information source, and codes it in an optimum way for further transmission — either by removing redundant bits, or compressing it in some other fashion. The encryption block re-codes the data in order to enhance transmission security. The channel encoding performs a variety of transformations on the input data to minimize the overall degradation due to channel impairments. Modulation impresses the encoded data onto the radio frequency carrier, which is then combined with other signals in a multiple access scheme, and finally delivered to the transmit antenna. Each block in the receiver chain performs the inverse operation to that of the transmit chain.

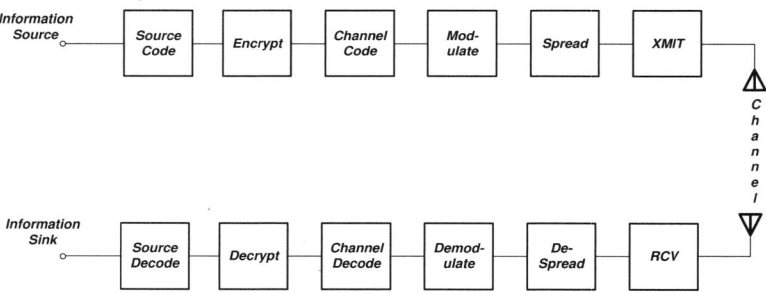

Fig. 0.1. Block diagram of modern digital communications system

The advantages of digital communications compared to traditional analog techniques can immediately be seen from this brief overview. First, the transmitted information is coded in such a way to make its reception insensitive to channel impairments, private, and free of unnecessary redundant information that would waste valuable spectrum. The data is then modulated onto a carrier in a manner that can predictably minimize the bandwidth and power requirements for a given desired data and error rate. The level of control over the security, bandwidth, and error rate that digital communications techniques allow is significantly greater than that of traditional analog techniques.

Nonlinear techniques can be applied in a straightforward manner to the encryption/decryption blocks of the system. In this manner, data can be "embedded" in a chaotic sequence, which is only known to the desired receiver – significantly enhancing security. Nonlinear techniques can also be potentially applied to channel encoding/decoding functions, where there may be some benefit to chaotic channel coding techniques for greater immunity to channel fading problems. Chaotic modulation and spreading techniques may allow for improved multiple channel access approaches and improved immunity to potential jamming and fading conditions. Chaotic modulation of digital data may be less sensitive to electronic nonlinearities in the transmit and receive portions of the device.

The spectra of chaotic signals make them very attractive for use as carriers in spread spectrum communications. Because chaotic signals are generated by deterministic dynamical systems, two coupled chaotic systems can be synchronized to produce nearly identical chaotic oscillations. This insight provides the key to the recovery of information that is modulated onto a chaotic carrier. In addition, a chaos-based communications system could also improve privacy, security, and probability of intercept, because chaotic sequences, unlike pseudorandom sequences, can be made completely nonperiodic.

Optically-based chaotic communications, which are based on the transmission of messages encoded on a chaotic waveform, have attracted very extensive research activity. Most of the systems are based on synchronization of chaos between a transmitter and a receiver, which are linked by a transmission channel. For such systems, synchronization between the transmitter and receiver is mandatory, since the bit-error rate (BER) of the decoded message at the receiver depends on the accuracy and robustness of synchronization.

Many systems based on either semiconductor lasers or fiber lasers have been proposed and studied for chaotic optical communications based on nonlinear dynamics and chaos. Chaotic optical systems that can reach the rates typically employed by traditional optical communications systems, such as the OC-48 standard bit rate of 2.5 Gb/s and the OC-192 standard bit rate of 10 GB/s, are particularly attractive. In this book, we present three of the leading semiconductor laser systems that are most actively investigated and are most promising for high-bit-rate chaotic optical communications: optical injection system, optical feedback, and the optoelectronic feedback.

This research was supported by the Army Research Office (ARO) and the Multi-University Research Initiative Program. Dr. John Lavery was the ARO Program Manager guiding the research on the program, and we would like to his acknowledge his tireless support and creative insights throughout the project.

We would also like to acknowledge the technical leadership of Professor Henry Abarbanel of UCSD. As Director of the Institute for Nonlinear Sciences (INLS) at UCSD, Professor Abarbanel has pioneered the applications of nonlinear dynamics to a variety of fields — from communications to neurobiology. Many of the technical insights and ideas that are developed here were inspired by his creativity and significant contributions.

We would also like to acknowledge the contributions of Professor Kung Yao of UCLA, for his deep insights into the field of digital communications, and his creativity and contributions to this new field of nonlinear dynamics-based communications. We would also like to acknowledge the support of Drs. Lou Pecora and Tom Carroll of the Naval Research Laboratories for their many useful discussion and their continuous support.

Finally, we would like to acknowledge the support of the superb administrative staff at INLS and UCSD — Ms. Mary Jones, Ms. Terry Peters, Ms. Beryl Nosworthy, and Mr. James Thomas.

La Jolla,   *Lawrence Larson*
June, 2004   *Jia-Ming Liu*
                                                                                              Lev Tsimring

Contents

1 An Overview of Digital Communications Techniques Using Chaos and Nonlinear Dynamics
Lawrence E. Larson, Lev S. Tsimring, Henry D. I. Abarbanel,
Jia-Ming Liu, Kung Yao, Alexander R. Volkovskii, Nikolai F. Rulkov,
Mikhail M. Sushchik 1
1.1 Introduction 1
1.2 Wireless Communications Based on Nonlinear Dynamics and Chaos 3
1.3 Optical Communications Based on Nonlinear Dynamics 15
1.4 Conclusions... 23
References 25

2 Digital Communication Using Self-Synchronizing Chaotic Pulse Position Modulation
Nikolai F. Rulkov, Alexander R. Volkovskii, Michail M. Sushchik,
Lev S. Tsimring, Lucas Illing 29
2.1 Introduction 29
2.2 CPPM Basics 32
2.3 CPPM implementation 36
2.4 Experimental Studies of CPPM with Channel Distortions 42
2.5 CPPM Performance and Features 46
2.6 Multiuser Extension of CPPM 53
2.7 Conclusions... 54
References 55

3 Spread Spectrum Communication with Chaotic Frequency Modulation
Alexander R. Volkovskii, Lev S. Tsimring, Nikolai F. Rulkov,
Ian Langmore, Stephen C. Young 59
3.1 Introduction 59
3.2 Phase and Frequency Modulation of Chaotic Carrier 61
3.3 Chaotic Frequency Modulation of Periodic Carrier 67

3.4 Communication Using CFM Signals 74
3.5 Conclusions... 86
References 87

4 Ultra-Wideband Communications Using Pseudo-Chaotic Time Hopping
David C. Laney, Gian Mario Maggio 91
4.1 Background 91
4.2 Single-User Pseudo-Chaotic Time Hopping 96
4.3 Multiple Access for Pseudo-Chaotic Time Hopping 113
4.4 Conclusions.. 127
References 130

5 Optimum Spreading Sequences for Asynchronous CDMA System Based on Nonlinear Dynamical and Ergodic Theories
Kung Yao, Chi-Chung Chen 133
5.1 Introduction 133
5.2 Introduction to CDMA Communication System 135
5.3 Chaotic CDMA Communication System........................ 138
5.4 CDMA System Models...................................... 140
5.5 Derivation of Optimal Sequences 141
5.6 Ergodic Dynamical Systems 143
5.7 Chaotic Optimal Spreading Sequences Design 146
5.8 Performance Comparisons of CDMA Systems 148
5.9 Construction of Optimal Spreading Sequences from Gold Codes 152
5.10 Acquisition Time of Optimal Spreading Sequences 154
5.11 Conclusions.. 157
References 159

6 Nonlinear Phenomena in Turbo Decoding Algorithms
Ljupco Kocarev 163
6.1 Introduction 163
6.2 Dynamics of Iterative Decoding Algorithms 165
6.3 Nonlinear Dynamical Systems................................ 169
6.4 Fixed Points in the Turbo-Decoding Algorithm................... 175
6.5 Bifurcation Analysis of Turbo-Decoding Algorithm 182
6.6 Control of Transient Chaos 185
6.7 Conclusions.. 188
References 189

7 Security of Chaos-Based Communication and Encryption
Roy Tenny, Lev S. Tsimring, Henry D.I. Abarbanel, Lawrence E. Larson 191
7.1 Introduction 191
7.2 Chaos-Based Encryption Schemes 192
7.3 Cryptanalysis Attacks on Chaos-Based Encryption Schemes 200

7.4 Security of Chaotic Encryption Schemes Based on Active/Passive
    Decomposition 203
7.5 Public Key Encryption Using Distributed Dynamics 212
7.6 Conclusions 226
References 228

8 Random Finite Approximations of Chaotic Maps
Jesús Urías, Eric Campos, Nikolai F. Rulkov 231
8.1 Introduction 231
8.2 Random Finite Approximations 232
8.3 Maps with a Generating Partition 235
8.4 Approximations for the Tent Maps 238
8.5 Conclusions 241
References 241

**9 Numerical Methods for the Analysis of
Dynamics and Synchronization of Stochastic Nonlinear
Systems**
How-Foo Chen, Jia-Ming Liu 243
9.1 Introduction 244
9.2 Numerical Simulation of Stochastic Differential Equations 246
9.3 Characterization of Chaos 259
9.4 Robustness of Chaos Synchronization 268
9.5 Chaotic Communications 275
9.6 Conclusions 281
References 283

**10 Dynamics and Synchronization of
Semiconductor Lasers for Chaotic Optical Communications**
Jia-Ming Liu, How-Foo Chen, Shuo Tang 285
10.1 Introduction 286
10.2 Basic Concepts of Laser Dynamics 287
10.3 Single-Mode Semiconductor Lasers 291
10.4 Nonlinear Dynamics of Single-Mode Semiconductor Lasers 298
10.5 Basic Concept of Chaos Synchronization 311
10.6 Chaos Synchronization of Single-Mode Semiconductor Lasers 315
10.7 Synchronization in the Presence of Message Encoding 330
10.8 Conclusions 335
References 337

**11 Performance of Synchronized Chaotic Optical
Communication Systems**
Shuo Tang, How-Foo Chen, Jia-Ming Liu 341
11.1 Introduction 341
11.2 General Issues on Chaotic Optical Communications 345
11.3 Experiment of Chaotic Optical Communication at 2.5 Gb/s 350

11.4 Comparison of Different Encoding and Decoding Schemes 354
11.5 Chaotic Optical Communications at 10 Gb/s 363
11.6 Conclusions 375
References 376

Index 379

List of Contributors

Henry D. I. Abarbanel
Department of Physics and
Marine Physical Laboratory (Scripps
Institution of Oceanography)
University of California, San Diego,
La Jolla,CA 92093-0402
habarbanel@ucsd.edu

Eric Campos
IICO–UASLP, 78210 San Luis
Potosí, Mexico

Chi-Chung Chen
Electrical Engineering Department,
University of California, Los Angeles
Los Angeles, CA 90095-1594
ccchen@seas.ucla.edu

How-Foo Chen
Electrical Engineering Department,
University of California, Los Angeles
Los Angeles, CA 90095-1594
howfoo@seas.ucla.edu

Lucas Illing
Dept. of Physics, Duke University,
Durham, NC 27708
illing@phy.duke.edu

Ljupco Kocarev
Institute for Nonlinear Science,
University
of California, San Diego, La Jolla,
CA 92093-0402
lkocarev@ucsd.edu

David C. Laney
Center for Wireless Communications,
University of California, San Diego,
La Jolla CA 92093-0407
david.laney@cubic.com

Ian Langmore
Dept. of Electrical and Computer
Engineering,
University of California, San Diego,
CA 92093-0354
ilangmor@ucsd.edu

Lawrence E. Larson
Dept. of Electrical and Computer
Engineering, University of California,
San
Diego, La Jolla, CA 92093-0354
larson@ece.ucsd.edu

Jia-Ming Liu
Electrical Engineering Department,
University of California, Los Angeles
Los Angeles, CA 90095-1594
liu@ee.ucla.edu

Gian Mario Maggio
Center for Wireless Communications,
University of California, San Diego,
CA 92093-0407
gmaggio@ucsd.edu

Nikolai F. Rulkov
Institute for Nonlinear Science,
University
of California, San Diego, La Jolla,
CA 92093-0402
nrulkov@ucsd.edu

Mikhail M. Sushchik
Therma-Wave Inc., Fremont, CA
94539
msushchi@thermawave.com

Shuo Tang
Electrical Engineering Department,
University of California, Los Angeles
Los Angeles, CA 90095-1594
tangs@ee.ucla.edu

Roy Tenny
Dept. of Electrical and Computer
Engineering, University of California,
San Diego, La Jolla, CA 92093-0354
rtenny@ucsd.edu

Lev S. Tsimring
Institute for Nonlinear Science,
University
of California, San Diego, La Jolla,
CA 92093-0402
ltsimring@ucsd.edu

Jesús Urías
IICO–UASLP, 78210 San Luis
Potosí, Mexico

Alexander R. Volkovskii
Institute for Nonlinear Science,
University
of California, San Diego, La Jolla,
CA 92093-0402
avolkovsk@ucsd.edu

Kung Yao
Electrical Engineering Department,
University of California, Los Angeles
Los Angeles, CA 90095-1594
yao@ee.ucla.edu

Stephen C. Young
Dept. of Physics, University of
Southern California, Los Angeles,
CA 90089-0484

1

An Overview of Digital Communications Techniques Using Chaos and Nonlinear Dynamics

Lawrence E. Larson, Lev S. Tsimring, Henry D. I. Abarbanel, Jia-Ming Liu,
Kung Yao, Alexander R. Volkovskii, Nikolai F. Rulkov, and
Mikhail M. Sushchik

Summary. This chapter provides a brief overview of some of the digital communications techniques that have been proposed recently employing nonlinear dynamics, along with a comparison to traditional approaches. Both wireless modulation techniques as well as optical communications approaches are be presented. [1]

1.1 Introduction

Digital communications techniques have been continuously developed and refined for the past fifty years to the point where today they form the heart of a multi-hundred billion dollar per year industry employing hundreds of thousands of people on a worldwide basis. There is a continuing need for transmission and reception of digital signals at higher and higher data rates. There are a variety of physical limits that place an upper limit on these data rates, and so the question naturally arises: are there alternative communication techniques that can circumvent these natural limits?

For example, most digital communication today is carried out using electronic devices that are essentially "linear," and linear system theory has been used to continually refine their performance. In many cases, inherently nonlinear devices are linearized in order to achieve a certain level of linear system performance. However, as device technology reaches its fundamental limits, the natural question arises: can the intrinsic nonlinearity of electronic devices be exploited in some fundamental way to improve communications system performance?

One example of the type of improvement that can potentially be achieved with the judicious application of the intrinsic nonlinearity of an electronic device is the well-known use of *solitons* in fiber-optic transmission systems. The inherent nonlinearity of an optical fiber ensures that a digital pulse — a soliton — retains a constant shape over a large distance, and that the pulses

---

[1] Portions of this chapter were taken from publications [1–3].

do not diffuse or disperse during transmission. In this case, the nonlinearity of the fiber compensates for its dispersion, and vice versa. Solitons have many interesting and useful properties, including essentially distortionless propagation over enormous distances and "damage-free" soliton-soliton collision.

The potential advantages of operation of nonlinear devices for generation of digital communication signals include improved efficiency, lower dc power, lower probability of intercept, and lower probability of detection. These potential advantages have to be balanced against the need for efficient spectrum management, and the cost of this new technology.

In order to understand the full impact that nonlinear techniques may have on digital communications, we begin with a brief overview of modern digital communications techniques. Figure 1.1 shows a block diagram of a typical digital communications system [4]. Each block in the transmission chain performs a unique function. The source encoding block takes the data provided by the information source, and codes it in an optimum way for further transmission — either by removing redundant bits, or compressing it in some other fashion. The encryption block re-codes the data in order to enhance transmission security. The channel encoding performs a variety of transformations on the input data to minimize the overall degradation due to channel impairments. Modulation impresses the encoded data onto the radio frequency carrier, which is then combined with other signals in a multiple access scheme, and finally delivered to the transmit antenna. Each block in the receiver chain performs the inverse operation to that of the transmit chain.

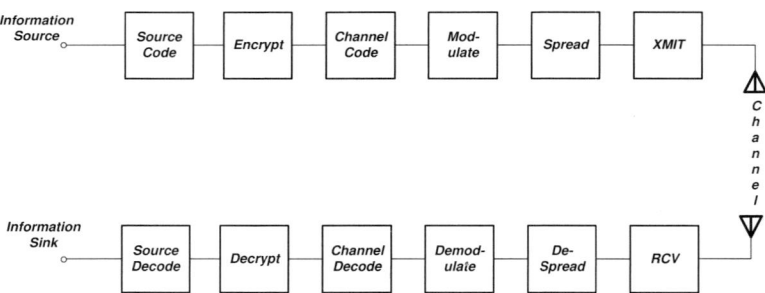

Fig. 1.1. Block diagram of modern digital communications system.

The advantages of digital communications compared to traditional analog techniques can immediately be seen from this brief overview. First, the transmitted information is coded in such a way to make its reception insensitive to channel impairments, private, and free of unnecessary redundant information that would waste valuable spectrum. The data is then modulated onto a carrier in a manner that can predictably minimize the bandwidth and power requirements for a given desired data and error rate. The level of control over the security, bandwidth, and error rate that digital communications techniques allow is significantly greater than that of traditional analog techniques.

Nonlinear techniques can be applied in a straightforward manner to the encryption/decryption blocks of the system. In this manner, data can be "embedded" in a chaotic sequence, which is only known to the desired receiver - significantly enhancing security. Nonlinear techniques can also be potentially applied to channel encoding/decoding functions, where there may be some benefit to chaotic channel coding techniques for greater immunity to channel fading problems. Chaotic modulation and spreading techniques may allow for improved multiple channel access approaches and improved immunity to potential jamming and fading conditions. Chaotic modulation of digital data may be less sensitive to electronic nonlinearities in the transmit and receive portions of the device.

This chapter provides a brief overview of some of the digital communications techniques that have been proposed recently employing nonlinear dynamics, along with a comparison to traditional approaches.

1.2 Wireless Communications Based on Nonlinear Dynamics and Chaos

Wireless digital communications devices typically have a variety of requirements, such as data transmission rate, bit-error rate (BER), bandwidth, complexity, and cost. However, in hostile environments (e.g., multipath propagation, interference from other such devices, potential eavesdroppers, etc.), additional desirable features may include security, low probability of intercept, spread-spectrum, and efficient battery usage.

These various features can be traded off against each other in a communications system using spread-spectrum techniques. Spread-spectrum communication systems modulate a relatively narrowband message signal onto a wideband carrier. Common spread-spectrum technologies use correlation techniques to match the received signal with a certain known pattern. As a result, the desired signal is accumulated coherently, and the channel noise and interference are averaged out. This desirable property of spread-spectrum systems—to suppress interference through utilization of a wider bandwidth— is called the processing gain of the system.

One new approach to these various challenges makes use of chaotic dynamical systems. Chaotic signals exhibit a wide spectrum and have been studied in connection with spread-spectrum applications [5]. Due to their irregular nature, they can be used to efficiently encode the information in a number of ways. Because chaotic signals are generated by deterministic dynamical systems, two coupled chaotic systems can be synchronized to produce identical chaotic oscillations. This provides the key recovery of information that is modulated onto a chaotic carrier [6].

The broad continuous spectra of chaotic signals make them very attractive for use as carriers in spread spectrum communications [5, 7–13]. Because chaotic signals are generated by deterministic dynamical systems, two coupled chaotic systems can be synchronized to produce identical chaotic oscillations. This insight provides the key to the recovery of information that is modulated onto a chaotic carrier. A number of chaos-based communication schemes have been suggested, but many of these systems are very sensitive to distortion, filtering, and noise. However, a chaos-based communications system could also improve privacy, security, and probability of intercept, inasmuch as chaotic sequences, unlike pseudorandom sequences, can be made completely nonperiodic.

1.2.1 Wireless Communications Based on Chaotic Carriers

Several differing chaos-based modulation schemes are now described. The operation of a Differential Chaos Shift Keying (DCSK) [9, 14] modulator and demodulator is illustrated in Figure 1.2. For each transmitted bit, the transmitter outputs a chaotic sequence x_i of length M followed by the same sequence multiplied by the information signal $b_l = \pm 1$, where l is the bit counter. As a result, the transmitted signal s_i is given by

$$s_i = \begin{array}{ll} x_i, & 0 < i \leq M \\ b_l x_{i-M}, & M < i \leq 2M \end{array} \tag{1.1}$$

The receiver takes the received signal r_i and multiplies it by the received signal, delayed by M (r_{i-M}). The result is then averaged over the spreading sequence length M. Thus the output of the correlator can be written as

$$S = \sum_{i=1}^{M} r_i r_{i+M}. \tag{1.2}$$

If we make the standard assumptions that the received signal r_i is given by $r_i = s_i + \xi_i$, where ξ_i is a stationary random process with $< \xi_i >= 0$, that ξ_i and ξ_j are statistically independent for any $i \neq j$, and that we can maintain perfect bit synchronization. Then the correlator output can be written as

$$S = \sum_{i=1}^{M} (s_i + \xi_i)(s_{i+M} + \xi_{i+M})$$
$$= \sum_{i=1}^{M} \left(b_l x_i^2 + x_i(\xi_{i+M} + b_l \xi_i) + \xi_i \xi_{i+M} \right) \tag{1.3}$$
$$= b_l \sum_{i=1}^{M} x_i^2 + \sum_{i=1}^{M} \left(x_i(\xi_{i+M} + b_l \xi_i) + \xi_i \xi_{i+M} \right)$$

where the first term is the desired signal and the second is a zero mean random quantity representing the noise and interference terms.

One shortcoming of this method is the need to transmit the same chaotic sequence twice, which makes this system prone to interception and wasteful of power. Also, the transmitter requires a delay element and a switch, or a generator capable of reproducing the same chaotic sequence.

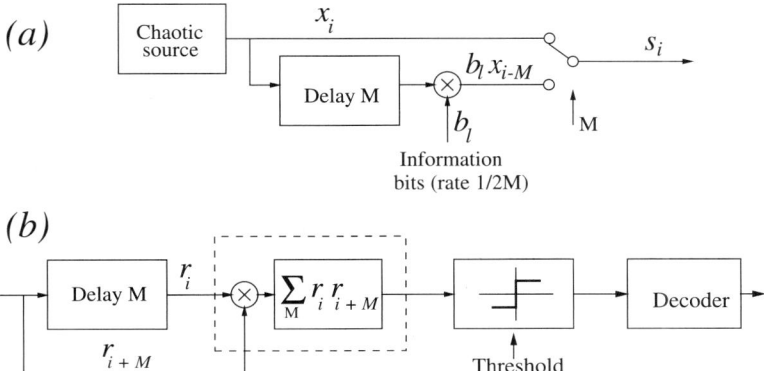

Fig. 1.2. DCSK operation: (a) transmitter, (b) receiver. (Reprinted with permission from [2]. ©2000 IEEE.)

An alternative technique is Correlation Delay Shift Keying (CDSK) [2] . In the CDSK modulator (Figure 1.3), the transmitted signal is the sum of a chaotic sequence and the same delayed chaotic sequence multiplied by the information signal $b_l = \pm 1$. As a result, CDSK overcomes the problems of DCSK: the switch in the transmitter is now replaced by an adder, and the transmitted signal is never repeated. The CDSK receiver (Figure 1.3b) is the same as for DCSK, except the delay now does not have to be equal to the spreading sequence length. The correlator output S is given by the sum

$$S = \sum_{i=1}^{M} (x_i + b_l x_{i-L} + \xi_i)(x_{i-L} + b_{l-1} x_{i-2L} + \xi_{i-L})$$

$$= b_l \sum_{i=1}^{M} x_{i-L}^2 + \sum_{i=1}^{M} \eta_i, \qquad (1.4)$$

where

$$\eta_i = x_i x_{i-L} + b_{l-1} x_i x_{i-2L} + b_l b_{l-1} x_{i-L} x_{i-2L}$$
$$+ x_i \xi_{i-L} + b_l x_{i-L} \xi_{i-L} + x_{i-L} \xi_i + b_{l-1} x_{i-2L} \xi_i$$
$$+ \xi_i \xi_{i-L}.$$

The first term in (1.4) is the desired signal and the second comes not only from noise part of the correlator input, but also from correlating chaotic segments over finite time. This additional interference leads to degraded performance of CDSK, compared to DCSK.

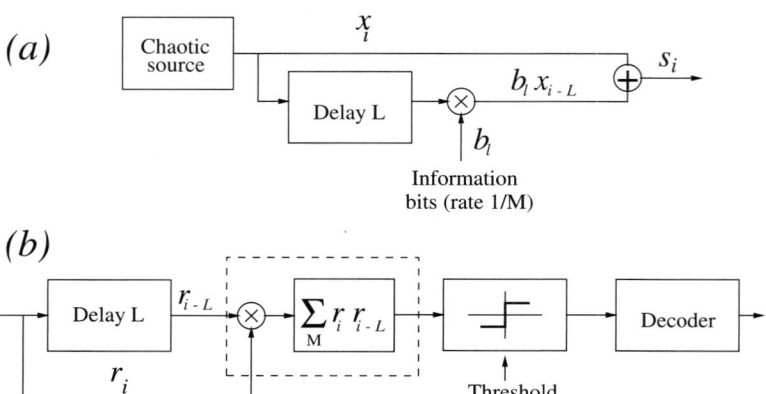

Fig. 1.3. CDSK operation: (a) transmitter, (b) receiver. (Reprinted with permission from [2], ©2000 IEEE.)

An alternative to including the reference signal in the transmitted signal involves re-creating the reference signal in the receiver. This approach is taken in the design of Symmetric Chaos Shift Keying (SCSK) [2] whose operation is illustrated in Figure 1.4. The central element of a SCSK transmitter is the chaotic map

$$\mathbf{x}_{i+1} = \mathbf{F}(\mathbf{x_i}), \tag{1.5}$$

where $\mathbf{x}_i$ is the internal state vector. The first component of this vector multiplied by the information signal $b_l = \pm 1$ is the transmitted signal: $s_i = b_l x_i$. In the receiver this signal is driving a matched chaotic system:

$$\mathbf{y}_{i+1} = \mathbf{G}(|s_i|, \mathbf{y_i}). \tag{1.6}$$

$\mathbf{F}(\bullet)$ and $\mathbf{G}(\bullet)$ are chosen such that the drive-response system that they form has a stable identically synchronous regime with respect to the first components of vectors $\mathbf{x_i}$ and $\mathbf{y_i}$: $x_i^1 = y_i^1$. The simplest example of such drive-response system is a two one-dimensional map

$$x_{i+1} = F(x_i),$$
$$y_{i+1} = F(s_i),$$

where $F(\bullet)$ is even, $F(x) = F(-x)$. In the noise-free case, the output of the chaotic system in the receiver is the same as the output of the chaotic system in the transmitter, and the same as the signal in the channel, except for information-dependent polarity modulation. The sign of b_l can therefore be determined by taking the product of the received signal and the output of the chaotic system in the receiver. The product can then be averaged over the length of the spreading sequence in order to reduce the effects of channel noise.

In general, the correlator output for SCSK can be written as

$$S = \sum_{i=1}^{M} y_i^1 (b_l x_i^1 + \xi_i), \tag{1.7}$$

where y_i^1 is the output of the chaotic system in the receiver. In the case when the chaotic map is one-dimensional

$$S = \sum_{i=1}^{M} F(b_l x_{i-1} + \xi_{i-1})(b_l x_i + \xi_i). \tag{1.8}$$

We can introduce $\tilde{\xi}_i = \xi_i / b_l$ and rewrite this in the form

$$\begin{aligned} S &= b_l \sum_{i=1}^{M} F(x_{i-1} + \tilde{\xi}_{i-1})(F(x_{i-1}) + \tilde{\xi}_i) \\ &= b_l \sum_{i=1}^{M} x_i^2 + b_l \sum_{i=1}^{M} F(x_{i-1} + \tilde{\xi}_{i-1})\tilde{\xi}_i \\ &\quad + b_l \sum_{i=1}^{M} \left(F(x_{i-1} + \tilde{\xi}_{i-1}) - F(x_{i-1}) \right) F(x_{i-1}). \end{aligned} \tag{1.9}$$

The first sum in this expression is the desired signal, and the second is the interference.

The SCSK approach has advantages over both DCSK and CDSK. The transmitter design is simpler and the SCSK transmitted sequence is non-repeating, leading to a lower probability of intercept. Additionally, demodulation of the SCSK signal requires a matched nonlinear system in the receiver, thus offering better protection against an unauthorized reception. These advantages come at the expense of some performance loss and of the more restricted choice of nonlinear systems used for chaos generation.

Extensive analysis has been carried out on the performance of these systems in [2], along with a comparison to more traditional periodic carrier communications approaches. What follows are some representative performance comparisons of these new chaotic modulation approaches compared to existing techniques.

The output of the correlator for DCSK is given by (1.3). It can be written in the form

$$S = b_l A + b_l \zeta + \eta, \qquad A > 0. \tag{1.10}$$

Here $A = <x_i^2> M$, $\zeta = \sum_{i=1}^{M} x_i^2 - A$ and

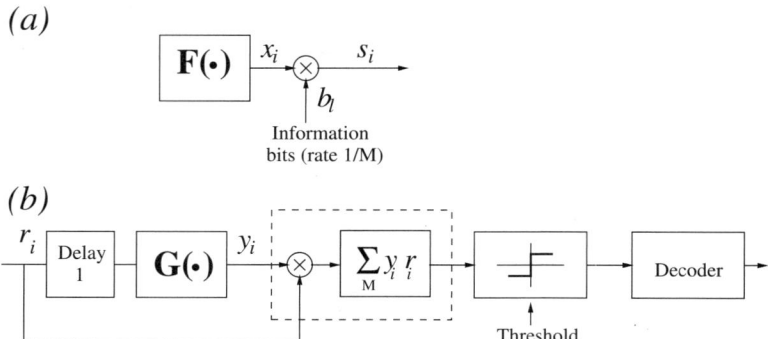

Fig. 1.4. SCSK operation: (a) transmitter, (b) receiver. (Reprinted with permission from [2]. ©2000 IEEE.)

$$\eta = \sum_{i=1}^{M} x_i \xi_{i+M} + b_l \sum_{i=1}^{M} x_i \xi_i + \sum_{i=1}^{M} \xi_i \xi_{i+M}.$$

For DCSK $A = E_b/2$. We require that x_i is stationary, and that the correlation between x_i and x_{i+k} decay quickly as $|k|$ increases (which is standard for chaotic systems). We further assume that M is much larger than the characteristic correlation decay times. Under these assumptions, as M increases, the distributions of ζ and η approach Gaussian distributions [15].

Thus the bit error rate for the DCSK is given by

$$BER = \frac{1}{2} \text{erfc} \left(\sqrt{ \frac{E_b}{4N_0} \left(1 + \frac{2}{5M} \frac{E_b}{N_0} + \frac{N_0}{2E_b} M \right)^{-1} } \right). \qquad (1.11)$$

Figure 1.5a presents the results of numerical simulations with different values of M. Channel noise ξ_i was taken to be Gaussian. The bit error rate for conventional binary phase shift keying (BPSK) $BER = \text{erfc} \left(\sqrt{E_b/N_0} \right) / 2$ is also shown for comparison. In Figure 1.6 we observe excellent agreement between the analytical prediction and the results of numerical simulations for $M = 100$.

From Figure 1.5 we also see that at large M the performance degrades with increasing M, which is consistent with (1.11). This trend occurs due to increasing contribution of noise-noise cross terms $\xi_i \xi_{i-M}$ in (1.3) and is typical for correlation decoding of TR signals. As we increase M keeping E_b/N_0 constant at a fixed signal amplitude, we increase N_0 proportionally to M. Thus, although the useful signal in (1.3) increases linearly with M, and so does the standard deviation of $\sum_{i=1}^{M} x_i(b_l \xi_{i-M} + \xi_i)$, $\sim \sqrt{MN_0} \sim M$, the standard deviation of $\sum_{i=1}^{M} \xi_i \xi_{i-M}$, $\sim \sqrt{MN_0^2} \sim M^{3/2}$, grows faster.

The bit error rate for the CDSK system is given by

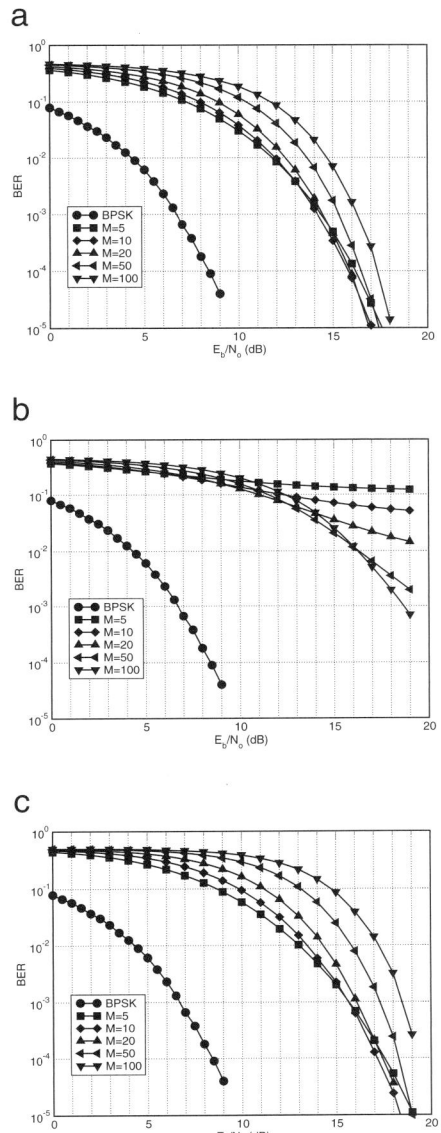

Fig. 1.5. Performance of correlation-based detection methods: DCSK (a), CDSK (b), and SCSK (c). A comparison to BPSK is included for comparison. (Reprinted with permission from [2]. ©2000 IEEE.)

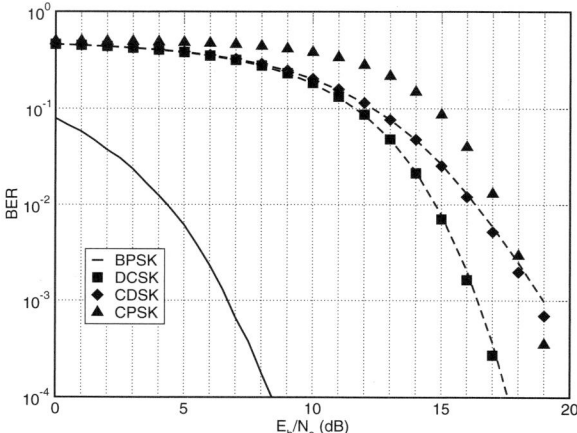

Fig. 1.6. Performance of DCSK, CDSK, and SCSK with $M = 100$. Numerical data and analytical estimates are shown with symbols and lines, respectively. (Reprinted with permission from [2]. ©2000 IEEE.)

$$BER = \qquad\qquad\qquad\qquad\qquad\qquad\qquad (1.12)$$

$$\tfrac{1}{2}\mathrm{erfc}\left(\sqrt{\tfrac{E_b}{8N_0}\left(1 + \tfrac{19}{20M}\tfrac{E_b}{N_0} + \tfrac{M}{4}\tfrac{N_0}{E_b}\right)^{-1}}\right).$$

The results of numerical simulations with $L = 200$ are shown in Figure 1.5*b*. The comparison between the analytical and numerical results is given in Figure 1.6. Considering that x_i and x_j can be considered statistically independent only approximately at large M, the analytical and the simulation curves at $M = 100$ match reasonably well.

In Figure 1.6 we also see the CDSK system performs between two and three dB worse than the DCSK system. This is due to two factors. First, due to the nature of the transmitted signal, there are four signal-noise cross terms in (1.5), compared to only two such terms in (1.3) for DCSK. Second, in addition to interference terms due to noise (noise-signal and noise-noise terms) there are three interference terms due to incomplete orthogonality of chaotic segments on two consecutive time intervals. These terms are present even when noise amplitude is zero, thus, the bit error rate saturates at large E_b/N_0 at the value $BER_{sat} = \mathrm{erfc}(\sqrt{5M/38})/2$. This saturation is visible in Figure 1.5*b* which shows the bit error rate curves computed numerically for different values of M. Here we also see that, as in the case of DCSK, increasing M at constant E_b/N_0 leads to performance degradation.

In general the correlator output (1.9) for this system can be written in the form (1.10) with $A = E_b + M\Delta E_b$, where

$$\Delta E_b = \left\langle \left(F(x_{i-1} + \tilde{\xi}_{i-1}) - F(x_{i-1}\right) F(x_{i-1})\right\rangle. \tag{1.13}$$

η and ζ can be defined as in the previous two cases. When M is large, ζ is a zero mean Gaussian variable with the variance in the case of the tent map $\sigma_\zeta^2 = 4E_b^2/(5M)$. η for SCSK is defined as

$$\eta = \sum_{i=1}^{M} F(x_{i-1} + \tilde{\xi}_{i-1})\tilde{\xi}_i$$

$$+ \sum_{i=1}^{M} \left(F(x_{i-1} + \tilde{\xi}_{i-1}) - F(x_{i-1}))F(x_{i-1}) - \Delta E_b\right),$$

and is also in the limit of large M a zero mean Gaussian variable. Because of the nonlinearity of F, it is difficult to find analytically the explicit formula for the bit error rate; however, we may expect to see the same general features in its performance as in DCSK and CDSK. Figure 1.5c shows the numerical performance curves for SCSK. This figure confirms our expectation that the performance of SCSK should follow the same trends as that of DCSK or CDSK. In particular we again observe the degradation of performance at large values of M.

1.2.2 Wireless Chaotic Pulse Position Modulation

An impulsive -based transmission system, where the information is modulated on the time intervals between the pulses, has a number of potential advantages over those described in the previous section. The negative effects of filtering and channel distortions, which typically severely impair the ability of chaotic systems to synchronize, are substantially reduced by using impulse signals.

One approach to these impulsive-based communications systems is to use chaotically timed pulse sequences rather than continuous chaotic waveforms [16]. Each pulse has an identical shape to all the others, but the time delay between them varies chaotically. Because the information about the state of the chaotic system is contained entirely in the timing between pulses, the distortions that affect the pulse shape will not significantly influence the ability of the chaotic pulse generators to synchronize and be utilized in communications.

This proposed system is similar to other ultra-wide bandwidth impulse radios [17], which are very promising communication platforms, especially in severe multipath environments or where they are required to co-exist with a large number of other radio systems. Chaotically varying the spacing between the narrow pulses enhances the spread spectrum characteristics of the system by removing any periodicity from the transmitted signal. Because of the absence of characteristic frequencies, chaotically positioned pulses are extremely hard to observe and detect illegitimately. Thus, one expects transmission based on chaotic pulse sequences to have a very low probability of intercept. Additional security can be accomplished by using one of the coding schemes suggested for covert communications with chaotic systems [18].

The particular chaotic encoding method, Chaotic Pulse Position Modulation (CPPM) [3], is related to the dynamical feedback modulation method [16]. The communication scheme is built around a Chaotic Pulse Regenerator, CPRG (see Figure1.7). Given a pulse train with interpulse intervals T_n, the CPRG produces a new pulse sequence with intervals $T_n + \Delta T_n$ where ΔT depends on the input sequence: $\Delta T_n = F(T_n, ..., T_{n-k})$; $F(\bullet)$ is such that with no input and with the feedback loop closed, the transmitter generates a pulse train with chaotic interpulse intervals.

The binary information is applied to the pulse train leaving the CPRG by adding an extra block in the feedback loop that leaves the signal unchanged if "0" is being transmitted, or delays the pulse by a fixed time if "1" is being transmitted. This modulated pulse sequence is the transmitted signal. Because an unauthorized receiver has no information on the spacing between the pulses leaving the CPRG, it cannot determine whether a particular received pulse was delayed, and thus whether "0" or "1" was transmitted. This provides a certain degree of privacy. At the receiver side, the signal is applied to the input of an identical CPRG.

The outputs from the CPRGs in the transmitter and the receiver are identical. Thus the signal at the output of the receiver CPRG is identical to the signal in the channel, except some pulses in the transmitted signal are delayed. By evaluating the relative pulse timings in the received signal and in the signal at the output of the CPRG, the receiver can decode the digital message. When the CPRGs are not matched with sufficient precision, a large decoding error results. Thus the parameters of the CPRGs act like a privacy key.

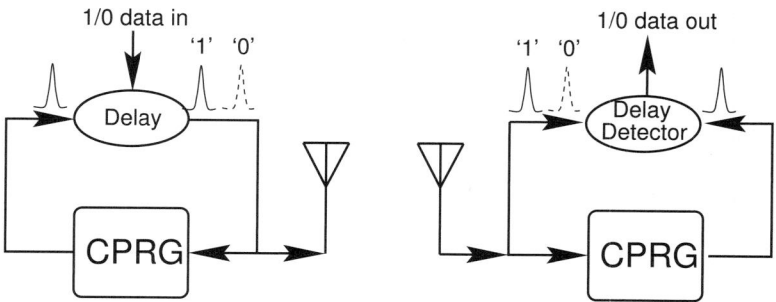

Fig. 1.7. Illustration of the basic CPPM approach. (Reprinted with permission from [3]. ©1999 IEEE.)

When synchronized, the receiver "knows" the time interval or a window where it can expect a pulse corresponding to a "1" or a "0". This allows the input to be blocked at all times except when a pulse is expected. The time intervals when the input to a particular receiver is blocked can be utilized by other users. To decode a bit of information we must determine whether a pulse from the transmitter falls into the window corresponding to "0" or that corresponding to "1", which can be done by integrating the input signal within the windows around the expected locations of pulses corresponding to "0" and to "1". If all pulses have the same polarity and synchronization is perfect, the CPPM performance is equivalent to that of OOK—3 dB worse than BPSK.

This is shown in the "PPM" curve in Figure 1.8. This performance is achieved with the window size equal to the pulse duration. In the case of imperfect synchronization the window cannot be so narrow, and with the larger window size and the same detection method we shall have an additional loss of performance equal to $10 \log T/\tau$ where T is the window size and τ is the pulse duration.

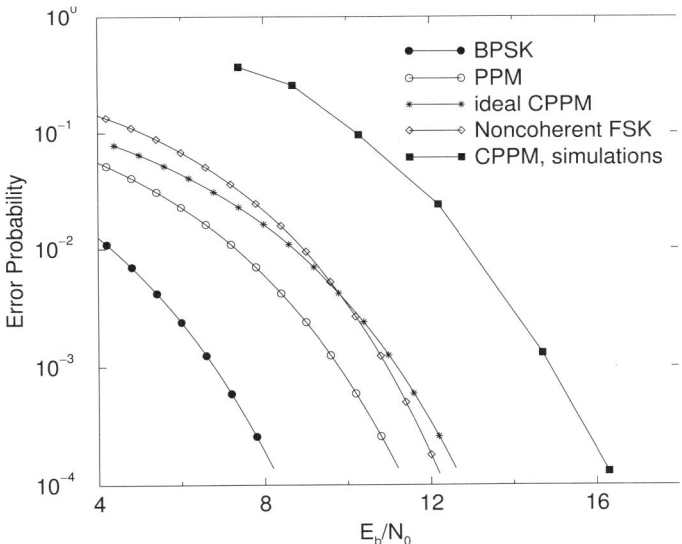

Fig. 1.8. Performance of the PPM schemes: PPM (the synchronous binary pulse position modulation scheme) is offset by 3 dB from BPSK; ideal CPPM includes the additional shift derived from KSE. (Reprinted with permission from [3]. ©1999 IEEE.)

There is another degradation factor common to chaos-based communication schemes. Most traditional schemes are based on periodic signals and systems where the carrier is generated by a stable system. All such systems are characterized by zero Kolmogorov-Sinai entropy (KSE): without any input the average speed of non-redundant information generation is zero. Chaotic systems have positive KSE and continuously generate non-redundant information. Even in an ideal environment, in order to perfectly synchronize two chaotic systems, one must transmit the amount of information per unit time that is equal to or larger than the KSE [19]. Consider a tent map $x_{n+1} = \alpha |0.5 - |0.5 - x_n||$ with $\alpha = 1.3$. The KSE for chaotic iterations of this map is $E_{KS} = \log_2 \alpha = 0.38$. Thus, to synchronize two tent maps and transmit one bit of information per iteration one would have to actually send on average 1.38 bits per iteration. This correction shifts the ideal performance curve for the CPPM scheme by 1.39 dB to the right, shown by the "ideal CPPM" curve in Figure 1.8.

The receiver that provides the best results is shown in Figure 1.9. Based on the state of the synchronized CPRG, the input is blocked at all times except the time windows around the expected locations of the pulses corresponding to "1" and "0". The signals within these windows are applied to peak detectors (PD). Based on which window contained the peak of the maximum height, we decide whether "1" or "0" was transmitted and the signal within the corresponding time window is passed to the receiver SPRG.

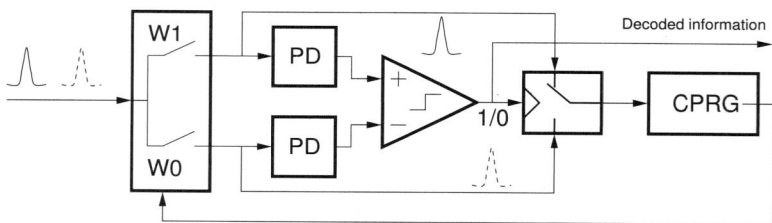

Fig. 1.9. Diagram of an optimized CPPM receiver. (Reprinted with permission from [3]. ©1999 IEEE.)

The channel is modeled by adding WGN to the output of the transmitter and then LP-filtering the signal with a FIR filter, which models the transmitter and receiver antennas. We measure SNR and find E_b/N_0 using the following formula [4]: $E_b/N_0 = S/N\,(W/R)$, where S/N is the SNR, W is the channel bandwidth, and R is the bit rate.

The performance curve corresponding to these parameters is shown in Figure 1.8. One can see that CPPM performs 4 dB worse than the ideal CPPM system in simulation. Most of this difference is attributed to the imperfect synchronization in the receiver. Although CPPM seems to perform worse than even non-coherent FSK, we should emphasize that (*i*) this spread spectrum system provides low intercept probability and covertness through a very simple design; (*ii*) to our knowledge, this system performs exceptionally well compared to other chaos-based covert communication schemes; (*iii*) there exists a multiplexing strategy that can be used with CPPM [20] and (*iv*) the system can be improved at lower bit rates, with narrower bandwidth. All this makes CPPM a prime base for development of chaos-based covert spread spectrum systems.

1.3 Optical Communications Based on Nonlinear Dynamics

Optically based chaotic communications, which are based on the transmission of messages encoded on a chaotic waveform, have attracted very extensive research activity [21]. Most of the systems are based on synchronization of chaos between a transmitter and a receiver, which are linked by a transmission channel. For such systems, synchronization between the transmitter and receiver is mandatory, because the bit-error rate (BER) of the decoded message at the receiver depends on the accuracy and robustness of synchronization.

Many systems based on either semiconductor lasers [22–34] or fiber lasers [35–37] have been proposed and studied for chaotic optical communications. We are especially interested in chaotic optical systems that can reach the rates typically employed by traditional optical communications systems, such as the OC-48 standard bit rate of 2.5 Gb/s and the OC-192 standard bit rate of 10 GB/s. We present three of the leading semiconductor laser systems that are most actively investigated and are most promising for high-bit-rate chaotic optical communications. They are the optical injection system [24, 25], optical feedback, and the optoelectronic feedback system [29, 30], which are schematically shown in Figures (1.10a-c), respectively.

Synchronization between the transmitter laser diode and the receiver laser diode can be either unidirectional or bidirectional, but unidirectionally coupled systems are typically used for high-bit-rate communications. The receiver can be operated in open-loop or closed-loop mode, but an open-loop receiver can be more stably synchronized to the transmitter than a closed-loop receiver [31]. Therefore, we consider only unidirectional systems with open-loop receivers as shown in Figure 1.10.

Information can be impressed onto these optoelectronic systems in several different ways, as shown in Figure 1.10. Several encryption methods have been considered and demonstrated for chaotic communication systems. The most important ones include chaos shift keying (CSK) [38], chaos masking (CMS) [39], chaos modulation [6], [40], and chaotic pulse position modulation (CPPM) [2], [3]. In the case of chaos modulation, both additive chaos modulation (ACM) [6] and multiplicative chaos modulation (MCM) have been considered [40].

The CSK encryption method is implemented by encoding the message through direct current modulation of the transmitter. Decoding of the message is done by subtracting the output of the receiver from the signal that is transmitted to the receiver. True synchronization cannot be expected in the

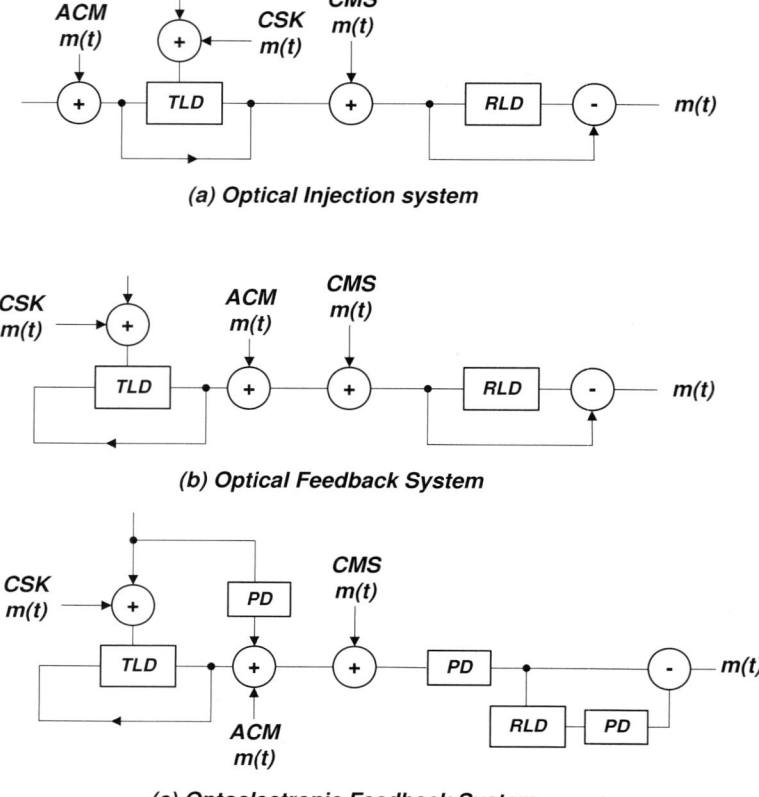

Fig. 1.10. Schematic diagrams for three synchronized chaotic optical communications systems using semiconductor lasers. Also shown are the message encoding and decoding schemes for the three methods: (a) optical injection, (b) optical feedback, and (c) optoelectronic feedback.

process of message encoding with CSK because the transmitter is current-modulated with the message but the receiver is not. The chaotic state of the transmitter is influenced by the message in the CSK encryption.

In the CMS encryption method, a message is encoded on the chaotic output of the transmitter by simply adding the message to the chaotic waveform being transmitted to the receiver. The message is decoded by subtracting the output of the receiver from the signal that is transmitted to the receiver. Because information of the message is injected into the receiver but no information of the message is sent to the transmitter, the symmetry between the receiver and the transmitter cannot be maintained. Consequently, true synchronization cannot be expected when a message is encoded with CMS.

However, the chaotic state of the transmitter is not influenced by the message in the CMS encryption.

In the ACM encryption method, the message is encoded by adding it to the chaotic output of the transmitter, but, different from CMS, the information of the message is sent equally to both the transmitter and the receiver. Decoding is performed again by subtracting the output of the receiver from the signal that is transmitted to the receiver. True synchronization is possible when a message is encoded with ACM because the symmetry between the transmitter and the receiver is not broken by the message-encoding process of this encryption method. The chaotic state of the transmitter, as well as the complexity of chaos, varies with the message in the ACM encryption.

For the optical injection system shown in Figure 1.10(a), the nonlinear dynamics of the transmitter is generated by optical injection from another laser. When synchronization is achieved, the optical frequency, phase, and amplitude of the two lasers are all locked together synchronously [34].

According to the configuration in Figure 1.10(a), the transmitter can be modeled by the following coupled equations in terms of the complex laser field amplitude and the carrier density [22, 23]

$$
\begin{aligned}
\frac{dA^T}{dt} = &-\left(\frac{\gamma_c^T}{2} + \eta\alpha\right)A^T + i(\omega_0 - \omega_c)A^T \\
&+ \frac{\Gamma}{2}(1 - ib^T)gA^T + F_{sp}^T + \eta\left\{\alpha A_i e^{-i\Omega t}\left[1 + m_{ACM}(t)\right]\right\}
\end{aligned}
\tag{1.14}
$$

and

$$
\frac{dN^T}{dt} = \frac{J\left[1 + m_{CSK}(t)\right]}{ed} - \gamma_s N^T - \frac{2\varepsilon_0 n^2}{\hbar\omega_0}|A^T|^2
\tag{1.15}
$$

and the receiver can be described by

$$
\begin{aligned}
\frac{dA^R}{dt} = &-\left(\frac{\gamma_c^R}{2}\right)A^R + i(\omega_0 - \omega_c)A^R \\
&+ \frac{\Gamma}{2}(1 - ib^R)gA^R + F_{sp}^T + \eta s(t)
\end{aligned}
\tag{1.16}
$$

and

$$
\frac{dN^R}{dt} = \frac{J}{ed} - \gamma_s N^R - \frac{2\varepsilon_0 n^2}{\hbar\omega_0}|A^R|^2.
\tag{1.17}
$$

where A is the complex intracavity field amplitude, γ_c^R is the cavity decay rate, ω_c the longitudinal mode frequency of the cavity, Γ the confinement factor, b^T the linewidth enhancement, g the gain coefficient including second-order effects, F_{sp}^T the spontaneous emission noise source, N the carrier density, J the injection current density, d the active layer thickness, γ_s the spontaneous carrier decay rate, n the refractive index, and F the carrier noise.

In this model, $A_i e^{-i\Omega t}$ is the optical injection field with a detuning frequency Ω, η is the injection rate of an optical field into the laser, and α defines the coupling strength between the transmitter and the receiver. The subscript of the encoding message in (1.14) and (1.15) indicates the type of encoding scheme used. When a particular scheme is used, $M(t)$ with a subscript of other schemes should be set to zero. As we can see from the mathematical model, when CMS is applied, the encoding message is injected into the receiver but not into the transmitter. When CSK is applied, only the transmitter, but not the receiver, is current modulated with the encoding message. By comparison of the rate equations above for different encoding schemes, we can see that the transmitter and the receiver are never mathematically identical when a message is encoded with the CMS or the CSK scheme. They can be identical only when a message is encoded with the ACM scheme and when their parameters are properly matched.

A general characteristic of chaotic communications based on synchronization, is that the recovered message is contaminated by the channel noise as well as the synchronization error. Therefore, the performance of this system is determined by an experimental comparison of the results in the time domain (Figure 1.11). In order to reveal the effect of the channel noise on the message recovery, the data shown in this figure are obtained when the laser noise is not considered. The value of SNR is set at 36 dB. It is observed that message recovery is almost impossible for the CSK scheme because message encoding with CSK causes frequent desynchronization bursts and the success of message decoding for this encryption scheme is determined by the resynchronization time. At a bit rate of 10 Gb/s, resynchronization is difficult to achieve within the short time of the bit duration of 0.1 ns. The performance can be improved at low bit rate when the bit duration gets longer than the resynchronization time.

In the case of CMS, the synchronization error mainly arises from the breaking of the mathematical identity between the transmitter and the receiver by the encoded message. Because the encoding message used is small in comparison to the transmitter output, the encoded message acts only as a perturbation on the synchronization. Therefore, the recovered message shows some resemblance to the pattern of the encoding message. Better message recovery can be expected if a low-pass filter is used.

The performance of ACM is the best among the three encryption methods because message encoding by ACM does not break the mathematical identity between the transmitter and the receiver. The error bits are contributed by synchronization error caused by the channel noise, as well as by the laser noise when it is considered. Whether synchronization deviation or desynchronization bursts dominate in the generation of error bits depends on the amount of noise present.We can see that the single error bit seen in the ACM decoded message in Figure 1.11 is generated by the occurrence of a desynchronization burst.

The system performance measured by the BER as a function of channel SNR for the optical injection system is shown in Figure 1.12 for each of the three encryption schemes.We observe that CSK and CMS have similar performance when the laser noise is not considered. The performance of CMS is barely affected by the laser noise because the breaking of the mathematical identity between the transmitter and the receiver caused by the encoded

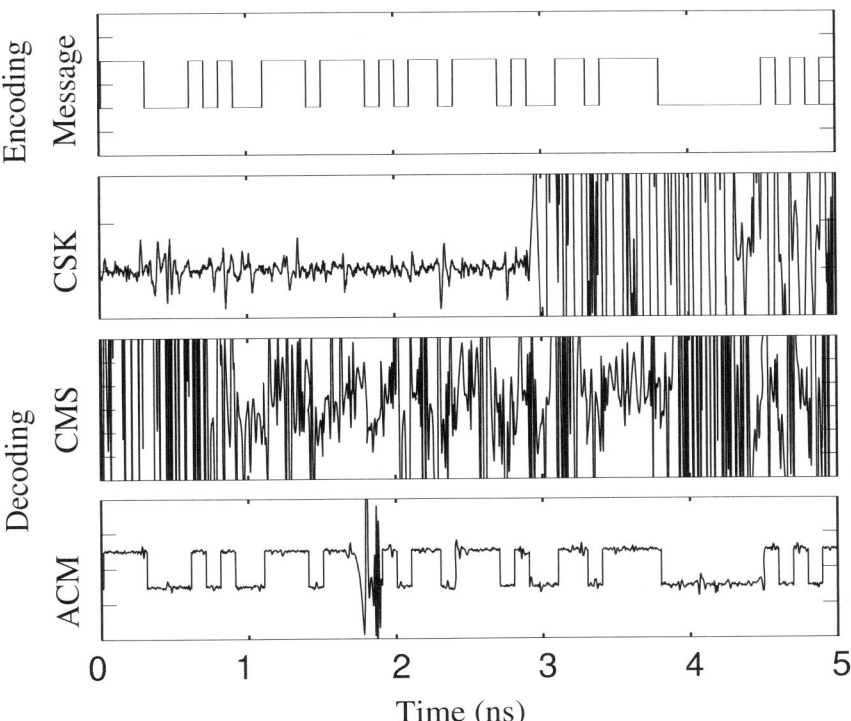

Fig. 1.11. Time series of the decoded messages of the three different encryption schemes in the optical injection system.

message has a much more significant effect on creating synchronization error than the perturbation of the laser noise to the system. The performance of CSK is, however, deteriorated by introducing the laser noise because the noise further increases the desynchronization probability and the resynchronization time. As for the performance of the ACM scheme, a BER lower than 10^{-5} can be obtained for an SNR larger than 60 dB under the condition that the laser noise is absent. However, in the presence of the laser noise at a level corresponding to a laser linewidth of around 100 kHz, for example, the BER saturates at a value higher than 10^{-3}, and the BER saturates at a higher value for a larger laser noise.

For the optical feedback system shown in Figure 1.10(b), the nonlinear dynamics of the transmitter is generated by optical feedback of the laser output. According to the configuration in Figure 1.10(b), the transmitter can be modeled by the following coupled equations

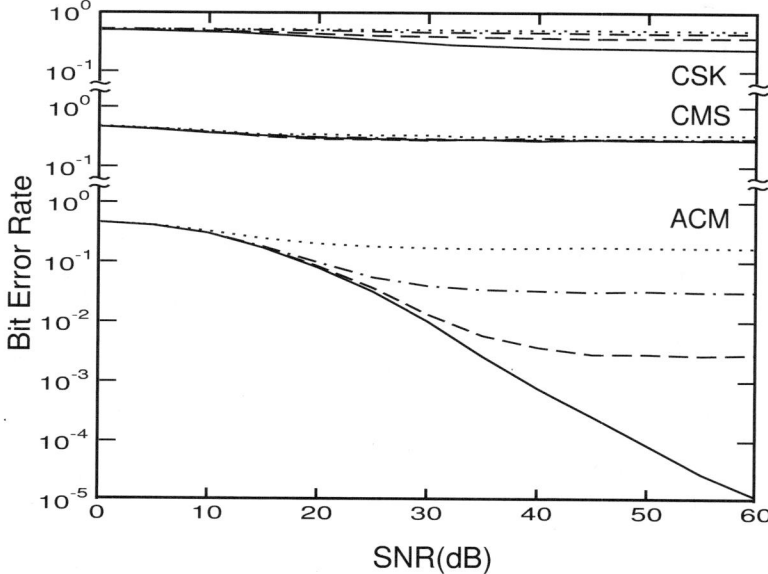

Fig. 1.12. BER versus SNR for the three different encryption schemes in the optical injection system. Solid line: obtained when laser noise is absent, dashed line: obtained when the laser noise level of $\Delta\nu = 100kHz$ for both the TLD and RLDs. Dot-dashed line obtained when $\Delta\nu = 1MHz$ and dotted line obtained when $\Delta\nu = 10MHz$.

$$\frac{dA^T}{dt} = -\frac{\gamma_c^T}{2}A^T + i(\omega_0 - \omega_c)A^T$$
$$+\frac{\Gamma}{2}(1 - ib^T)gA^T + F_{sp}^T + \eta\left\{\alpha A^T(t - \tau) + m_{ACM}(t)\right\} \tag{1.18}$$

and

$$\frac{dN^T}{dt} = \frac{J\left[1 + m_{CSK}(t)\right]}{ed} - \gamma_s N^T - \frac{2\varepsilon_0 n^2}{\hbar\omega_0}|A^T|^2 \tag{1.19}$$

and the receiver can be characterized by

$$\frac{dA^R}{dt} = -\left(\frac{\gamma_c^R}{2}\right)A^R + i(\omega_0 - \omega_c)A^R$$
$$+\frac{\Gamma}{2}(1 - ib^R)gA^R + F_{sp}^R + \eta s(t) \tag{1.20}$$

and

$$\frac{dN^R}{dt} = \frac{J}{ed} - \gamma_s N^R - \frac{2\varepsilon_0 n^2}{\hbar\omega_0}|A^R|^2 \tag{1.21}$$

In this case, τ is the feedback delay time and η is the injection rate. Because the feedback strength has to be equal to the coupling strength between the transmitter and the receiver for the existence of perfect chaos synchronization, the parameter α is used for both quantities. The transmitted signal has the form $s(t) = \alpha A^T(t - \tau) + m(t)$ for both the ACM and CMS schemes, and the form $s(t) = \alpha A^T(t - \tau)$ for the CSK scheme. The subscript of the encoding message in (1.18) and (1.21) indicates the type encoding scheme used.

When CMS is applied, the encoding message is sent to the receiver but is not fed back to the transmitter. When CSK is applied, only the transmitter, but not the receiver, is current modulated with the encoding message. By comparison of the rate equations above for different encoding schemes, we can see that the transmitter and the receiver can be identical in the presence of a message *only* when ACM is applied and when their parameters are well matched.

The system performance, measured by the BER as a function of channel SNR for the optical feedback system is shown in Figure 1.13. From Figure 1.13, we find that message recovery for the CSK scheme is not possible at the high bit rate studied here because the resynchronization time after a desynchronization burst has to be shorter than the bit duration for a following bit to be recoverable. The performance of CMS in this system is similar to that of CMS in the optical injection system, and it is barely affected by the laser noise for the same reason as that mentioned above for the optical injection system. As for the performance of the ACM scheme, a BER lower than 10^{-3} cannot be obtained even when the channel SNR is as large as 120 dB. This is caused by the frequent occurrence of desynchronization bursts in this system even at an extremely low level of channel noise. In this system, the effect of the laser noise of each laser is about the same as that of the equivalent channel noise on the system BER performance, and the transmitter noise has as equal an effect as the receiver noise on the BER performance.

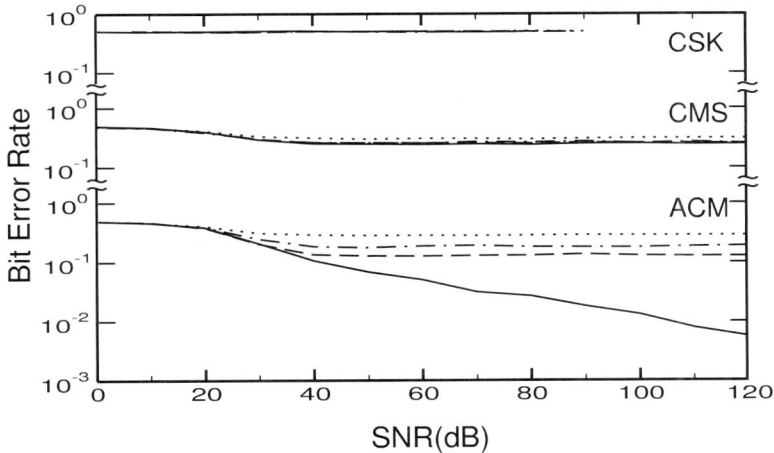

Fig. 1.13. BER versus SNR for the three different encryption schemes in the optical feedback system Each curve corresponds to the same curve in Figure 1.11.

The final synchronization approach employs optoelectronic feedback According to the configuration in Figure 1.10(c), the transmitter can be modeled by the following coupled equations in terms of the photon density S^T and the carrier density N^T

$$\frac{dS^T}{dt} = -\gamma_c^T S^T + \Gamma g S^T + 2\sqrt{S_0 S^T} F_S^T \qquad (1.22)$$

and

$$\frac{dN^T}{dt} = \frac{J\left[1 + m_{CSK}(t)\right]}{ed}\left[1 + \xi y^T(t - \tau)\right] - \gamma_s N^T - g S^T \qquad (1.23)$$

and the receiver can be described by

$$\frac{dS^R}{dt} = -\gamma_c S^R + \Gamma g S^R + 2\sqrt{S_0 S^R} F_S^R \qquad (1.24)$$

and

$$\frac{dN^R}{dt} = \frac{J\left[1 + m_{CSK}(t)\right]}{ed}\left[1 + \xi y^R(t - \tau)\right] - \gamma_s N^R - g S^R \qquad (1.25)$$

where ξ is the feedback strength, τ is the feedback delay time, and $f(t)$ is the normalized response function of the feedback loop, including the photodetector and the amplifier. The transmitted signal has the form $s(t) = S^T(t) + m(t)$ for both the ACM and CMS schemes, and the form $s(t) = S^T(t)$ for the CSK scheme. The subscript of the encoding message $m(t)$ in 1.23 indicates the encoding scheme used. The transmitter and the receiver can be mathematically identical only when a message is encoded with the ACM scheme and when their parameters are well matched.

The system performance as measured by the BER as a function of channel SNR for the optoelectronic feedback system is shown in Figure 1.14. From Figure 1.14, we find again that message recovery at a high bit rate is not possible for the CSK scheme. The CMS scheme has better performance in this system than it does in both the optical injection and the optical feedback systems. The performance of CMS is barely affected by the laser noise for the same reason as that mentioned above for the other two systems. As for the performance of ACM, a lower BER can be obtained when the SNR is larger than 38 dB, which is much better than the performance of ACM in the other two systems discussed above.

The reason for this improved performance is that the channel noise in the optoelectronic feedback system is converted into electronic noise to be electrically injected into the receiver, whereas the channel noise in both the optical injection and the optical feedback systems is optically injected into the receiver. Because the carrier decay rate is much smaller than the cavity decay rate, the carrier density fluctuation caused by the channel noise that is electrically injected into the receiver in the optoelectronic feedback system is

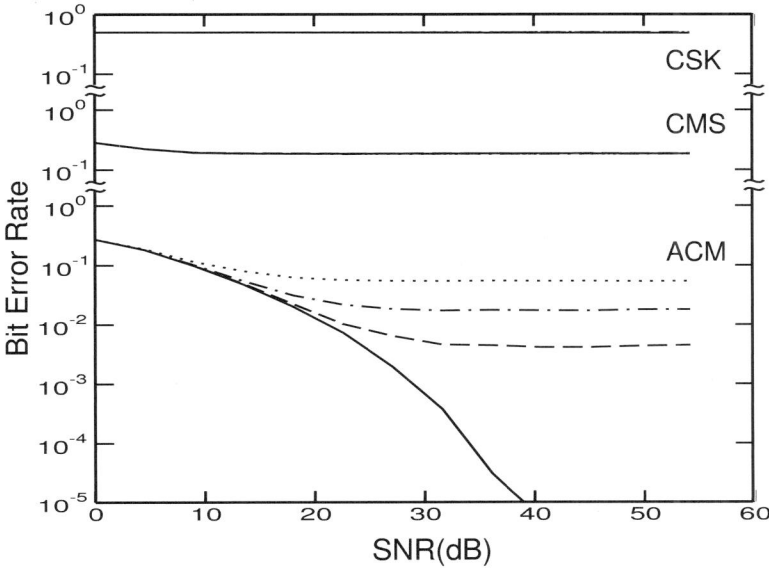

Fig. 1.14. BER versus SNR for the three different encryption schemes in the optoelectronic feedback system. Each curve corresponds to the same curve in Figure 1.11.

much lower than the amplitude and phase fluctuations of the laser field caused by the channel noise that is optically injected into the receiver in the other two systems. However, because the laser noise directly causes fluctuations in the intracavity laser field, the laser noise in this system saturates the BER to a value higher than as it does in the optical injection system. Therefore, the effect of the laser noise on the BER performance of this system is very different from that of the channel noise. Note that the problem mentioned above that makes the use of the CSK scheme not possible for high-bit-rate communications with this system does not exist for the ACM scheme because no desynchronization bursts are observed when ACM is applied to this system.

1.4 Conclusions

Communications systems based on chaos and nonlinear dynamics are still in their infancy, although astonishing progress has been made in recent years. In the wireless arena, the technology promises a high degree of security because of the essentially nonperiodic nature of the carrier. In addition, the simplicity and ease of control of these approaches makes them attractive candidates for low-cost secure communications.

A number of chaos-based communication schemes have been suggested, but many of these systems are very sensitive to distortion, filtering, and noise. We have examined several of the most promising approaches in this chapter, including DCSK, CDSK, SCSK, and CPPM. At their best, these systems have comparable performance to traditional OOK systems, but with far greater security and low- observability.

Three semiconductor laser systems— the optical injection system, the optical feedback system, and the optoelectronic feedback system — which are capable of generating broadband, high-frequency chaos for high-bit-rate communications have been considered in this chapter. The inherent advantage of any optical communication system is its ability to handle high-bit-rate communications. Optoelectronic-feedback laser systems with ACM message encoding-decoding at 2.5 Gb/s has demonstrated chaotic optical communication at a bit rate matching the requirement of the OC-48 standard. Chaotic optical communications at the OC-192 standard bit rate of 10 Gb/s are possible when high-speed semiconductor lasers are used.

The performance of each system at 10 Gb/s is numerically studied for the three encryption schemes of CSK, CMS, and ACM. Noise sources have very significant effects on the system performance at high bit rates, primarily because they cause synchronization error in the forms of synchronization deviation and desynchronization bursts in these synchronized chaotic communication systems. Among the three laser systems, the optoelectronic feedback system is least susceptible to noise-induced desynchronization and thus has the best performance, whereas the optical feedback system is most susceptible to noise-induced desynchronization and, thus, has the worst performance.

Among the three encryption schemes, it is found that, at high bit rates, only the performance of ACM with low-noise lasers is acceptable because ACM allows true synchronization in the process of message encoding by maintaining the mathematical identity between the transmitter and the receiver. Both CSK and CMS cause significant bursts or synchronization deviation in the systems because they break the identity between the transmitter and the receiver in the process of message encoding. The possibility of designing stably synchronized systems and optimized filters to improve the performance of synchronized chaotic optical communication systems is an important subject to be addressed in the future.

Acknowledgments

The authors would like to acknowledge many valuable discussions with Drs. L. Pecora and T. Carroll of the Naval Research Laboratories, Dr. G. Maggio of ST Microelectronics, Dr. M. Kennel, D. Laney, and L. Illing of UCSD, and Dr. S. Tang and H. Chen of UCLA. This work was partially supported by the Army Research Office under MURI grant DAAG55-98-1-0269 under the guidance of Dr. John Lavery.

References

1. J. M. Liu, H. F. Chen, and S. Tang, Synchronized Chaotic Optical Communications at High Bit Rates, *IEEE J. Quantum Electron.*, vol. 38, pp. 1184-1196, 2002.
2. M. Sushchik, L. S. Tsimring, and A. R. Volkovskii, Performance analysis of correlation-based communications systems using chaos, *IEEE Trans. Circuits Sys.– I*, vol. 47, no. 12, pp. 1684-1691, 2000.
3. M. Sushchik, N. Rulkov, L. Larson, L. Tsimring, H. Abarbanel, K. Yao, and A. Volkovskii, Chaotic pulse position modulation: a robust method of communicating with chaos *IEEE Comm. Lett.*, vol.4, pp. 128-130, 2000.
4. B. Sklar *Digital Communications*, Prentice-Hall Englewood Cliffs, NJ, 1988.
5. G. Mazzini, G. Setti, and R. Rovatti, Chaotic complex spreading sequences for asynchronous DS-CDMA - part I: System modeling and results (vol 44, p. 937, 1997), *IEEE Trans. Circuit. Syst. I*, vol. 45, no. 4, pp. 515–516, 1998.
6. C.W. Wu and L.O. Chua, A simple way to synchronize chaotic systems with applications to secure communication systems, *Int. J. Bifur. Chaos*, vol. 3, no. 6, pp. 1619–1627, 1993.
7. M. Hasler, Synchronization of chaotic systems and transmission of information, *Int. J. Bifur. Chaos*, vol. 8, no. 4, pp. 647–659, 1998.
8. R. Rovatti, G. Setti, and G. Mazzini, Chaotic complex spreading sequences for asynchronous DS-CDMA - part II: Some theoretical performance bounds, *IEEE Trans. Circuit. Syst. I*, vol. 45, no. 4, pp. 496–506, 1998.
9. G. Kolumbán, M.P. Kennedy, and L.O. Chua, The role of synchronization in digital communications using chaos. ii. chaotic modulation and chaotic synchronization, *IEEE Trans. Circuits Sys.– I*, vol. 45, no. 11, pp. 1129–1140, 1998.
10. T. L. Carroll and L. M. Pecora, Synchronizing hyperchaotic volume-preserving maps and circuits, *IEEE Trans. Circuit. Syst. I*, vol. 45, no. 6, pp. 656–659, 1998.
11. T. Yang and L. O. Chua, Chaotic digital code-division multiple access (CDMA) communication systems., *Int. J. Bifur. Chaos*, vol. 7, no. 12, pp. 2789–2805, 1997.
12. H. Dedieu, M. P. Kennedy, and M. Hasler, Chaos shift keying: Modulation and demodulation of a chaotic carrier using self-synchronizing Chua's circuits, *IEEE Trans. Circ. and Systems - II*, vol. 40, no. 10, pp. 634–642, 1993.
13. U. Parlitz and S. Ergezinger, Robust communication based on chaotic spreading sequences, *Phys. Lett. A*, vol. 188, pp. 146–150, 1994.
14. G. Kolumbán, B. Vizvári, W. Schwarz, and A. Abel, Differential chaos shift keying: a robust coding for chaos communication, in *Proceedings of the NDES*, Seville, pp. 87–92, 1996.
15. N. I. Chernov, Limit theorems and Markov approximations for chaotic dynamical systems, *Prob. Theory and Related Fields*, v.101, no.3, pp. 321-362, 1995.
16. A. R. Volkovskii and N. F. Rulkov, Synchronous chaotic response of a nonlinear oscillator system as a principle for the detection of the information component of chaos, *Sov. Tech. Phys. Lett.*, vol. 19, pp. 97–99, 1993.
17. M. Z. Win and R. A. Scholtz, Impulse radio: how it works. *IEEE Communications Letters*, vol.2(2), pp.36–38, 1998.
18. M. Hasler, Synchronization of chaotic systems and transmission of information. *Int. J. Bifur. Chaos*, 8(4), pp.647–659, 1998.

19. T. Stojanovski, L. Kocarev, and R. Harris, Applications of symbolic dynamics in chaos synchronization. *IEEE Trans. Circuits Syst. I*, vol. 44(10), pp. 1014–1017, 1997.

20. H. Torikai, T. Saito, and W. Schwarz, Synchronization via multiplex pulse-train. *IEEE Trans. Circuits Syst. I*, vol.46, pp. 1072–1085, 1999.

21. Special issue on applications of chaos in modern communication systems, *IEEE Trans. Circuits Syst. I*, vol. 48, 2001.

22. J. M. Liu, H. F. Chen, and S. Tang, Optical communication systems based on chaos in semiconductor lasers, *IEEE Trans. Circuits Syst. I*, vol. 48, pp. 1475–1483, 2001.

23. H.F. Chen and J.M. Liu, Open-loop chaotic synchronization of injection- locked semiconductor lasers with gigahertz range modulation, *IEEE J. Quantum Electron.*, vol. 36, pp. 27–34 2000.

24. Y. Liu, H. F. Chen, J. M. Liu, P. Davis, and T. Aida, Communication using synchronization of optical-feedback-induced chaos in semiconductor lasers, *IEEE Trans. Circuits Syst. I,* vol. 48, pp. 1484–1489, 2001.

25. A. Sanchez-Diaz, C. R. Mirasso, P. Colet, and P. Garcia-Fernandez, Encoded Gbit/s digital communications with synchronized chaotic semiconductor lasers, *IEEE J. Quantum Electron.*, vol. 35, pp. 292–297, 1999.

26. J. Paul, S. Sivaprakasam, P. S. Spencer, R. Rees, and K. A. Shore, GHz bandwidth message transmission using chaotic diode lasers, *Electron. Lett.*, vol. 38, pp. 28–29, 2002.

27. S. Sivaprakasam, E. M. Shahverdiev, and K. A. Shore, Experimental verification of synchronization condition for chaotic external cavity diode laser, *Phys. Rev. E*, vol. 62, pp. 7505–7507, 2000.

28. F. Rogister, A. Locquet, D. Pieroux, M. Sciamanna, O. Deparis, P. Megret, and M. Blondel, Secure communication scheme using chaotic laser diodes subject to incoherent optical feedback and incoherent optical injection, *Opt. Lett.*, vol. 26, pp. 1486–1488, 2001.

29. S. Tang and J.M. Liu, Message encoding-decoding at 2.5 Gbits/s through synchronization of chaotic pulsing semiconductor lasers, *Opt. Lett.*, vol. 26, pp. 1843–1845, 2001.

30. S. Tang, H.F. Chen, S.K. Hwang, and J.M. Liu, Message encoding and decoding through chaos modulation in chaotic optical communications, *IEEE Trans. Circuits Syst. I,* vol. 49, pp. 163–169, 2002.

31. H. D. I. Abarbanel, M. B. Kennel, L. Illing, S. Tang, H. F. Chen, and J. M. Liu, Synchronization and communication using semiconductor lasers with optoelectronic feedback, *IEEE J. Quantum Electron.*, vol. 37, pp. 1301–1311, 2001.

32. L. Larger, J.P. Goedgebuer, and F. Delorme, Optical encryption system using hyperchaos generated by an optoelectronic wavelength oscillator, *Phys. Rev. E,* vol. 57, pp. 6618–6624, 1998.

33. J.B. Cuenot, L. Larger, J.P. Goedgebuer, and W.T. Rhodes, Chaos shift keying with an optoelectronic encryption system using chaos in wavelength, *IEEE J. Quantum Electron.*, vol. 37, pp. 849–855, 2001.

34. Y. Liu, P. Davis, and T. Aida, Synchronized chaotic mode hopping in DBR lasers with delayed optoelectric feedback, *IEEE J. Quantum Electron.*, vol. 37, pp. 337–352, 2001.

35. G.D. VanWiggeren and R. Roy, Chaotic communication using time delayed optical systems, *Int. J. Bifur. Chaos*, vol. 9, pp. 2129–2156, 1999.

36. C.T. Lewis, H.D.I. Abarbanel, M.B. Kennel, M. Buhl, and L. Illing, Synchronization of chaotic oscillations in doped fiber ring lasers, *Phys. Rev. E*, vol. 63, pp. 1–15, 2000.
37. L.G. Luo, P.L. Chu, and H.F. Liu, 1-GHz optical communication system using chaos in erbium-doped fiber lasers, *IEEE Photon. Technol. Lett.*, vol. 12, pp. 269–271, 2000.
38. U. Parlitz, L.O. Chua, L.Kocarev, K.S. Halle, and A. Shang, Transmission of digital signals by chaotic synchronization, *Int. J. Bifur. Chaos*, vol. 2, pp. 973–977, 1992.
39. L. Kocarev, K.S. Halle, K. Eckert, L.O. Chua, and U. Parlitz, Experimental demonstration of secure communications via chaotic synchronization, *Int. J. Bifur. Chaos*, vol. 2, pp. 709–713, 1992.
40. K.S. Halle, C.W. Wu, M. Itoh, and L.O. Chua, Spread spectrum communication through modulation of chaos, *Int. J. Bifur. Chaos*, vol. 3, pp. 469–477, 1993.

2

Digital Communication Using Self-Synchronizing Chaotic Pulse Position Modulation

Nikolai F. Rulkov, Alexander R. Volkovskii, Michail M. Sushchik, Lev S. Tsimring, and Lucas Illing

Summary. We review a new approach to communication with chaotic signals based upon chaotic signals in the form of pulse trains where intervals between the pulses are determined by chaotic dynamics of a pulse nary information is modulated onto this carrier by the pulse position modulation method, such that each pulse is either left unchanged or delayed by a certain time, depending on whether 0 or 1 is transmitted. By synchronizing the receiver to the chaotic-pulse train we can anticipate the timing of pulses corresponding to 0 and 1 and thus can decode the transmitted information. Based on the results of theoretical and experimental studies we discuss the basic design principles for the chaotic-pulse generator, its synchronization, and the performance of the chaotic-pulse communication scheme in the presence of channel noise and filtering.

2.1 Introduction

In the last two decades intensive studies of chaotic behavior have produced various methods for controlling chaos and ideas for its possible applications. Chaos found in nonlinear electrical circuits [1–4] and lasers [5–11] provides means for generation of chaotic signals that can potentially be used as carriers for information transmission. The simplicity of chaos generators and the rich structure of chaotic signals are the two most attractive features of chaos that have caused a significant interest in possible utilization of chaos for communication.

Because the chaotic signal is nonperiodic, it cannot be stored in the receiver as a reference in order to achieve coherent detection of the transmitted signal. To overcome this problem, in some of the proposed communication schemes, the original chaotic waveform is transmitted along with the modulated signal (transmitted reference scheme) either using a separate channel or using time division [12]. Thus, a reliable detection can be achieved at the expense of at least 3 dB of the signal-to-noise ratio. In another approach, a chaotic reference is regenerated at the receiver using the phenomenon of chaos synchronization [13]. It was shown in many experiments and theoretical studies that two coupled chaotic systems can be synchronized in a sense that a chaotic system at the receiver can follow the time evolution of the identical system located in the transmitter. Therefore, a chaotic signal generated in the transmitter can be replicated in the receiver in a stable manner [14, 15]. The regime of the transmitter-receiver synchronization was proposed as a possible mechanism for the recovery of information encoded in the received chaotic signal [16–19].

A number of chaos-based covert communication methods that mix an information signal with a chaotic one and then recover the information using synchronization of chaos have been suggested [20]. In one class of these methods, the information signal $m(t)$ is added to the chaotic output $x_e(t)$ generated by a chaotic encoder whose oscillations do not depend on $m(t)$. The mixture $x_e(t) + m(t)$ is transmitted to the decoder where it is used as a driving signal for the matched response system. Various implementations of the matched response systems were proposed; see, for example [1, 2, 21–24] and references therein. The common shortcoming of such methods of communication is that the driving signal which is "distorted" by the message $m(t)$, does not perfectly fit the decoder. As a result, the recovered message $m_r(t)$ will always contain some traces of chaotic waveforms no matter how perfectly the parameters of the decoder match those of the encoder.

A different approach to the problem of chaotic encoding and decoding was suggested in a number of papers [3, 16–18, 25–27]. The main idea of this approach is that information signal $m(t)$ is injected into one of the feedback loops of the chaotic system; see Figure 2.1. The feedback should be selected in such a way that the remaining subsystem (CT) is conditionally stable. In this case the distorted chaotic feedback signal $x_e(t) + m(t)$ returns back to CT and drives the encoder oscillations. When the same signal, $x_e(t) + m(t)$, is applied to the decoder, it excites oscillations of the response system which are identical to the oscillations in the encoder. As the result the message $m(t)$ can be recovered in the receiver using the open feedback loop of the response system. In this case (in the absence of noise), after initial transients, the information can be restored exactly.

All practical communication channels introduce signal distortions that alter the chaotic waveform, and as a result, the received chaotic oscillations do not precisely match the transmitter oscillations $x_e(t) + m(t)$. Channel noise, filtering, attenuation variability, and other distortions in the channel corrupt the chaotic carrier and information signal. The presence of these channel distortions significantly hamper the onset of identical synchronization of the chaotic systems [29, 30].

The strong sensitivity to the distortions of the chaotic signal and the resulting problems with chaos synchronization is the major obstacle for practical implementation of chaos-based communications systems. In order to overcome the problems of channel distortions, a number of special chaotic communication methods have been proposed [31–35]. At least in theory and nu-

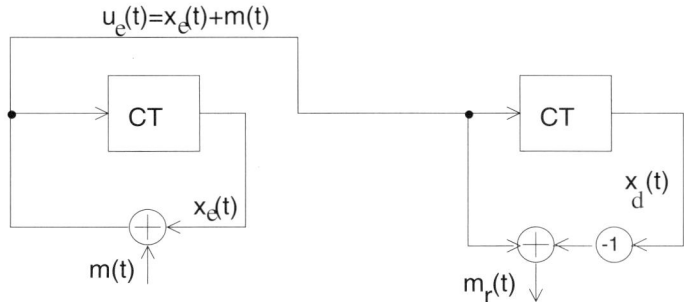

Fig. 2.1. Block-diagram of signal transmission with chaos suggested in [17]. (Reprinted with permission from [28], ©John Wiley & Sons, Ltd.)

merical simulations, it appears that the regime of identical synchronization in these specially designed systems is significantly less sensitive to channel noise and waveform distortions caused by the limited bandwidth of the channel [28, 36, 37].

One of the ways to minimize the effects of channel distortions was suggested in [38], where a chaotically timed pulse sequences substituted continuous chaotic waveforms. Each pulse in the sequence has identical shape, but the time delay between them varies chaotically. Because the information about the state of the chaotic system is contained entirely in the timing between pulses, the distortions that affect the pulse shape will not significantly influence the ability of the chaotic pulse generators to synchronize. Therefore synchronizing chaotic impulse generators can be utilized for communication via realistic wideband channels and at the same time allow the use of bandpass filters for noise reduction. The information can be encoded in the pulse train by alteration of time position of pulses with respect to chaotic carrier. This is the essence of the Chaotic Pulse Position Modulation (CPPM) system [39].

This proposed system belongs to the general class of ultra-wide bandwidth wireless communication systems. These systems received significant attention recently [40] (see also Chapter 4) because they offer a very promising alternative communication possibilities, especially in severe multipath environments or where they have to co-exist with a large number of other wireless systems. Chaotically varying spacing between narrow pulses enhances the spectral characteristics of the system by removing any periodicity from the transmitted signal. Because of the absence of characteristic frequencies, chaotically positioned pulses are difficult to observe and detect for the unauthorized user. Thus one expects that transmission based on chaotic pulse sequences can be designed to have a very low probability of intercept. A secure information transmission based on chaotic pulse trains with different methods of information encoding have been studied in [41].

The chapter is organized as follows. Section 2.2 gives a detailed overview of the CPPM system. In Section 2.3 we describe our implementation of the CPPM system and present the results of the experimental performance analysis in communication through a model channel with noise, filtering, and attenuation. We also consider the limitations in its performance caused by parameter mismatch between transmitter and receiver. The experimental setups we used for the analysis of CPPM performance in realistic noisy channels are described in Section 2.4. The channels we studied include a model bandlimited channel with white Gaussian noise, a low-power wireless link, and a free-space laser communication link. The last channel was characterized by the presence of severe communication signal distortions caused by atmospheric turbulence. Results of the experimental analysis are discussed in Section 2.5. Section 2.6 describes a modification of the CPPM which improves the performance of CPPM in multiuser environments.

2.2 CPPM Basics

In this section we describe the Chaotic Pulse Position Modulation system (CPPM). The CPPM method was suggested as a possible modification of the chaos-based communication approach shown in Figure 2.1 which significantly reduces the sensitivity of this communication to the channel distortions [39].

2.2.1 CPPM Principle and Operation

Consider a chaotic pulse generator that produces the chaotic pulse signal

$$U(t) = \sum_{j=0}^{\infty} w(t - t_j), \qquad (2.1)$$

where $w(t - t_j)$ represents the waveform of a pulse generated at time $t_j = t_0 + \sum_{n=0}^{j} T_n$, and T_n is the time interval between the nth and $(n-1)$th pulses. We assume that the sequence of the time intervals, T_i, represents iterations of a chaotic process. For simplicity we consider the case where chaos is produced by a one-dimensional map $T_n = F(T_{n-1})$, where $F(\ )$ is a nonlinear function. Some studies of such chaotic pulse generators can be found in [38, 42].

The information is encoded within the chaotic pulse signal by using additional delays in the interpulse intervals, T_n. As a result, the generated pulse sequence is given by a new map

$$T_n = F(T_{n-1}) + d + mS_n, \qquad (2.2)$$

where S_n is the information-bearing signal. Here we consider only the case of binary information, and therefore, S_n equals zero or one. The parameter m characterizes the amplitude of modulation. The parameter d is a constant time delay which is needed for practical implementation of our modulation and demodulation method. The role of this parameter is discussed later. In the design of the chaotic pulse generator, the nonlinear function $F(\ )$, and parameters d and m are selected to guarantee chaotic behavior of the map.

The modulated chaotic pulse signal $U(t) = \sum_{j=0}^{\infty} w(t - t_0 - \sum_{n=0}^{j} T_n)$, where T_n is generated by Eq. (2.2), is the transmitted signal. The duration of each pulse $w(t)$ in the pulse train is assumed to be much shorter than the minimal value of the interpulse intervals, T_n. To detect information at the receiver end, the decoder is triggered by the received pulses, $U(t)$. The consecutive time intervals T_{n-1} and T_n are measured and the information signal is recovered from the chaotic iterations T_n with the formula

$$S_n = (T_n - F(T_{n-1}) - d)/m. \qquad (2.3)$$

If the nonlinear function, $F(\ )$, and parameters d and m in the receiver are the same as in the transmitter, then the encoded information S_n can be easily recovered. When the nonlinear functions are not matched with sufficient precision, a large decoding error results. Therefore, an unauthorized receiver who has no information about the dynamical system producing chaotic pulses in the transmitter cannot determine whether a particular received pulse was delayed with respect to its original (chaotic) position, and thus whether S_n was "0" or "1".

Because the chaotic map of the decoder in the authorized receiver is matched to the map of the encoder in the corresponding transmitter, the time of the next arriving pulse can be predicted. In this case the input of the synchronized receiver can be blocked up to the moment of time when the next pulse is expected. The time intervals when the input to a particular receiver is blocked can be utilized by other users, thus providing a multiplexing strategy. Such selectivity based on the synchronization between the transmitter and the receiver can substantially improve the performance of the system by reducing the probability of false triggering of the decoder by channel noise.

The method described above is easy to implement in analog circuitry, and we used it in our experimental studies. However, we expect that various modifications of the detection method can be suggested to improve the system performance or simplify theoretical performance analysis. Consider one of them. When the demodulator is synchronized to the modulator, in order to decode a single bit of transmitted information the demodulator must simply determine whether a pulse from the transmitter was delayed relative to its anticipated position. If the ideal synchronization is established, but the signal is corrupted by noise, the optimal detection scheme operates as follows. Integrate the signal over the pulse duration inside the windows where pulses corresponding to "1" and "0" are expected to occur. The decision on whether "1" or "0" is received is made based upon whether the integral over "1"-window is larger or smaller than that over "0"-window. Such a detection scheme is employed in the ideal case of perfect synchronization in conventional (nonchaotic) Pulse Position Modulation (PPM) scheme. The performance of this scheme is known to be 3 dB worse than the BPSK system. Although in the case of perfect synchronization this detection scheme is ideal, according to our numerical simulations, its performance quickly degrades when synchronization errors due to the channel noise are taken into account. For this reason and for the sake of design simplicity we use a threshold detector in all our experiments and analysis.

In a noise-free environment, the arrival time of chaotic pulses can be easily registered by a variety of methods. However, in a noisy environment, the receiver can mistake a large noise fluctuation for an incoming pulse, and detect the wrong information bit. Furthermore, this false pulse can destroy chaotic synchronization, and thus prompt a sequence of errors until the receiver re-synchronizes with the transmitter. In fact, one of the advantages of using chaotic pulse generators is that the system re-acquires synchronization automatically, without any specific "hand-shaking" protocol. The decoder only needs to detect two correct consecutive pulses in order to re-establish synchronization. We studied the bit-error performance in a noisy environment both theoretically and experimentally. The results of these studies are presented below.

2.2.2 CPPM BER Performance Evaluation

We characterize the performance of our system by studying the dependence of the bit-error rate on the ratio of energy per one transmitted bit to the spectral density of noise, E_b/N_0. This dependence is shown in Figure 2.3, where it is compared to the performance of more traditional communication schemes, BPSK, PPM, and noncoherent FSK.

We can obtain a rough analytical estimate of the CPPM BER performance of our detection scheme in the case of rectangular pulse shape. In order to do so, let us consider a simplified model of our detection method. In the detector the signal that is a sum of the transmitted pulse signal and WGN is low-pass filtered and is applied to a threshold element. Let us assume that the low-pass filter can be approximated by the running average filter:

$$y(t) = \frac{1}{\tau} \int_{t-\tau}^{t} x(\xi)d\xi.$$

Let the windows where the pulses corresponding to "1" and "0" have the same duration, T; see Figure 2.2. We assume that the receiver maintains synchronization at all times, so that every "0"-pulse is within the "0"-window, $0 < t \le T$, and every "1"-pulse is within the "1"-window, $T < t \le 2T$. Let us divide the interval where "0"-pulse is expected into bins of duration $1/f$, where f is the filter cut-off frequency. We assume that $f = 1/\tau$, τ being the pulse duration, and that when a pulse arrives, it is contained entirely within one bin. We sample the output $y(t)$ from the filter once at the end of every bin.

In our model detection scheme the threshold element is set off when the
the output from one of the bins is larger than the threshold. If the threshold
is crossed where a pulse corresponding to "0" is expected, a "0" is detected,
otherwise a "1" is detected by default.

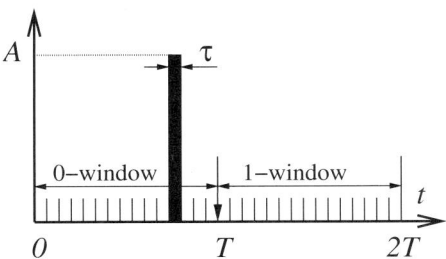

Fig. 2.2. The illustration of the detection scheme. (Reprinted with permission from
[43], ©2001 IEEE.)

Let A be the pulse amplitude, H, the threshold value, and σ^2, the noise
variance at the filter output.

First, we evaluate the error probability when "1" is transmitted, $P_{0|1}$. This
probability can be found from $P_{1|1} + P_{0|1} = 1$, where $P_{1|1}$, the probability to
correctly detect "1" can be easily found. It is the probability that the filter
output, y_i, from any bin in the "0" does not exceed the threshold. Using the
statistical independence of the measurements for each window in the case of
white noise, we can write:

$$P_{1|1} = \prod_{i=1}^{T/\tau} p_i(y_i < H) = [p(y < H)]^{T/\tau} = \left[\frac{1}{\sqrt{2\pi\sigma^2}} \int_{-\infty}^{H} \exp\left(-\frac{x^2}{2\sigma^2}\right) dx \right]^{T/\tau}$$

$$= \left[\frac{1}{2}\left(1 + \mathrm{erf}\left(\frac{H}{\sqrt{2\sigma^2}}\right)\right) \right]^{T/\tau} = \left[\frac{1}{2}\left(1 + \mathrm{erf}\left(h\sqrt{\frac{E_b}{N_0}}\right)\right) \right]^{T/\tau}.$$

Here we introduced the relative threshold value, $h = H/A$, the energy per
bit, $E_b = A^2\tau$, and the spectral power density of noise, $N_0 = 2\sigma^2\tau$.

The probability to detect "0" when "1" is transmitted is then:

$$P_{0|1} = 1 - P_{1|1} = 1 - \left[\frac{1}{2}\left(1 + \mathrm{erf}\left(h\sqrt{\frac{E_b}{N_0}}\right)\right) \right]^{T/\tau}.$$

The error probability in the case when "0" is transmitted can be found
similarly. The error occurs when the output from all bins in the "0" window
remains lower than the threshold, despite the fact that the transmitted signal
is nonzero within one of them:

$$P_{1|0} = \left[\frac{1}{\sqrt{2\pi\sigma^2}} \int_{-\infty}^{H} \exp\left(-\frac{x^2}{2\sigma^2}\right) dx \right]^{T/\tau - 1}$$

$$\times \left[\frac{1}{\sqrt{2\pi\sigma^2}} \int_{-\infty}^{H} \exp\left(-\frac{(x-A)^2}{2\sigma^2}\right) dx \right]$$

$$= \left[\frac{1}{2}\left(1 + \mathrm{erf}\left(\frac{H}{\sqrt{2\sigma^2}}\right)\right) \right]^{T/\tau - 1} \times \left[\frac{1}{2}\left(1 + \mathrm{erf}\left(\frac{H-A}{\sqrt{2\sigma^2}}\right)\right) \right].$$

Rewriting the last part of the equation in terms of the relative threshold value h, the energy per bit E_b, and the spectral power density of noise N_0 one obtains

$$P_{1|0} = \left[\frac{1}{2}\left(1 + \mathrm{erf}\left(h\sqrt{\frac{E_b}{N_0}}\right)\right) \right]^{T/\tau - 1} \times \left[\frac{1}{2}\mathrm{erfc}\left((1-h)\sqrt{\frac{E_b}{N_0}}\right) \right].$$

The overall error probability is the combination of $P_{1|0}$ and $P_{0|1}$: $BER = p_1 P_{0|1} + (1 - p_1)P_{1|0}$ where p_1 is the characteristic of the data stream which is the ratio of "1"s in it. In our experiment both p_1 and h were equal to $1/2$. In this case the expression for the BER can be written in a shorter form:

$$BER = \frac{1}{2}\left(1 - \mathrm{erf}\left(\frac{\sqrt{\epsilon}}{2}\right)\left(\frac{1}{2}\left(1 + \mathrm{erf}\left(\frac{\sqrt{\epsilon}}{2}\right)\right)\right)^{T/\tau - 1}\right),$$

where $\epsilon = E_b/N_0$

Figure 2.3 shows the BER performance of the CPPM as estimated analytically, computed in numerical simulations with $h = 0.5$, $p_1 = 0.5$, and $T/\tau = 10$ and measured in the experiment. The difference of approximately 1 dB between the analytical and numerical curves can be largely attributed to burst errors arising from the loss of synchronization, which is not taken into account in our model. Details on the experimental setup used to obtain the experimental BER data are given in Section 2.4.

2.3 CPPM implementation

2.3.1 CPPM Implementation

The implementation of the chaotic pulse modulator used in our experiments is illustrated in Figure 2.4. The Integrator produces a linearly increasing voltage, $V(t) = \beta^{-1}(t - t_n)$, at its output. At the Comparator this voltage is compared with the threshold voltage produced at the output of the nonlinear converter $F(x)$. The threshold level $F(V_n)$ is formed by a nonlinear conversion of voltage $V_n = V(t_n)$ which was acquired and saved from the previous iteration using sample and hold (S&H) circuits. When voltage $V(t)$ reaches this threshold

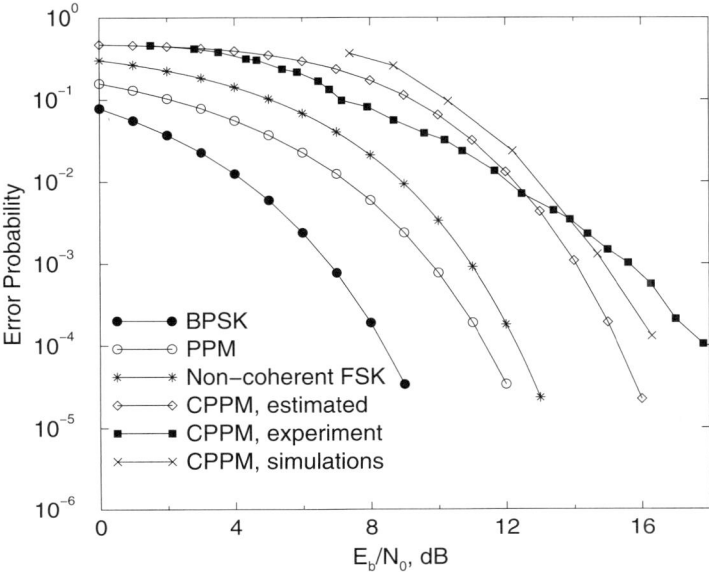

Fig. 2.3. Error probabilities of ideal BPSK, noncoherent FSK, and ideal PPM systems compared to the performance of the CPPM system. (Reprinted with permission from [43], ©2001 IEEE.)

level, the comparator triggers the Pulse Generator I. It happens at the moment of time $t'_{n+1} = t_n + \beta F(V_n)$. The generated pulse (Chaotic Clock Signal) causes the Data Generator to update the transmitted information bit. Depending on the information bit transmitted, S_{n+1}, the Delay Modulator delays the pulse produced by the Pulse Generator by the time $d + mS_{n+1}$. Therefore the delayed pulse is generated at the moment of time $t_{n+1} = t_n + \beta F(V_n) + d + mS_{n+1}$. Through the sample and hold circuit this pulse first resets the threshold to the new iteration value of the chaotic map $V(t_{n+1}) \rightarrow F(V(t_{n+1}))$, and then resets the integrator output to zero, $V(t) = 0$. The dynamics of the threshold value is determined by the shape nonlinear function $F(\ )$. The spacing between the nth and $(n+1)$th pulses is proportional to the threshold value V_n, which is generated according to the map

$$T_{n+1} = \beta F(\beta^{-1} T_n) + d + mS_{n+1}, \tag{2.4}$$

where $T_n = t_{n-1} - t_n$, and S_n is the binary information signal. In the experimental setup the shape of the nonlinear function was built to have the following form

$$F(x) \equiv \alpha f(x) = \begin{cases} \alpha x & \text{if } x < 5V, \\ \alpha(10V - x) & \text{if } x \geq 5V. \end{cases} \tag{2.5}$$

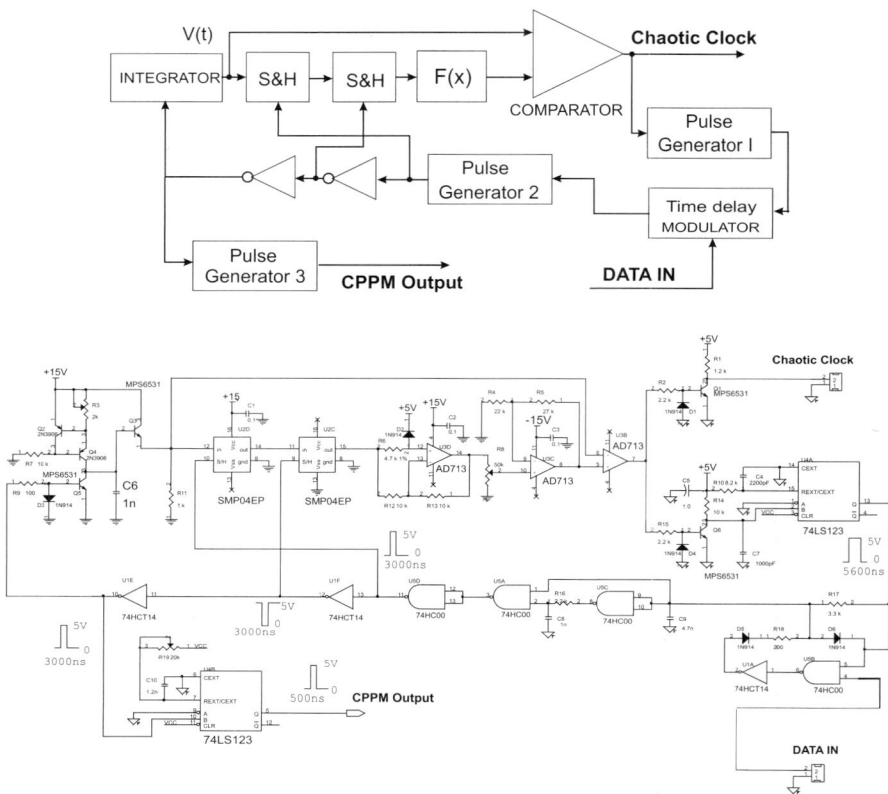

Fig. 2.4. Block diagram (top panel) and schematics (bottom panel) of the chaotic pulse modulator.

The selection of the nonlinearity in the form of a piecewise linear function helps to ensure the robust regimes of chaos generation for rather broad ranges of parameters of the chaotic pulse position modulator.

The position-modulated pulses, $w(t - t_j)$, are shaped in the Pulse Generator II. These pulses form the output signal $U(t) = \sum_{j=0}^{\infty} w(t - t_j)$, which is transmitted to the receiver.

The demodulator scheme is illustrated in Figure 2.5. In the receiver the Integrator, S&H circuits, and the nonlinear function block generating the threshold values are reset or triggered by the pulse received from the transmitter rather than by the pulse from the internal feedback loop. To be more precise, they are triggered when the input signal, $U(t)$, from the channel exceeds a certain input threshold. The time difference between the anticipated location of the pulse without modulation, $t'_{n+1} = t_n + \beta F(V_n)$, and the actual arrival time t_{n+1} translates into the difference between the threshold value,

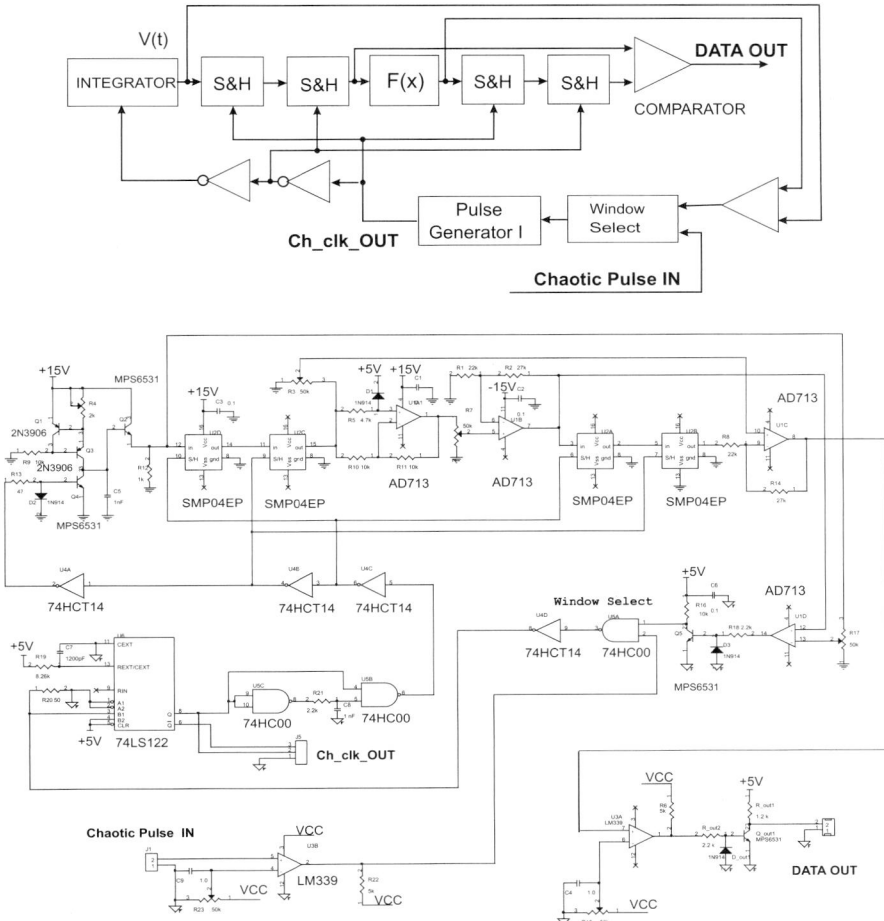

Fig. 2.5. Block diagram (top panel) and schematics (bottom panel) of the chaotic pulse demodulator.

$F(V_n)$ generated by the nonlinear function and the voltage, $V(t_{n+1})$ at the Integrator at the moment when the input signal $U(t)$ exceeds the input threshold. For each received pulse the difference $V(t_{n+1}) - F(V_n)$ is computed and is used for deciding whether the pulse was delayed. If this difference is less than the reference value $\beta(d + m/2)$, the detected data bit S_{n+1} is "0"; otherwise it is "1".

Another important part of the receiver is the Window Selection block. Once the receiver correctly observes two consecutive pulses, it can predict the earliest moment of time when it can expect to receive the next pulse. This means that we can block the input to the demodulator circuit until shortly before such a moment. This is done by the Window Select block. In the experiment, this circuit opens the receiver input at the time $t'_{n+1} = t_n + \beta F(V_n)$ by Window Control pulses generated by the Comparator (see Figure 2.5). The input stays open until the decoder is triggered by the first pulse received. Using such windowing greatly reduces the chance of the receiver being triggered by noise, interference, or pulses belonging to other users, however, it may increase the time necessary to re-acquire synchronization after a string of erroneous pulses.

2.3.2 Parameters Mismatch Limitations

It is known that because the synchronization-based chaotic communication schemes rely on the identity of synchronous chaotic oscillations, they are susceptible to negative effects of parameter mismatches. Here we evaluate how precisely the parameters of our modulator and demodulator have to be tuned in order to ensure errorless communication over a distortion-free channel.

The information detection in our case is based on the measurements of time delays, therefore it is important that the modulator and the demodulator can maintain synchronous time reference points. The reference point in the modulator is the front edge of the chaotic clock pulse. The reference point in the demodulator is the front edge of the window control pulse. Ideally, if the parameters of the modulator and the demodulator were exactly the same and the systems were synchronized, then both reference points would be always at the times $t'_{n+1} = t_n + \beta F(V_n)$, and the received pulse would be delayed by the time d for $S_{n+1} = 0$ and $d + m$ for $S_{n+1} = 1$. In this case, setting the bit separator at the delay $d + m/2$ would guarantee errorless detection in a noise-free environment.

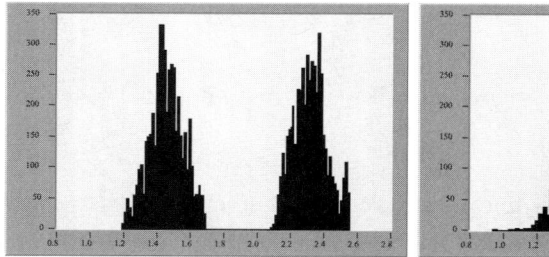 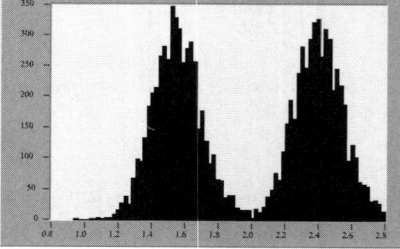

Fig. 2.6. Histograms of the fluctuations of the received pulse positions with respect to the receiver reference point: noise-free channel (left) and channel with WGN $E_b/N_o \sim 18$ dB (right). (Reprinted with permission from [43], ©2001 IEEE.)

In an analog implementation of a chaotic pulse position modulator and demodulator system, the parameters of the circuits are never exactly the same.

Therefore, the time positions $t_n'^{(M)}$ and $t_n'^{(D)}$ of the reference points in the modulator and the demodulator chaotically fluctuate with respect to each other. Due to these fluctuations the position of the received pulse, $t_n = t_n'^{(M)} + d + S_n$, is shifted from the arrival time predicted in the demodulator, $t_n'^{(D)} + d + S_n$. The errors are caused by the following two factors. First, when the amplitude of fluctuations of the position shift is larger than $m/2$, some delays for "0"s and "1"s overlap and cannot be separated. Second, when the fluctuations are such that a pulse arrives before the demodulator opens the receiver input ($t_n < t_n'^{(D)}$), the demodulator skips the pulse, loses synchronization and cannot recover the information until it resynchronizes. In our experimental setup the parameters $\beta_{M,D}$ were tuned to be as close as possible, and the nonlinear converters were built using 1% components. The fluctuations of the positions of the received pulses with respect to the window control pulse were studied experimentally by measuring time delay histograms. Figure 2.6 presents typical histograms measured for the case of a noise-free channel and for a channel with noise when $E_b/N_o \sim 18$ dB.

Assuming that systems were synchronized up to the $(n-1)$ pulse in the train, the fluctuations of the separation between the reference time positions equal

$$\Delta_n \equiv t_n'^{(D)} - t_n'^{(M)} =$$
$$\beta_D F_D(\beta_D^{-1} T_{n-1}) - \beta_M F_M(\beta_M^{-1} T_{n-1}), \tag{2.6}$$

where indices D and M stand for demodulator and modulator, respectively. As discussed above, in order to achieve errorless detection, two conditions should be satisfied for all time intervals in the chaotic pulse train produced by the modulator. These conditions are the synchronization condition, $\{\Delta_n\}_{max} < d$, and the detection condition $\{|\Delta_n|\}_{max} < m/2$. As an example we consider the simplest case where all parameters of the systems are the same except for the mismatch of the parameter α in the nonlinear function converter; see Eq. (2.5). Using Eq. (2.5) and Eq. (2.6) the expression for the separation time can be rewritten in the form

$$\Delta_n = (\alpha_D - \alpha_M)\beta f(\beta^{-1} T_{n-1}). \tag{2.7}$$

It is easy to show that the largest possible value of the nonlinearity output $f(\ )$, which can appear in the chaotic iterations of the map, equals 5 V. Note that in the chaotic regime only positive values of $f(\ )$ are realized. Therefore, if conditions

$$\beta(\alpha_D - \alpha_M) < d/5V \quad \text{and} \quad 2\beta|\alpha_D - \alpha_M| < m/5V. \tag{2.8}$$

are satisfied and there is no noise in the channel, then information can be recovered from the chaotic pulse train without errors.

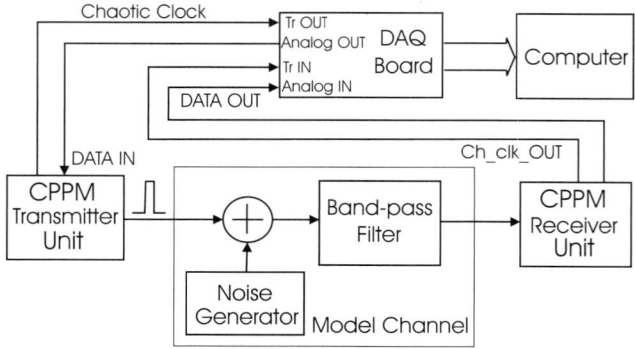

Fig. 2.7. Diagram of the experiment.

2.4 Experimental Studies of CPPM with Channel Distortions

2.4.1 CPPM Experiment with Model Channel

For the experimental evaluation of BER performance in a bandlimited noisy channel we transmitted the CPPM signal mixed with white Gaussian noise over a channel modeled with an active bandpass filter (Figure 2.7). We used a computer with a data acquisition board as the data source. Each update of the information bit loaded to the CPPM encoder was triggered by the chaotic clock generated in the transmitter. The computer calculated the displacement of the received chaotic pulse from the demodulator subtractor and detected the received information bit using a threshold element.

The model channel circuit consisted of a WGN generator and a bandpass filter with the pass band 1 kHz-500 kHz. The pulse duration was 500 ns. The distance between the pulses varied chaotically between 12 μs and 25 μs. This chaotic pulse train carried the information flow with the average bit rate ~60 kb/sec. The amplitude of pulse position modulation m was 2 μs. More details on the spectral characteristics of CPPM signals in passing through this model channel can be found elsewhere [43].

2.4.2 Wireless Implementation of CPPM

We have implemented and tested a low-power prototype of a wireless digital communication link using the chaotic pulse modulator and demodulator described above. The circuit diagrams of the transmitter and receiver units, which were added to the CPPM modulator and demodulator blocks to provide the wireless link, are shown in Figures 2.8 through 2.11.

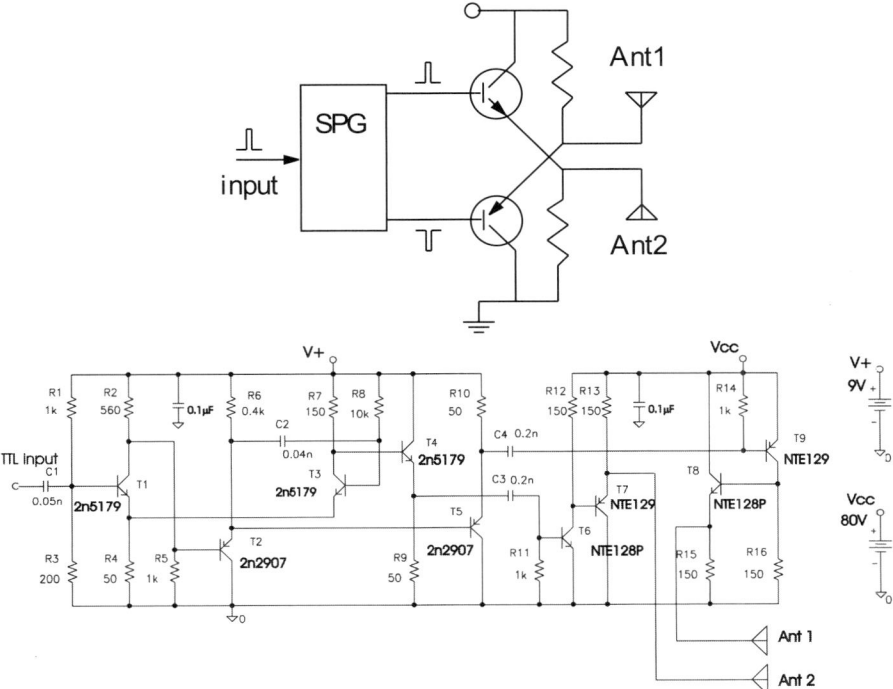

Fig. 2.8. Block-diagram (top panel) and schematics (bottom panel) of the ultra-wideband transmitter used in the wireless CPPM communication system.

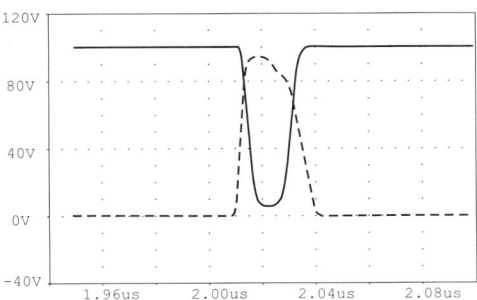

Fig. 2.9. Voltage pulse on the transmitter antennas Ant1 (solid line) and Ant2 (dotted line) in the exciter circuit, shown in Figure 2.8 with Vcc = 100 V. Plots are calculated using a MicroSim circuit simulator.

Transmitter

The transmitter contains the generator of a pair of symmetric short pulses (SPG) and the symmetric exciter for the two-pole antenna; see Figure 2.8. SPG is triggered by a TTL pulse (500 ns) from the output of the CPPM modulator (see Figure 2.4) and generates a short pulse (20 ns measured at a half of the amplitude). This short pulse is applied to the symmetric exciter that has two outputs; see Figure 2.9. The first output produces a high voltage positive pulse. The second output produces a high voltage negative pulse. These two outputs are applied to the corresponding poles of the antenna.

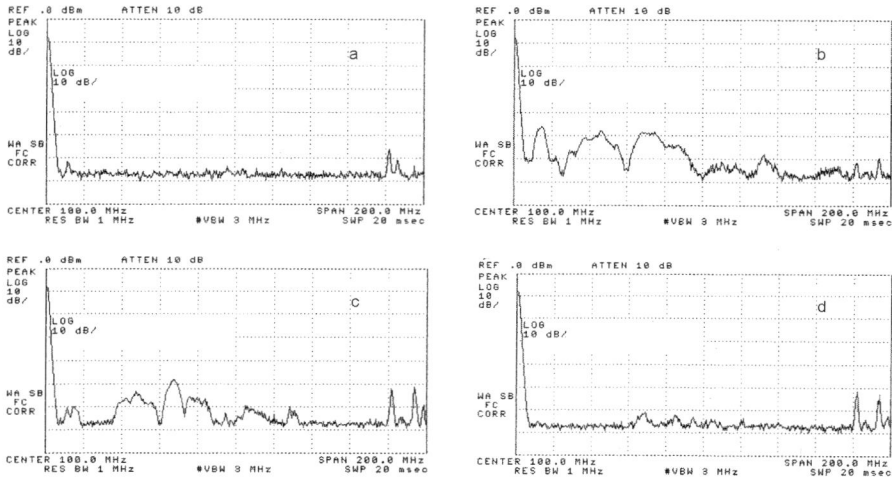

Fig. 2.10. Power spectral density of signals measured in the frequency range from 0 to 200 MHZ. The background noise measured with the CPPM transmitter turned off is shown in (a). Panels (b), (c), and (d) present the spectra of the CPPM signal measured at the distance L from the CPPM transmitting antenna: L = 1 m (b); L = 3 m (c) and L = 10 m (d). (Reprinted with permission from [43], ©2001 IEEE.)

Receiver

The block diagram of the receiver is presented in Figure 2.11. The electromagnetic pulses received by the two-pole antenna (Ant1 and Ant2) are applied to the differential amplifier whose output is then amplified in the two-stage amplifier (Amp) and sent to the symmetric detector circuit. The output of the symmetric detector is then amplified by the second differential amplifier to generate enough voltage for triggering the pulse generator (PG) which produces the TTL pulse of the duration about 500 ns. The TTL pulse is then applied to the CPPM demodulator block shown in Figure 2.5.

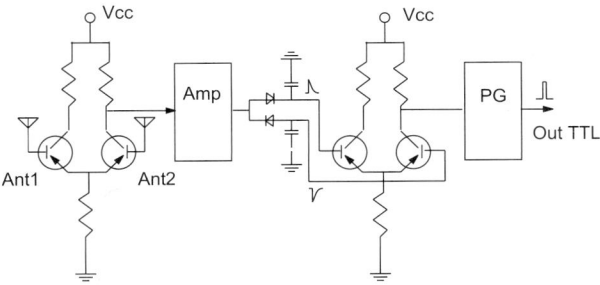

Fig. 2.11. Block diagram of the pulse receiver. (Reprinted with permission from [43], ©2001 IEEE .)

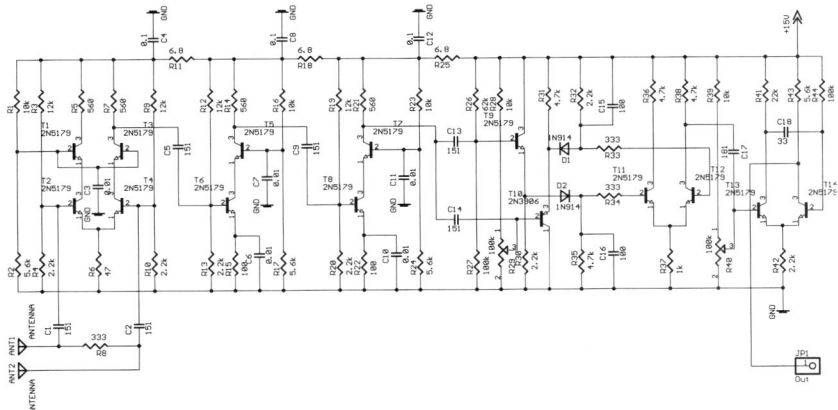

Fig. 2.12. Schematics of the ultra-wideband receiver used in the wireless CPPM system.

We tested our prototype wireless CPPM system in indoor experiments with separation between transmitter and receiver of up to 15 m and were able to achieve a stable regime of error-free communication.

2.4.3 Optical Chaos Communication Through Turbulent Atmosphere

The ability of the self-synchronizing CPPM method to communicate in the presence of significant nonstationary signal distortions in the channel has been studied experimentally using a free-space laser communication link [44]. Here the communication carrier signal consists of a sequence of optical pulses that travel through air. The characteristics of the communication channel are thus determined by the optical properties of air and the severe communication signal distortions in this experiment are a result of atmospheric turbulence.

A schematic representation of the chaotic free-space laser communication system is shown in Figure 2.13. In this experiment the CPPM modulator circuit described in Section 2.3.1 was used to modulate the output-intensity of a semiconductor laser ($\lambda = 690$ nm). The resulting intensity-modulated 10 mW semiconductor laser beam was coupled into a single-mode fiber. The beam emanating from the fiber was first expanded to a 4 in. diameter using a lens relay system (lenses L_1, L_2) and a Celestron transmitter-telescope and was then directed through a 26 ft long vertical air-locked pipe to a 45° mirror placed inside a shed on the roof of the building. From there the light propagated over an atmospheric path of length $L \simeq 2.5$ km to a 4 in. corner cube reflector placed on top of a water tower. After reflection the laser beam propagated from the water tower back to a communication receiver telescope on the roof-mounted shed. The receiver system used the same Celestron telescope and lens relay system (lenses L_1 and L_3) as did the transmitter system. The total double-pass atmospheric laser beam propagation distance was approximately $2L \simeq 5$ km long. The received light power was registered by the PIN photo-detector (PDA55) placed in the lens L_3 focal plane, amplified by the low-noise preamplifier (SR560 with a gain of 20), and the resulting signal served as input to the CPPM demodulator circuit (see Section 2.3.1).

2.5 CPPM Performance and Features

2.5.1 Experimental BER Performance Evaluation

We experimentally evaluated the BER performance of CPPM system using the model channel setup shown in Figure 2.7. We measured the dependence of BER on the ratio E_b/N_0, where E_b is the energy of CPPM signal per bit and N_0 is the spectral density of noise. This dependence is plotted in Figure 2.3 along with the analytical and numerical estimates that we discussed in Section 2.2.

The slightly better than expected performance of the experimental system at high levels of noise can be explained by the observed significant deviations of the noise distribution from Gaussian at high noise amplitudes. The quicker than expected degradation of performance at low levels of noise is primarily due to the small mismatch of the parameters in the transmitter and the receiver. Still, considering the crudeness of the analytical model and the experimental difficulties, all three plots agree reasonably well.

2.5.2 Effects of Atmospheric Turbulence in an Optical Chaos Communication Experiment

We tested the self-synchronizing CPPM communication scheme in an experiment where information was transmitted over a ~5 km laser link through

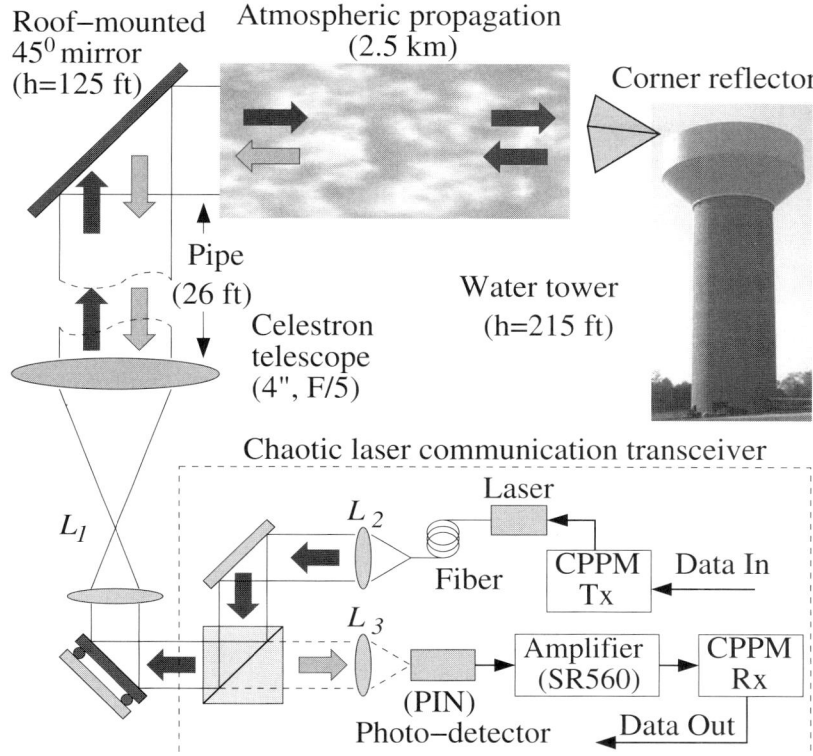

Roof–mounted
45^0 mirror
(h=125 ft)

Atmospheric propagation
(2.5 km)

Corner reflector

Pipe
(26 ft)

Celestron
telescope
(4", F/5)

Water tower
(h=215 ft)

Chaotic laser communication transceiver

L_1

Laser

L_2

Fiber

CPPM
Tx

Data In

L_3

(PIN)
Photo–detector

Amplifier
(SR560)

CPPM
Rx

Data Out

Fig. 2.13. Schematic for the free-space laser communication system based on the CPPM transceiver. Details are given in the text. (Reprinted with permission from [44], ©2002 American Physical Society.)

turbulent air. Turbulence in the atmospheric communication channel leads to severe laser beam intensity scintillations that result in deep fluctuations of the received communication signal (received laser beam power). The most dele-terious effects from receiver plane scintillations are the loss of signal-to-noise ratio and drop-outs (information loss).

To demonstrate the level of intensity scintillations in the atmospheric chan-nel, we measured the received signal from a continuously running laser with constant output intensity. Figure 2.14a shows the fluctuations of the normal-ized received power, measured by the PIN photodetector placed in the lens L_3 focal plane (Figure 2.13), amplified by the low noise preamplifier (SR500 with gain of 20), and then acquired with sampling rate 1000 samples/sec. The corresponding ensemble-averaged received signal power spectrum $S(f)$ is shown in Figure 2.14b. The severeness of the signal distortions is clearly visible in Figure 2.14 and is quantified by an estimated standard deviation of the normalized received signal as high as 0.8-0.9, which is indicative of a

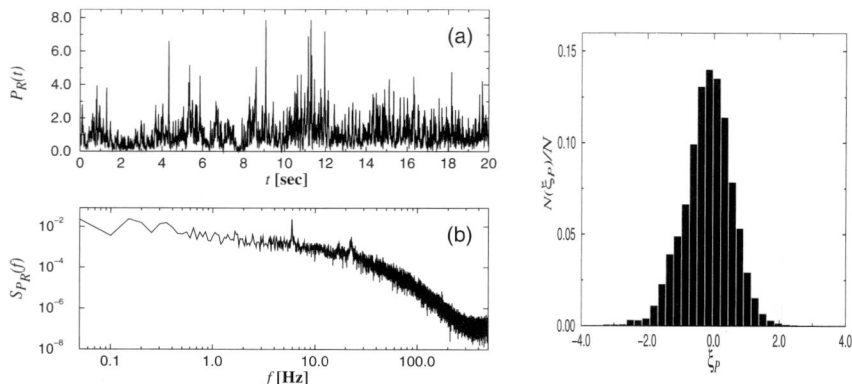

Fig. 2.14. Fluctuations of the received power $P(t)$ in the experiment with a non-modulated laser generating a constant output intensity (10 mW). The normalized received power $P(t)/\langle P(t) \rangle$ measured at the photo-detector output (a), and the corresponding averaged power spectrum (b) illustrate the presence of strong laser beam intensity scintillations. (c) Histogram of the probability distribution for the random variable $\ln(P/\langle P \rangle)$. (Reprinted with permission from [44], ©2002 American Physical Society.)

strong scintillation regime. In atmospheric optics the laser beam scintillations are traditionally described in terms of the distribution of the logarithm of the received power (for finite receiver telescope): $\ln(P/\langle P \rangle)$, where $\langle P \rangle$ is the ensemble (time) averaged value [45]. The histogram of the random variable $\ln(P/\langle P \rangle)$ is shown in Figure 2.14c. Representing an approximation to the probability distribution of the received power the histogram closely matches the log-normal distribution that theory predicts for turbulent atmosphere [45].

In the communication experiment the CPPM modulator described in Section 2.3.1 (denoted by CPPM Tx in Figure 2.13) generates a sequence of TTL pulse signals that are used to trigger the laser resulting in a chaotic sequence of short-term ($\sim$1.0 μs) pulses of light intensity. The interpulse intervals $\{T_n\}$ fluctuated chaotically ranging from 10 μsec to 25 μsec and supported a $\sim$60 kbit per sec bit-rate. After the double-pass propagation through air the distorted light pulses are detected by the PIN photo diode, the output of which is applied to the CPPM demodulator circuit (CPPM Rx in Figure 2.13). If the output exceeds a certain input threshold, which in our experiment was set to $\sim$200 mV, the timer circuit in CPPM Rx is triggered and the information signal is recovered from the chaotic iterations of interpulse intervals, $\{T_n\}$, using formula (2.3).

The CPPM communication method encodes information in the interpulse intervals $\{T_n\}$ and can therefore tolerate strong signal distortions and amplitude variations like the ones caused by atmospheric turbulence as long as fluctuations of the propagation time in the turbulent channel remain small. Figure 2.15a illustrates the severe pulse amplitude fluctuations of the signal

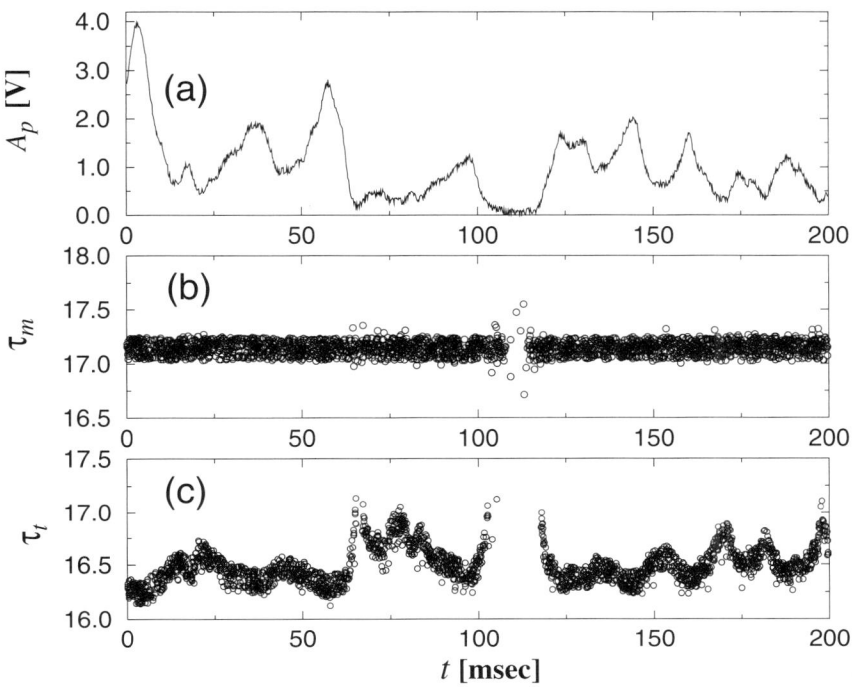

Fig. 2.15. Fluctuations of the CPPM pulses of light intensity after traveling through atmospheric turbulence. Pulse amplitude A_p measured in volts at the output of amplifier (a). Propagation times τ_m (b) and τ_t (c) in μsec. The pulse propagation times are computed from data acquired simultaneously at the output of CPPM Tx and output of Amplifier (SR560) at a sampling rate of 5×10^6 samples per sec. (Reprinted with permission from [44], ©2002 American Physical Society.)

entering the CPPM demodulator. Figure 2.15b shows the pulse propagation time τ_m, which is measured between the leading front of the TTL pulse applied to the laser and the maximal point of the received pulse. It varied only within a 0.2 μsec time interval. However, in order to trigger CPPM Rx the received pulse amplitude has to exceed a certain threshold level. This level ($\sim$200 mV) was selected to minimize instances of receiver controller triggering caused by noise, or by pulses originating from local pulse reflections off nearby optical surfaces. Therefore, the actual delay time τ_t, measured between the leading front of TTL pulses generated by CPPM Tx and the moments of CPPM Rx triggering, depends on the amplitude of the received pulses and fluctuates; see Figure 2.15c. Although τ_t changes with the amplitude variation these changes remain less than the modulation amplitude $m \sim 1.5\mu$sec; see Eq. (2.2). Thus, the variations of the pulse propagation time are small enough for the CPPM controller to self-synchronize and to maintain the stability of the communication link.

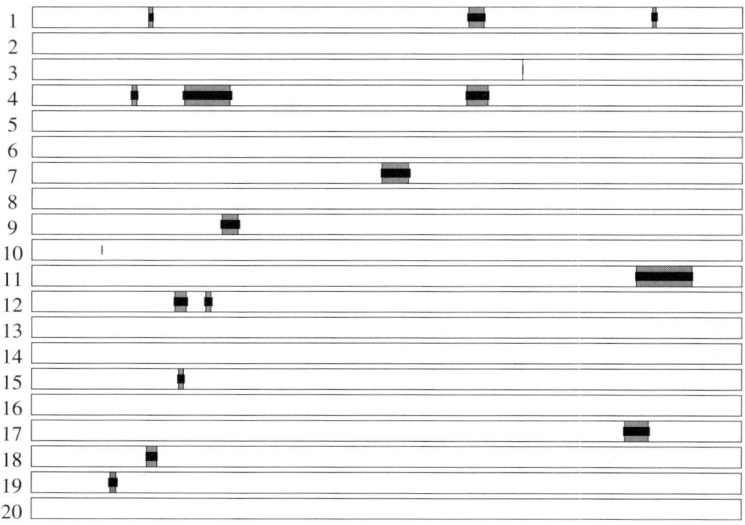

Fig. 2.16. Typical structure of errors shown in 20 consecutive measured data streams each of length ∼170 msec transmitted at ∼2 min intervals. Each strip presents 10,000 bits which are transmitted with the CPPM method. White intervals of the strips mark blocks of data received without errors. Narrow black ribbons in the middle of strips mark the blocks of the data received with errors. The gray background shows the blocks of the dropped-out data caused by the loss of CPPM pulses due to fading instances.

The gaps visible in Figures 2.15b and 2.15c are caused by pulse amplitude fading when the pulse amplitude falls to the photo-receiver noise level and below threshold level, respectively. The gaps are audible as occasional clicks, when using the free-space laser communication system described here for real-time voice communication, which was implemented by digitizing the output of a microphone with a delta-modulator and transmitting the binary signal. Except for these short-term (less than 50 msec) drop-offs, the voice communication was stable and clear.

Figure 2.16 shows an example of a map of communication errors that resulted from the transmission of binary pseudo-random data. The total BER of 1.92×10^{-2} measured in this experiment has three contributions. First, the loss of bits carried by the pulses that did not trigger the CPPM receiver due to the fading in the channel contributes $\sim 1.78 \times 10^{-2}$ to the BER ($\sim 92.7\%$). The fading moments occur randomly during the communication and cause the drop-outs of blocks of data up to 1000 consecutive bits. Second, errors that occur within time intervals immediately before and after the failure of communication by fading contribute $\sim 1.4 \times 10^{-3}$ to the BER ($\sim 7.3\%$). In these time intervals the amplitude of the received pulses is still close to the threshold and, as a consequence, even small noise in the channel can result in significant fluctuation of the interpulse intervals (see Figure 2.15c). Third, errors that are not related to the complete failure of the channel because of fading instances contributed to the BER only $\sim 5.5 \times 10^{-5}$.

This structure of errors indicates that the CPPM communication method supports robust communication over a turbulent channel except for instances when the communication link fails due to fading. Thanks to the self-synchronizing feature of the CPPM method and the fact that the CPPM receiver needs to obtain just two consecutive correct pulses to re-establish the regime of chaos synchronization, the communication after drop-out events is re-established almost immediately.

2.5.3 Wireless CPPM and Low Probability of Detection

In the experiments with wireless CPPM, we studied spectral characteristics of the chaotic pulse signal radiated by the wireless CPPM. The Power Spectral Density (PSD) was measured with HP 8590A Portable RF Spectrum Analyzer using a two-pole receiving antenna. The results of the measurements are presented in Figure 2.10. The PSD of the signal received by the RF spectrum analyzer, when the CPPM transmitter is turned off, is shown in Figure 2.10a. The spectrum was measured in the range of frequencies from 1 MHz to 200 MHz with the video bandwidth filter of 3 MHz. The same measurements were done when the CPPM transmitter was turned on. The amplitude of the pulse measured across the poles of the transmitting antenna is about 130 V. Duration of the pulse is about 30 nsec. Figure 2.10 presents measured PSD of the received RF signal at two distances from the transmitting antenna (1 and 10 m). As one can see, already at 10 m the signal PSD is at the level or below the background RF noise. This shows a potential for using CPPM systems in applications requiring low probability of detection.

Here we use the results of the experimental analysis of radiated pulse signals to illustrate the advantages of CPPM over more conventional communications schemes in the area of low probability of detection. The existing communications schemes rely on digitally generated pseudo-random sequences to eliminate from transmitted signals the features that allow an adversary to detect and intercept the transmission. This approach has two intrinsic shortcomings.

First, the pseudo-random sequences eventually repeat, and second, the digital character of the generation algorithm introduces into the signal features associated with the corresponding quantization. These two points are illustrated in Figure 2.17, where we show the spectrum of a pseudo-random pulse sequence transmitted by the same method as in the CPPM transmission. The transmitted signal consisted of a periodically repeated sequence of 128 pulses

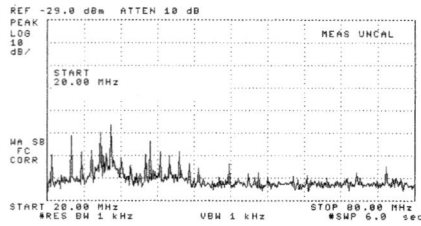

Fig. 2.17. The spectrum of a transmission of a pulse train with pseudo-random pulse timing (left panel). The reference spectrum of the background (right panel). The start frequency is 20 MHz and the stop frequency is 80 MHz. (Reprinted with permission from [43], ©2001 IEEE.)

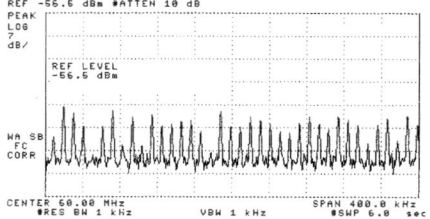
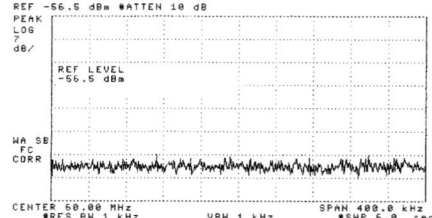

Fig. 2.18. The spectrum of a transmission of a pulse train with pseudo-random pulse timing (the same as in Figure 2.17), but measured at a finer frequency resolution (left panel). A typical spectrum of a CPPM transmission measured with the same resolution (right panel). (Reprinted with permission from [43], ©2001 IEEE.)

with the timing determined by the raising edges of a pseudo-random sequence of the length of 511 bits.

In Figure 2.17 (left panel) one can clearly see the periodicity due to the quantization in the pseudo-random timing sequence, with the chip rate ~ 5.4 MHz. The peaks in Figure 2.18 (left panel) are due to the periodic repetition of the pseudo-random sequence, with the characteristic frequency ~ 10.5 kHz. Either of these features can be exploited in order to detect such transmission. For comparison, in Figure 2.18 (right panel) we see the spectrum of a chaotic pulse train, with the same frequency resolution as in Figure 2.18 (left panel). CPPM avoids both sources of periodicity present in the pseudo-random pulse train transmission, and, as expected, its spectrum does not show the corresponding peaks.

More complicated methods[1] used for detection of pseudo-random pulse transmissions can be applied in order to discover chaotic pulse transmission.

---

[1] We should point out that such simple a and apparent scheme as observing a pulse sequence on an oscilloscope is not very efficient in a noisy environment, with band-pass filtering due to antennas. Many (or all, if there is no transmission) pulses will appear due to noise.

These methods, however, become much more efficient if one can recover a time reference, which in the case of CPPM transmission is more difficult.

2.6 Multiuser Extension of CPPM

Direct application of CPPM in a multiuser environment leads to a significant performance degradation. If multiple transmitters are operating at the same time, receiver "A" can capture pulses from other transmitters which would occasionally fit into the reception windows of the receiver, thus creating errors in the bit detection, and moreover, causing synchronization breakdowns. To reduce the probability of these events, we propose to send a fixed group of pulses instead of a single pulse. The structure of the pulse train should be unique for a given user, and the transmission time for the train as before is determined by the chaotic map. The detection of the pulse train arrival is achieved by the matched filter (correlation detector), thus providing selectivity and processing gain. The output of the correlator is then processed in the receiver in the same way as a single pulse is processed in the original scheme shown in Figure 2.5.

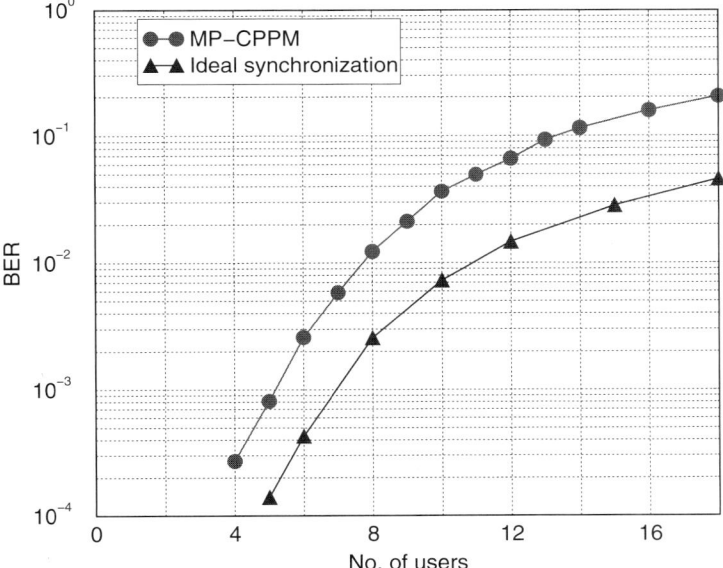

Fig. 2.19. Bit-error rate as a function of the number of users in a multipulse CPPM scheme. Circles correspond to the MP-CPPM, and triangles correspond to the standard (periodic) PPM scheme. (Reprinted with permission from [43], ©2001 IEEE.)

We tested this scheme in numerical simulations with up to 20 users. The pulse train patterns were chosen to minimize the maximum cross-correlation between different pulse trains. Chaotic intervals between the pulse trains were generated by the tent map with the slope 1.3. Then at the transmitter, the pulse train was produced either at time $T_n - \delta$ or $T_n + \delta$, depending on the value (0 or 1, respectively) of bit b_n being transmitted. Unlike the single-pulse CPPM, the detection scheme is based on the position of the maximum output of the correlator within a certain window with respect to the nominal (determined by the chaotic clock) position T_n. If the pulse train arrival time is closer to $T_n - \delta$ than to $T_n + \delta$, bit 0 is registered, and otherwise bit 1 is registered. We also employed an adjustable window size depending on the magnitude of the output from the matched filter, so when the signal is weak (synchronization is lost), the window becomes large in order to re-establish synchronization. Figure 2.19 shows the bit-error rate as a function of the number of users for 50-pulse trains. As was mentioned above, the performance of this system is degraded by occasional desynchronization events. For comparison, a corresponding plot for an ideally synchronized chaos oscillator at the receiver and transmitter is shown. As can be seen, the difference between these graphs is approximately 25% in terms of the number of users.

2.7 Conclusions

Discussing chaos-based communication systems, one may notice a potential disadvantage common to all such schemes utilizing synchronization. Most traditional schemes are based on periodic signals and carrier waveforms stored at the receiver and not transmitted through the channel. All such systems are characterized by zero Kolmogorov-Sinai entropy h_{KS} [46]: in these systems without any input the average rate of nonredundant information generation is zero. Chaotic systems have positive h_{KS} and continuously generate information. In the ideal environment, in order to perfectly synchronize two chaotic systems, one must transmit an amount of information per unit time that is equal to or larger than h_{KS} [46]. Although our detection method allows some tolerance in the synchronization precision, the need to transmit extra information to maintain the synchronization results in an additional shift of the actual CPPM performance curve relative to the case when ideal synchronization is assumed. Because the numerical and experimental curves in Figure 2.3 pass rather close to the analytical curve that assumes synchronization, the degradation caused by nonzero Kolmogorov-Sinai entropy does not seem to be significant.

Although CPPM performs slightly worse than BPSK, noncoherent FSK, and ideal PPM, we should emphasize that (i) this wideband system provides low probability of intercept and low probability of detection; (ii) improves the privacy while adding little circuit complexity; (iii) to our knowledge, this system performs exceptionally well compared to most other chaos-based covert communication schemes; (*iv*) there exists a multiplexing strategy described above that can be used with CPPM (see also [47, 48]); and (*v*) compared to other impulse systems, CPPM does not rely on a periodic clock, and thus can eliminate any trace of periodicity from the spectrum of the transmitted signal. All this makes CPPM attractive for development of chaos-based communications.

Acknowledgments

The authors are grateful to H.D.I. Abarbanel, L. Larson, L. Kocarev, and M.A. Vorontsov for helpful discussions. This work was supported in part by U.S. Department of Energy (grant DE-FG03-95ER14516), the U.S. Army Research Office (MURI grant DAAG55-98-1-0269). The authors also thank J. Gowens and J. Carrano for support in the development of the Atmospheric Laser Optics Testbed (A_LOT) at Adelpi, Maryland used in the laser experiments.

References

1. K. M. Cuomo and A. V. Oppenheim, Circuit implementation of synchronized chaos with applications to communications, *Phys. Rev. Lett.*, vol. 71(1), pp. 65–68, 1993.
2. L. Kocarev, K. S. Halle, K. Eckert, L. O. Chua and U.Parlitz, Experimental demonstration of secure communications via chaotic synchronization. *Int. J. Bifurcation Chaos*, vol. 2, pp. 709–713, 1992.
3. T. L. Carroll and L. M. Pecora, Synchronizing nonautonomous chaotic circuits, *IEEE Trans. Circuits Syst.*, vol. 40(10), pp. 646–650, 1993.
4. C. W. Wu and L. O. Chua, A simple way to synchronize chaotic systems with applications to secure communication systems, *Int. J. Bif. Chaos*, vol. 3(6), pp. 1619–1627, 1993.
5. P. Colet and R. Roy, Digital communication with synchronized chaotic lasers, *Optics Lett.*, vol. 19(24), pp. 2056–2058, 1994.
6. P. Celka, Chaotic synchronization and modulation of nonlinear time-delayed feedback optical systems. *IEEE Trans. Circuits Syst.*, vol. 42(8), pp. 455–463, 1995.
7. P. Celka, Synchronization of chaotic optical dynamical systems through 700 m of single mode fiber. *IEEE Trans. Circuits Syst.*, vol. 43(10), pp. 869–872, 1996.
8. C. R. Mirasso, P. Colet, and P. Garcia-Fernández, Photonics Technol. Lett **8**, 299 (1996)
9. G. D. VanWiggeren and R. Roy, Communication with chaotic lasers, *Science*, vol. 279(5354), pp. 1198–1200, 1998.
10. G. D. VanWiggeren and R. Roy, Optical communication with chaotic waveforms, *Phys. Rev. Lett.*, vol. 81(16), pp. 3547–3550, 1998.
11. H. D. I. Abarbanel and M. B. Kennel, Synchronizing high-dimensional chaotic optical ring dynamics, *Phys. Rev. Lett.*, vol. 80(14), pp. 3153–3156, 1998.

12. G. Kolumban, G. Kis, and Z. Jákó, M. P. Kennedy. FM-DCSK: a robust modulation scheme for chaotic communications. *IEICE Trans. Fundamentals of Electronics, Communication, and Computer Science*, vol.E81-A(9), pp. 1798–8002, 1998.

13. L. M. Pecora and T. L. Carroll. Synchronization in chaotic systems. *Phys. Rev. Lett.*, vol.64, pp. 821–824, 1990.

14. H. Fujisaka and T. Yamada, Stability theory of synchronized motion in coupled-oscillator systems, *Prog. Theor. Phys.*, vol. 69(1), pp. 32–47, 1984.

15. L. M. Pecora and T. L. Carroll, Synchronization in chaotic systems, *Phys. Rev. Lett.*, vol. 64(8), pp. 821–824, 1990.

16. D. R. Frey, Chaotic digital encoding: an approach to secure communication, *IEEE Trans. Circuits Syst.*, vol. 40(10), pp. 660–666, 1993.

17. A. R. Volkovskii and N. F. Rulkov, Synchronous chaotic response of a nonlinear oscillator system as a principle for the detection of the information component of chaos, *Tech. Phys. Lett.*, vol. 19(2), pp. 97–99, 1993.

18. U. Feldmann, M. Hasler, and W. Schwarz, Communication by chaotic signals: the inverse system approach, *Int. J. Ciruit Theory Appl.*, vol. 24(5), 1996, pp. 551–579.

19. L. Kocarev and U. Parlitz, General approach for chaotic synchronization with applications to communication. *Phys. Rev. Lett.*, vol. 74(25), pp. 5028–5031, 1995.

20. M. Hasler. Synchronization of chaotic systems and transmission of information. *Int. J. of Bifurcation and Chaos*, vol. 8(4), pp. 647–659, 1998.

21. T. L. Carroll and L. M. Pecora, Cascading synchronized chaotic systems. *Physica D*, vol. 67, pp. 126–140, 1993.

22. H. Dedieu, M. P. Kennedy, and M. Hasler, Chaos shift keying: modulation and demodulation of a chaotic carrier using self-synchronizing Chua's circuits. *IEEE Trans. Circuits Syst. - I*, vol. 40, pp. 634–642, 1993.

23. K. Murali and M. Lakshmanan, Transmission of signals by synchronization in a chaotic Van der Pol-Duffing oscillator. *Phys. Rev. E*, vol. 48, pp. R1624–1626, 1993.

24. Y. H. Yu, K. Kwak, and T. K. Lim, Secure communication using small time continuous feedback. *Physics Letters A*, vol. 197, pp. 311–316, 1995.

25. K. S. Halle, C. W. Wu, M. Itoh, and L. O. Chua, Spread spectrum communication through modulation of chaos. *Int. J. Bifurcation Chaos*, vol. 3, pp. 469–477, 1993.

26. U. Parlitz and L. Kocarev, General approach for chaotic synchronization with applications to communication. *Phys. Rev. Lett.*, vol. 74, pp. 5028–5031, 1995.

27. U. Parlitz, L. Kocarev, T. Stojanovski, and H. Preckel, Encoding messages using chaotic synchronization. *Phys. Rev. E*, vol. 53, pp. 4351–4361, 1996.

28. N. F. Rulkov and L. S. Tsimring. Synchronization methods for communications with chaos over band-limited channel *Int. J. Circuit Theory and Applications.*, vol. 27, pp. 555–567, 1999.

29. G. Kolumban, M. P. Kennedy and L. O. Chua, The role of synchronization in digital communications using chaos. II. Chaotic modulation and chaotic synchronization, *IEEE Trans. Circuits Syst. - I*, vol. 45(11), pp. 1129–1140, 1998.

30. C. Williams, Chaotic communication over radio channels, *IEEE Trans. Circuits Syst. - I*, vol. 48(12), pp. 1394–1404, 2001.

31. T. L. Carroll, Amplitude-independent chaotic synchronization. *Phys. Rev. E*, vol. 53(4A), pp. 3117–3122, 1996.

32. T. L. Carroll and G. A. Johnson, Synchronizing broadband chaotic systems to narrow-band signals, *Phys. Rev. E*, vol. 57(2), pp. 1555–1558, 1998.
33. E. Rosa, S. Hayes, and C. Grebogi, Noise filtering in communication with chaos. *Phys. Rev. Lett.*, vol. 78(7), pp. 1247–1250, 1997.
34. N. F. Rulkov and A. R. Volkovskii, Threshold synchronization of chaotic relaxation oscillations, *Physics Letters A*, vol. 179(4-5), pp. 332–336, 1993.
35. H. Torikai, T. Saito, and W. Schwartz, Synchronization via multiplex pulse trains, *IEEE Trans. Circuits Syst. - I*, vol. 46(9), 1072–1085, 1999.
36. T. L. Carroll, Communicating with use of filtered, synchronized, chaotic signals, *IEEE Trans. Circuits Syst. - I*, vol. 42(2), pp. 105–110, 1995.
37. T. L. Carroll, Noise-robust synchronized chaotic communication, *IEEE Trans. Circuits Syst. - I*, vol. 48(12), pp. 1519–1522, 2001.
38. N. F. Rulkov and A. R. Volkovskii. Synchronization of pulse-coupled chaotic oscillators. In *Proc. of the 2nd Experimental Chaos Conference*, pp. 106–115, World Scientific, 1993.
39. M. M. Sushchik *et al.* Chaotic Pulse Position Modulation: a Robust Method of Communicating with Chaos *IEEE Communications Letters*, vol.4, pp. 128–130, 2000.
40. M. Z. Win and R. A. Scholtz. Impulse radio: how it works. *IEEE Communications Letters*, vol.2(2), pp.36–38, 1998.
41. T. Yang and L. O. Chua. Chaotic impulse radio: A novel chaotic secure communication system. *Int. J. of Bifurcation and Chaos*, vol.10, pp. 345–357, 2000.
42. P. A. Bernhardt Coupling of the relaxation and resonant elements in the autonomous chaotic relaxation oscillator (ACRO) *Chaos* vol.2, 183–199, 1992.
43. N. F. Rulkov, M. M. Sushchik, L. S. Tsimring, and A. R. Volkovskii Digital Communication Using Chaotic-Pulse-Position Modulation. *IEEE Trans. Circuits Syst.* vol. 48 , pp. 1436–1444, 2001.
44. N. F. Rulkov, M. A. Vorontsov and L. Illing Chaotic Free-Space Laser Communication over Turbulent Channel. *Phys. Rev. Lett.* vol. 89 , 277905, 2002.
45. L. C. Andrews, R. L. Phillips, and C. Y. Hopen, Laser Beam Scintillation with Applications (SPIE Press, Bellingham, 2001).
46. T. Stojanovski, Lj. Kocarev, and R. Harris. Applications of symbolic dynamics in chaos synchronization. *IEEE Trans. Circuit. Syst. I*, vol.44(10), pp. 1014–1017, 1997.
47. H. Torikai, T. Saito, and W. Schwarz. Multiplex communication scheme based on synchronization via multiplex pulse-trains. In *Proceedings of the 1998 IEEE International Symposium on Circuits and Systems*, pp. 554–557, New York, 1998.
48. H. Torikai, T. Saito, and W. Schwarz. Synchronization via multiplex pulse-trains. *IEEE Trans. Circuit. Syst. I*, vol.46(9), pp.1072–1085, 1999.

3

Spread Spectrum Communication with Chaotic Frequency Modulation

Alexander R. Volkovskii, Lev S. Tsimring, Nikolai F. Rulkov, Ian Langmore, and Stephen C. Young

Summary. We describe two different approaches to employ chaotic signals in spread-spectrum (SS) communication systems with phase and frequency modulation. In the first one a chaotic signal is used as a carrier. We demonstrate that using a feedback loop controller, the local chaotic oscillator in the receiver can be synchronized to the transmitter. The information can be transmitted using phase or frequency modulation of the chaotic carrier signal. In the second system the chaotic signal is used for frequency modulation of a voltage controlled oscillator (VCO) to provide a SS signal similar to frequency hopping systems. We show that in a certain parameter range the receiver VCO can be synchronized to the transmitter VCO using a relatively simple phase lock loop (PLL) circuit. The same PLL is used for synchronization of the chaotic oscillators. The information signal can be transmitted using a binary phase shift key (BPSK) or frequency shift key (BFSK) modulation of the frequency modulated carrier signal. Using an experimental circuit operating at radio frequency band and a computer modeling we study the bit error rate (BER) performance in a noisy channel as well as multiuser capability of the system.

3.1 Introduction

Spread-Spectrum (SS) communications is one of the most rapidly developing fields in both theory and applications of electrical engineering. Spreading the spectrum can considerably improve the characteristics of communication systems. The larger frequency band and the lower spectral density of the signal complicate the signal detection problem for a surveillance receiver, which combined with other design attributes ensures the Low Probability of Intercept (LPI) for SS systems. In the SS receiver knowledge of parameters of the transmitted signal, such as the spreading function, is effectively used to reject the independent interference or jammer signal. This results in the capability of SS systems to operate in multiuser environments, where several users operate in the same frequency band and use Code Division Multiple Access (CDMA) to separate the different users. Not all interference signals are independent. In a multipath propagation environment the jammer is a replica of the transmitted signal shifted in time and frequency. SS systems are able to reject this type of jammer signal, which is a result of high resolution Time of Arrival (TOA) measurements typical for SS systems (see, for example [1, 2]).

In spite of fact that the main principles of SS have been known for more than 60 years, the SS systems were mostly used for military applications where such factors as reliability, security, accuracy, and jamming resistance usually prevail over complexity and manufacturing costs. Only recently, tremendous improvements in integrated circuit technology have made SS systems available as popular consumer products. These include mobile phones and Wireless Personal Area Networks (WPANs), micropower indoor radio communication devices connecting portable consumer electronics [3].

The key element of any SS system is the spreading technique. After decades of research many different methods (analog, digital, optomechanical, etc.) of generating high-frequency broadband noiselike signals have been developed. These signals can be generated directly from a high-frequency source, for example, by amplifying and filtering a thermal noise voltage on a resistor, or by modulating a high-frequency sinusoidal carrier by a low-frequency broadband signal. In modern SS communication systems Pseudo-Noise (PN) sequences are usually used as spreading functions for modulating a sinusoidal carrier. Because the performance of SS systems is primarily determined by the spreading, many new spreading techniques based on dynamical chaos have been intensively studied in last decade [4, 5].

Active research in this area was initiated by Pecora and Carroll [6] who first demonstrated the robust synchronization of one chaotic oscillator by another . Several different schemes implementing communication using chaos have been proposed in recent years. Some of them, such as Differential Chaos Shift Keying (DCSK) [7] , do not use the phenomenon of chaotic synchronization and rely on transmitting the chaotic reference signal along with the modulated signal. In other schemes (chaotic masking [8] , inverse systems [9], parameter modulation [10] , and generalized frequency modulation [11]) synchronization has been used. However, as subsequent research has shown [12], many of these methods are susceptible to channel distortions and typically yield poor Bit Error Rate (BER) performance. A better BER characteristic can be achieved with DCSK and it can be further improved by employing methods of Symbolic Dynamics (SD-DCSK) [13].

Studies of the communication systems with chaotic carriers show that most of these systems are highly susceptible to channel distortions: additive channel noise, amplitude, and frequency distortions can easily desynchronize the chaotic oscillators [14]. In order to eliminate the influence of channel distortions on the quality of chaotic synchronization it was suggested to use chaos for modulation of the temporal characteristics of carrier signal, the Chaotic Pulse Position Modulation (CPPM) [15]. This scheme has good BER performance; the transmitted power can occupy an Ultra-Wide frequency Band (UWB) with pulse coding and matched filters used to separate users in multiuser environment.

Chaotic waveforms can also be used in place of PN sequences as spreading functions in Direct Sequence (DS) and Frequency Hopping (FH) SS systems. Chaotic sequences for DS SS have been intensively studied in the last few years [16]. It has been shown that chaotic spreading sequences optimized with respect to the Multiple Access Interference (MAI) power can also improve the BER performance and allow about 15% more users in asynchronous CDMA systems than Gold codes [17, 18]. In FH SS systems the hopping sequences are usually optimized for better Hamming correlation properties to ensure small user-to-user interference. Analysis of spreading sequenced for FH SS generated by the chaotic 1-d maps shows that the chaotic sequences have as good correlation properties as PN sequences [19].

The rest of the chapter is organized as follows. In Section 3.2 we describe the synchronization algorithm and modulation-demodulation schemes for the communication system with frequency and phase modulation of a chaotic carrier, in Section 3.3 we present the general synchronization scheme for a communication system with Chaotic Frequency Modulation (CFM) of periodic oscillators, Section 3.4 is devoted to the performance characteristics of the CFM system with different types of modulation, and the conclusions are in Section 3.5.

3.2 Phase and Frequency Modulation of Chaotic Carrier

3.2.1 x- and τ-Loop Synchronization Scheme

Frequency modulation (FM) is widely used in communication (FM radio, TV, radar, etc.) because it is relatively simple and has some spread-spectrum capabilities, especially broadband FM . Because of the advantages of FM during the last decade several attempts have been made to use it with chaotic signals [11, 20, 21]. Although the modulation in many cases can be easily realized by variation of the time-scale parameter of the chaotic system in accordance with the information signal, to retrieve the information the receiver needs to keep tracking the changes of the received signal spectra, which is a much more complicated task. In conventional FM systems phase locked loops are commonly used as a simple and efficient FM demodulator [see, for example, [22]]. In a PLL FM demodulator circuit, the Local Oscillator (LO) frequency is controlled by the loop signal in such a way that it is equal to the frequency of the input (received) signal and the oscillators are phase-locked. Thus, when the modulation is applied to the transmitted signal, the loop signal follows the modulating signal .

To exploit the idea of phase lock-in for the chaotic systems:

$$\dot{\boldsymbol{x}}_{tr} = \tau_{tr}\boldsymbol{F}(\boldsymbol{x}_{tr}, \boldsymbol{p}_{tr}) \qquad \dot{\boldsymbol{x}}_{re} = \tau_{re}\boldsymbol{F}(\boldsymbol{x}_{re}, \boldsymbol{p}_{re}), \qquad (3.1)$$

where we use lower "tr" and "re" indexes for transmitter and receiver systems, correspondingly, $\boldsymbol{x}_{tr,re}$ represent the phase variables of the chaotic systems,

$\tau_{tr,re}$ are the time-scale parameters, and $\boldsymbol{p}_{tr,re}$ are parameters essential for the system dynamics; a feedback loop can be used to control the parameter τ_{re} in the receiver system in order to keep a constant time shift between the oscillations (τ-loop). If the trajectories are identical in the phase spaces, the time shift between them can be determined as a difference in time moments when the trajectories pass through the identical Poincaré cross-sections $\varGamma_{tr}$ and $\varGamma_{re}$. Then the τ-loop can use this information about the time shift to control the time-scale parameter in the receiver to keep the trajectories in-step with each other. To keep the chaotic trajectories identical in their respective phase spaces another feedback loop ($\boldsymbol{x}$-loop) can control the parameters $\boldsymbol{p}_{re}$ of the receiver system using the error signals detected on $\varGamma_{tr}$ and $\varGamma_{re}$. The block diagram of the control scheme is shown in Figure 3.1. Although one of the most

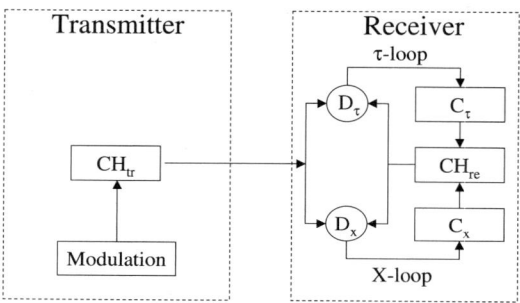

Fig. 3.1. Block-diagram of the chaotic PLL synchronization scheme. $Ch_{tr,re}$ are chaotic oscillators; D_τ is the time-shift detector; D_x is the phase space shift detector; C_τ and C_x are the controller devices for modulation parameters τ and p correspondingly. (Reprinted with permission from [21], ©1997 IEEE.)

general methods of such control is based on Ott-Grebogi-Yorke "controlling chaos" (OGY) [23, 24] , the Adaptive Parametric Control (APC) [25] is much simpler and more practical. We assume that in the $\boldsymbol{x}$-loop only the information about the space coordinates (regardless of the time) is used by the controller to modify only one of the receiver parameters p_{re}. In this case the $\boldsymbol{x}$-loop operates independently of the τ-loop and if the control method is stable the trajectories converge in space and become identical: $|\tilde{\boldsymbol{x}}_{tr}(n) - \tilde{\boldsymbol{x}}_{re}(n)| \to 0$, where $\tilde{\boldsymbol{x}}_{tr,re}(n)$ are trajectories on $\varGamma_{tr}$ and $\varGamma_{re}$. When the trajectories are identical in phase spaces but shifted in time the τ-loop performs similarly to a PLL with stable periodic oscillators and can synchronize the systems in time. The control equation for the τ-loop can be written as

$$\Delta\tau(n) = \Delta\tau(0) - \phi\left(\Delta t(n)\right), \tag{3.2}$$

where $\Delta\tau(n) = \tau_{re}(n) - \tau_{tr}$ is the current detuning of the time-scale parameters; $\Delta t(n) = t_{re}(n) - t_{tr}(n)$ is the time shift between the oscillations; and

$\phi(\Delta t)$ is the characteristic of phase discriminator. We assume it is a smooth function in some interval $\Delta t \in [\Delta t_{\min}, \Delta t_{\max}]$.

The time intervals between two intersections of Γ_{tr} and Γ_{re} surfaces $t_{tr}(n) - t_{tr}(n-1)$ and $t_{re}(n) - t_{re}(n-1)$ are functions of phase coordinates and system parameters:

$$t_{tr}(n) - t_{tr}(n-1) = \tau_{tr} T\left(\tilde{\boldsymbol{x}}_{tr}(n-1), \boldsymbol{p}_{tr}\right)$$
$$t_{re}(n) - t_{re}(n-1) = \tau_{re}(n-1) T\left(\tilde{\boldsymbol{x}}_{re}(n-1), \boldsymbol{p}_{re}(n-1)\right). \tag{3.3}$$

Using Equations (3.2) and (3.3) one can write the following linearized evolution equation for the time shift $\Delta t(n)$:

$$\Delta t(n) = \Delta t(n-1) + T\left(\tilde{\boldsymbol{x}}_{tr}(n-1), \boldsymbol{p}_{tr}\right)\Big(\Delta \tau(0) -$$
$$\phi(\Delta t(n-1))\Big) + \tau_{tr}\Big(T'_x \Delta X + T'_p \Delta P\Big) \tag{3.4}$$

where $|\tilde{\boldsymbol{x}}_{tr} - \tilde{\boldsymbol{x}}_{re}| << 1;$ $|\boldsymbol{p}_{re} - \boldsymbol{p}_{tr}| << 1;$ $T'_x \Delta X = \sum\limits_{k=1}^{m} \dfrac{dT}{d\tilde{x}^k}(\tilde{x}^k_{re}(n-1) -$
$\tilde{x}^k_{tr}(n-1));$ $T'_p \Delta P = \sum\limits_{i=1}^{m1} \dfrac{dT}{dp_i}(p^i_{re}(n-1) - P^i_{tr});$ m and $m1$ are the numbers of coordinates and parameters correspondingly. As mentioned above, the time-scale parameter does not have an effect on the system dynamics so the x-loop performs independently from the τ-loop. Suppose that the x-loop is designed to be stable in accordance with one of the known methods of controlling chaos and converges under $n \to \infty$ to $\tilde{x}_{re} = \tilde{x}_{tr}$ and $\boldsymbol{p}_{re} = \boldsymbol{p}_{tr}$ that leads to vanishing ΔX and ΔP in (3.4). The evolution equation for the time shift between the oscillations (3.4) therefore has the stationary solution:

$$\Delta t^* = \phi^{-1}\big(\Delta \tau(0)\big) = const. \tag{3.5}$$

The solution (3.5) corresponds to synchronous oscillations of the chaotic systems (3.1). This regime is stable (provided that the x-loop is stable) if:

$$\left|1 - T\big(\tilde{\boldsymbol{x}}_{tr}(n-1), \boldsymbol{p}_{tr}\big)\frac{d\phi}{d\Delta t}\bigg|_{\Delta t^*}\right| < 1, \tag{3.6}$$

Equations (3.4–3.6) describe the dynamics of the τ-loop, which is similar to the dynamics of PLL with periodic signals (for periodic oscillations $T(\tilde{\boldsymbol{x}}, \boldsymbol{p}) = const$). In particular, the main properties of the synchronous regime (3.5) are the same: (i) the synchronous regime exists in a certain interval of initial detunings $\phi_{\min} \leq \Delta \tau(0) \leq \phi_{\max}$, where $\phi_{\min}$ and $\phi_{\max}$ are the maximum and minimum output signals of the time shift detector D_τ, and (ii) the residual time shift Δt^* in the synchronous regime is determined by the initial detuning $\Delta \tau(0)$. By analogy with a PLL, the $x - \tau$ loop can be used in communication systems as a demodulator of frequency modulated signals. Modulation of the time-scale parameter τ_{tr} by the information signal leads to the frequency modulation of the broadband spectra of the chaotic signal in the transmitter, which corresponds to the modulation of $\Delta\tau(0)$ in (3.4–3.6). In the synchronous regime (3.5) the output of the time shift detector D_τ is a function of $\Delta\tau(0)$ and therefore the information can be recovered.

3.2.2 Demodulation of Phase- and Frequency-Modulated Chaotic Carriers

A communication scheme based on phase and frequency modulation of chaotic carrier signals has been described in [21], where ring oscillators [26] were used as chaotic systems. The mathematical model for a single oscillator (see [27] for circuit implementation) is the following

$$\begin{aligned} \tau\dot{x} &= y \\ \tau\dot{y} &= -x - \delta y + z \\ \tau\dot{z} &= \gamma(F(x) - z) - \sigma y \end{aligned} \qquad (3.7)$$

where:

$$F(x) = \begin{cases} 0.528, & x \le 1.2 \\ \alpha x(1 - x^2), & -1.2 < x < 1.2 \\ -0.528, & x \ge 1.2 \end{cases}$$

and τ is the time-scale parameter.

The system (3.7) demonstrates a variety of chaotic behaviors. The synchronization algorithm has been tested on two types of attractors shown in Figure 3.2a,b. The time shift detector in the τ-loop measured the difference be-

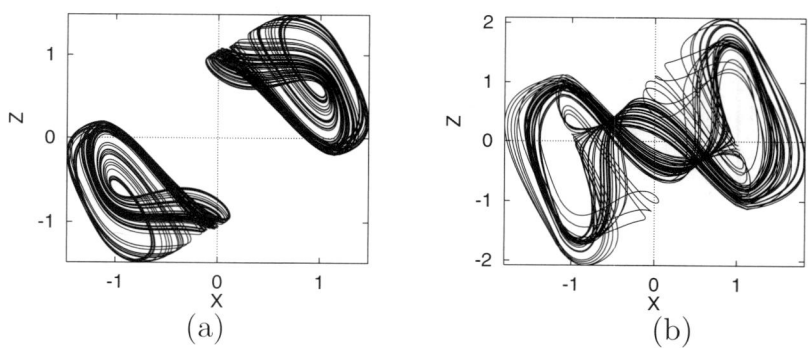

(a)   (b)

Fig. 3.2. Asymmetric (a) and symmetric (b) chaotic attractors of the system (3.7). (Reprinted with permission from [21], ©1997 IEEE.)

tween time moments $t_{tr}(n)$ when the trajectory of the transmitter system crosses the Poincaré surface Γ_{tr} and the time moments $t_{re}(n)$ corresponding to the receiver system trajectory crossing the Γ_{re}. In this case the time shift detector output function $\phi(\Delta t)$ is proportional to Δt inside the interval $\Delta t \in [0, t_{\min}]$, where $t_{\min} = \min\{t_{tr}(n) - t_{tr}(n-1), \quad t_{re}(n) - t_{re}(n-1)\}$, $\quad n = 1, 2, 3, \ldots$. Outside this interval $\phi(\Delta t) = \phi(\Delta t, n)$ is chaotic. The adaptive parametric control [25] was used in the x-loop for synchronization of the trajectories in the phase spaces. The transmitter chaotic oscillator was used as the *SYSTEM* in [25] and the receiver oscillator as the *MODEL*. Parameters of the *MODEL* were controlled instead of the *SYSTEM* ones. It was assumed that only the x_{tr} variable was available for the receiver. The planes $\dot{x}_{tr} = 0$ and $\dot{x}_{re} = 0$ were used as the Poincaré cross-sections. Parameter γ_{re} was used as the control parameter. Perturbations were made in accordance with the equation:

$$\gamma_{re}(n+1) = \gamma_{re}(n) + \mu\big(x_{re}(n) - x_{tr}(n)\big), \qquad (3.8)$$

where μ is the constant parameter of the adaptive control. Because the adaptive parametric control requires coupling between the *MODEL* and the *SYSTEM*, the x_{re} variable was adjusted after calculating $\gamma_{re}(n+1)$ so that $x_{re}(n) = x_{tr}(n)$.

The stability and robustness of synchronization have been studied in computer simulations. In-phase synchronization of chaotic oscillations corresponding to the attractors (Figure 3.2a) is illustrated in Figure 3.3a,b. In simulations

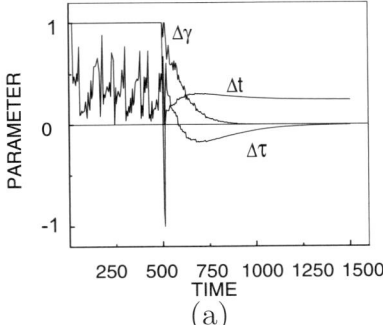
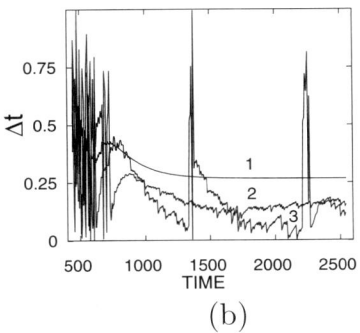

(a)   (b)

Fig. 3.3. (a) The normalized time shift between the oscillations $\Delta\tilde{t} = \frac{\Delta t}{\Delta t_{\max}}$ and the adjusted parameters: $\Delta\tilde{\tau} = \frac{\tau_{re}(n) - \tau_{tr}}{\Delta\tau_{\max}}$, $\Delta\tilde{\gamma} = \frac{\gamma_{re}(n) - \gamma_{tr}}{\Delta\gamma_{\max}}$ versus the time for $\alpha_{tr} = \alpha_{re} = 16, \delta_{tr} = \delta_{re} = 0.43, \sigma_{tr} = \sigma_{re} = 0.72, \gamma_{tr} = 0.1, \gamma_{re}(0) = 0.11$. (b) The influence of mismatch of parameters $\alpha_{tr,re}$ on the amplitude of fluctuations of the time shift Δt in the synchronous regime: $\delta_{tr} = \delta_{re} = 0.43, \sigma_{tr} = \sigma_{re} = 0.72, \gamma_{tr} = 0.1, \gamma_{re}(0) = 0.11, \alpha_{re} = 16$, and $\alpha_{tr} = 16$ — (1); 16.7 — (2); 16.9 — (3). (Reprinted from [21], ©1997 IEEE.)

shown in Figure 3.3a both parameters τ_{re} and γ_{re} were detuned at the starting point $t = 0$. The available ranges for the control parameters were set to the following

$$0.5 \leq \tau_{re} \leq 1.5 \qquad 0.09 \leq \gamma_{re} \leq 0.11$$

At the time moment $t = 500$ the control was turned on and the systems converged to the stationary synchronous regime. In spite of the constant time shift Δt the oscillations are synchronous and the values of the parameters in the transmitter and the receiver systems are equal: $\tau_{tr} = \tau_{re}$ and $\gamma_{tr} = \gamma_{re}$.

In practice however, some parameters could be inaccessible for the control and, if these parameters are not identical, the stationary solution (3.5) does not exist. However, if the parameters' mismatch is small enough the systems oscillate near the synchronous regime and the jitter (i.e., the deviation from synchrony) depends on the amount of mismatch . Figure 3.3b shows the transient for the case with uncontrolled parameter $\alpha_{re} \neq \alpha_{tr}$, whereas parameters γ_{re} and τ_{re} were controlled. When the detuning $|\alpha_{re} - \alpha_{tr}|$ increases, the amplitude of fluctuations becomes more significant and at some level of detuning synchronization can be lost.

As was mentioned above, the modulation of the time-scale parameter τ is analogous to frequency modulation and can be detected by the τ-loop in accordance with Equation (3.5). The characteristic time of the demodulator depends on the convergence speed of the control scheme and can be quite fast (see Figure 3.3). A similar control scheme can demodulate a Binary Phase Shift Keying (BPSK) modulated chaotic carrier signal (i.e., the chaotic signal $x_{tr}(t)$ multiplied by $+1$ or -1 depending on the binary information signal content) . As shown in Figure 3.4, the detectors D_τ and D_x receive the absolute value or square of the input signals $|x_{tr}|$ and $|x_{re}|$. Because $\dot{x}_{tr} = 0$

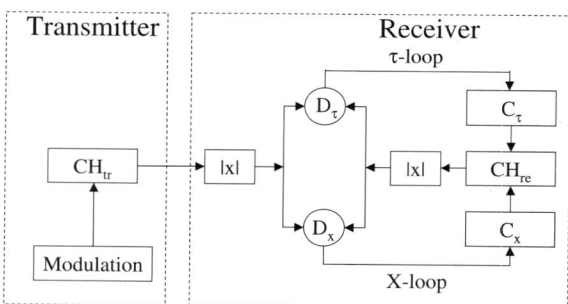

Fig. 3.4. Block-diagram of the chaotic PLL synchronization scheme for BPSK modulated chaotic carrier signals. (Reprinted with permission from [21], ©1997 IEEE.)

and $\dot{x}_{re} = 0$ surfaces were taken as Poincaré cross-sections, the τ-loop was equivalent to that shown in Figure 3.1, whereas the control equation for the x-loop (3.8) was changed in order to provide an appropriate direction of the parameter variation:

$$\gamma_{re}(n + 1) = \gamma_{re}(n) + \mu \, \text{sign}\Big(x_{re}(n)\Big) \Big(|x_{re}(n)| - |x_{tr}(n)| \Big)$$

As it is easy to see, this control scheme is invariant with respect to the substitution $x_{tr} \rightarrow -x_{tr}$ and can be used for in-phase and anti-phase synchronization. Because switchings of the input signal $x_{tr} \rightarrow -x_{tr}$ do not destroy the synchronization, this scheme can be used for demodulation of BPSK signals. Figure 3.5 shows signals in the transmitter and the receiver with BPSK modulation of the transmitted signal. The graphs in Figure 3.5 were plotted for pa-

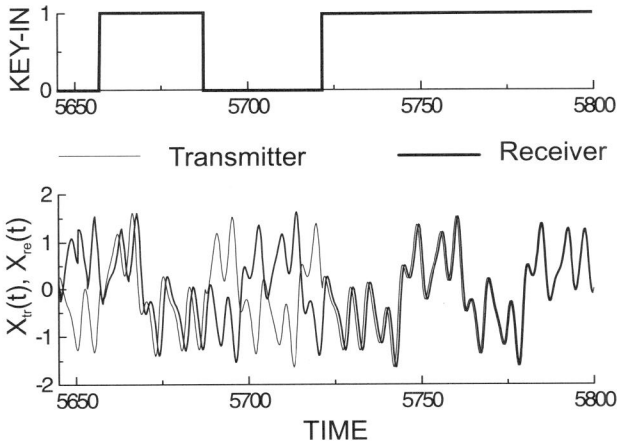

Fig. 3.5. Demodulation of BPSK chaotic signal. (Reprinted from [21], ©1997 IEEE.)

rameters of the systems corresponding to the attractors shown in Figure 3.2b. In accordance with some binary message the modulation: $x_{tr} \rightarrow -x_{tr}$ was applied to the transmitted signal $x_{tr}(t)$ at the time moments when $x_{tr} = 0$. Because the phase of oscillations in the receiver is not affected by the modulation, the message can be recovered by correlating the signals x_{tr} and x_{re}.

3.3 Chaotic Frequency Modulation of Periodic Carrier

3.3.1 Interpolated Frequency Hopping and Chaotic Frequency Modulation

Frequency hopping is one of the most efficient methods of spreading spectrum. In a typical scheme with noncoherent frequency hopping, the transmitter carrier signal frequency is modulated by a pseudo-random hopping code. It is assumed that the transmitter and the receiver share the same code and are synchronized (usually by exchanging special synchronizing codes). Then the frequency of the transmitter is shifted by the information signal and the receiver can extract the data by detecting this relative frequency shift.

In coherent FH systems, the receiver and transmitter maintain exact phase synchronization of the carrier, and the information is encoded in the phase variations of the transmitted signal at a current carrier frequency.

If the receiver is able to provide and maintain the exact carrier phase for demodulation (which is difficult because of discrete frequency hops), differential phase encoding can be used. Alternatively, as in the DCSK scheme [7], the first part of a time slot assigned for each binary symbol serves as a phase reference, and the subsequent part within the same slot is phase shifted with respect to it according to the information bit. In this method the carrier frequency is modulated by a chosen chaotic signal to spread the spectrum. It should be noted that this method does not exploit specific deterministic features of the chaotic signals. Indeed, the frequency-modulating signal does not have to be chaotic– any random or pseudo-random signal (as in conventional differentially coherent modulation techniques) would do as well.

FH and especially fast discrete frequency hopping, with at least several hops per information bit, provides a high level of redundancy and therefore high resistance against interference and fading. It is known, however, that FH systems have two main problems: spectra splatter and transient mismatch between the transmit and receive synthesizers. Besides the common solutions used to reduce these effects, such as voltage-controlled oscillator (VCO) pre-tuning, swapping ("ping-ponging") multiple synthesizers, and transient hop interval dwell and guard times (see [28] and references), the alternative Interpolated Frequency Hopping (IFH) technique was recently proposed in [28]. The hopping code in the IFH-transceiver is interpolated by the digital filter, so instead of abrupt hops, frequency varies smoothly in time, which results in better synchronization and therefore lower BER.

Analog chaotic oscillators provide a natural way of generating signals with smoothly varying Chaotic Frequency Modulation (CFM). An additional benefit of employing analog chaotic oscillators is that they may exhibit self-synchronization, thus eliminating the need for special synchronizing sequences. A traditional way of using frequency modulation consists of modulating the transmitter VCO by the signal from the chaos generator, and demodulating this signal at the receiver before feeding it into the receiver's chaos generator (see Figure 3.6). In many systems demodulation of FM signals is performed using a PLL [29]. The gain of the PLL must be chosen large enough to provide frequency variation of VCO_{re} within the entire frequency range of the transmitted signal. The problem with this approach is that any interfering signal with frequency close enough to the current carrier frequency of the transmitted signal will cause a large perturbation of the frequency of the receiver VCO. This may cause an instability and desynchronization of the PLL.

3.3.2 PLL-Based Synchronization Scheme for CFM Signals

In [30] it was proposed to use a different chaotic FM communication scheme based on including the phase lock loop and the chaotic generator in the feed-

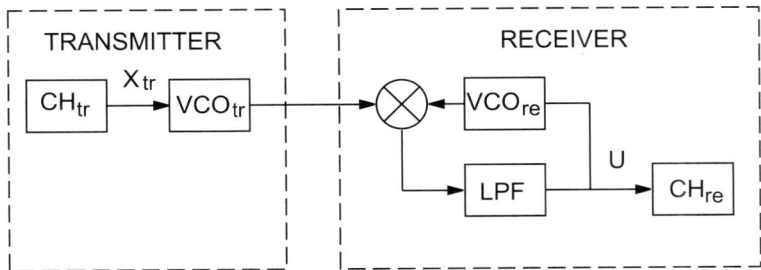

Fig. 3.6. Traditional PLL-based FM communication scheme. (Reprinted with permission from [30], ©1999 John Wiley & Sons, Ltd.)

back loop of the receiver VCO. It was shown that stable synchronization can be achieved for a certain class of chaotic FM systems and the synchronous regime is much more robust against additive disturbances in the channel than a simple FM scheme (Figure 3.6). It has also been demonstrated that within this scheme, a binary information signal can be transmitted via phase or frequency modulation with coherent or noncoherent detection at the receiver.

The block diagram of the proposed synchronization scheme is shown in Figure 3.7. In the transmitter, one of the state variables $x_{tr}(t)$ of the chaotic

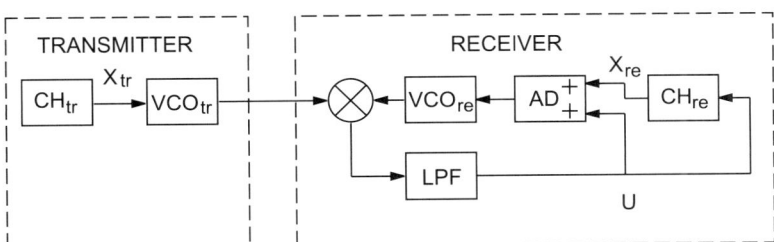

Fig. 3.7. PLL-based synchronization scheme for CFM signals. (Reprinted with permission from [30], ©1999 John Wiley & Sons, Ltd.)

oscillator (CH$_{tr}$) is used for the frequency modulation of the voltage-controlled oscillator (VCO$_{tr}$) to get a spread spectrum CFM signal. The receiver consists of the chaotic oscillator CH$_{re}$, voltage-controlled oscillator VCO$_{re}$, phase discriminator ($\otimes$), low-pass filter (LPF), and the adder (AD). The chaotic oscillators CH$_{tr,re}$ and voltage-controlled oscillators VCO$_{tr,re}$ are assumed to be identical. The sum of the state variable $x_r(t)$ and the output signal from the LPF $u(t)$ modulates the frequency of VCO$_{re}$. When the state variables $x_{tr}(t)$ and $x_{re}(t)$ are close to each other, the phase lock loop ($\otimes \rightarrow$ LPF $\rightarrow$ AD $\rightarrow$ VCO$_{re}$) locks on the current frequency of the transmitted signal.

In this regime the LPF output voltage u is proportional to the open loop frequency offset and therefore to the difference $x_{tr}(t) - x_{re}(t)$. The LPF output u feeds the CH_{re}, providing unidirectional dissipative coupling between the chaotic oscillators in the transmitter and in the receiver. This coupling leads to chaotic synchronization. When the chaotic oscillators are synchronized and the voltage-controlled oscillators are phase-locked, switching the phase or modulating the frequency of the CFM signal can be used for information transmission (see Section 3.4 below).

For the analysis of the proposed scheme suppose that $x_{tr}(t)$ and $x_{re}(t)$ are normalized state variables: $\max |x_{tr}(t)| = \max |x_{re}(t)| = 1$. The VCO frequencies are modulated by the chaotic signals:

$$\begin{aligned}
\dot{\varphi}_{tr} &\equiv \omega_{tr} = \omega_0(1 - m_1 x_{tr}) \\
\dot{\varphi}_{re} &\equiv \omega_{re} = \omega_0(1 - m_1 x_{re} - m_2 u(t)),
\end{aligned} \tag{3.9}$$

where ω_0 is the "natural" frequency of VCO, m_1 and m_2 are the modulation gain coefficients, and $u(t)$ is the LPF output signal normalized by the maximum output voltage of phase discriminator: $|u(t)| \le 1$.

Denote $\varphi = \varphi_{re} - \varphi_{tr}$ and consider the first-order low-pass filter with the transfer function $K(s) = 1/(1 + Ts)$. Then the equations for PLL can be written in the form:

$$\begin{aligned}
\dot{\varphi} &= \omega_0 m_1(x_{tr}(t) - x_{re}(t)) - \omega_0 m_2 u \\
T\dot{u} &= \phi(\varphi) - u,
\end{aligned} \tag{3.10}$$

where the function $\phi(\varphi)$ is the normalized characteristic of the phase discriminator $(\max(\phi(\varphi)) = 1)$. If a multiplier is used as a phase discriminator, $\phi(\varphi) \equiv \sin \varphi$.

The chaotic systems in the transmitter and the receiver can be described by the following equations

$$\begin{aligned}
\tau_{ch}\dot{x}_{tr} &= f(x_{tr}, \boldsymbol{x}_{tr}) & \tau_{ch}\dot{x}_{re} &= f(x_{re}, \boldsymbol{x}_{re}) + \epsilon u \\
\tau_{ch}\dot{\boldsymbol{x}}_{tr} &= \boldsymbol{F}(x_{tr}, \boldsymbol{x}_{tr}) & \tau_{ch}\dot{\boldsymbol{x}}_{re} &= \boldsymbol{F}(x_{re}, \boldsymbol{x}_{re}),
\end{aligned} \tag{3.11}$$

where:

x_{tr}, x_{re} are the state variables used for frequency modulation;
$\boldsymbol{x}_{tr}$, $\boldsymbol{x}_{re}$ are the vectors of other variables;
f and $\boldsymbol{F}$ are nonlinear functions;
τ_{ch} is a characteristic time of chaotic oscillations;
ϵ is a coupling parameter.

Combining (3.10) and (3.11) and changing time to a dimensionless form $\tau = t\sqrt{\omega_0 m_2/T}$, one can derive the following set of equations for both transmitter and receiver

$$\begin{aligned}
\dot{x}_{tr} &= \alpha f(x_{tr}, \boldsymbol{x}_{tr}), \\
\dot{\boldsymbol{x}}_{tr} &= \alpha \boldsymbol{F}(x_{tr}, \boldsymbol{x}_{tr}), \\
\dot{x}_{re} &= \alpha(f(x_{re}, \boldsymbol{x}_{re}) + \epsilon u), \\
\dot{\boldsymbol{x}}_{re} &= \alpha \boldsymbol{F}(x_{re}, \boldsymbol{x}_{re}), \\
\lambda \dot{\varphi} &= \beta(x_{tr}(t) - x_{re}(t)) - u, \\
\tfrac{1}{\lambda} \dot{u} &= \phi(\varphi) - u,
\end{aligned} \qquad (3.12)$$

where:
$\alpha = (\sqrt{T/\omega_0 m_2})/\tau_{ch} = \tau_{pll}/\tau_{ch}$ is the relative speed of chaotic frequency modulation;
$\beta = m_1/m_2$ is the relative depth of frequency modulation;
$\lambda = 1/\sqrt{\omega_0 m_2 T}$ is the PLL damping parameter.

If the modulation is slow: $\alpha \ll 1$, the dynamics of the system (3.12) can be separated into fast and slow motions. The phase lock loop dynamics (last two equations of (3.12)) is fast, and the dynamics of chaotic generators is slow. So, starting from arbitrary initial conditions, the system first approaches the manifold of slow motions in which the PLL dynamics is slaved by the dynamics of the chaotic oscillators (see Figure 3.8). On the slow motion manifold, one

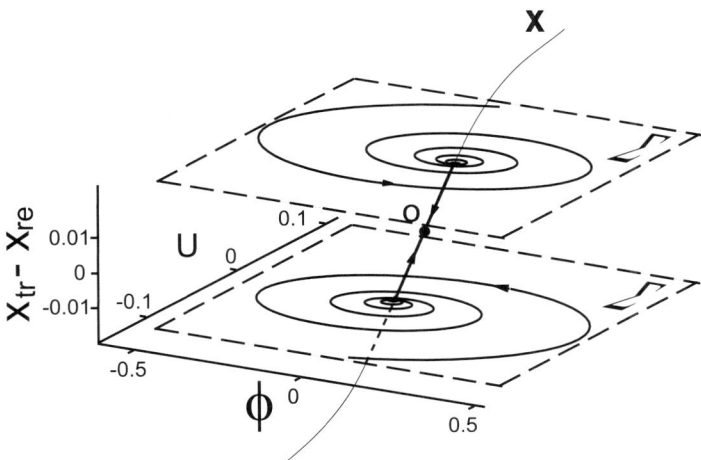

Fig. 3.8. Phase portrait of the system (3.12) for $\alpha \ll 1$. Σ indicates planes of fast locking of the phase-lock loop, and line ξ shows a slow synchronization of the chaotic generators; O indicates the steady state of the system (3.12), which corresponds to synchronization of both $CH_{tr,re}$ and $VCO_{tr,re}$. (Reprinted from [30], ©1999 John Wiley & Sons, Ltd.)

can neglect the time derivatives $\dot{\varphi}$ and $\dot{u}$, so $u = \beta(x_{tr}(t) - x_{re}(t))$, $\varphi = \phi^{-1}(u)$. Then the remaining equations read

$$\dot{x}_{tr} = \alpha f(x_{tr}, \boldsymbol{x}_{tr}),$$
$$\dot{\boldsymbol{x}}_{tr} = \alpha \boldsymbol{F}(x_{tr}, \boldsymbol{x}_{tr}),$$
$$\dot{x}_{re} = \alpha(f(x_{re}, \boldsymbol{x}_{re}) + \epsilon\beta(x_{tr}(t) - x_{re}(t))) \qquad (3.13)$$
$$\dot{\boldsymbol{x}}_{re} = \alpha \boldsymbol{F}(x_{re}, \boldsymbol{x}_{re}).$$

This system describes the dynamics of two dissipatively coupled chaotic oscillators. If the subsystem $\dot{\boldsymbol{x}} = \boldsymbol{F}(x, \boldsymbol{x})$ is stable (as determined by the conditional Lyapunov exponents [6]), then for large enough coupling strength ϵ, system (3.13) exhibits stable synchronization $|x_{tr}(t) - x_{re}(t)| \to 0$, $\|\boldsymbol{x}_{tr}(t) - \boldsymbol{x}_{re}(t)\| \to 0$ as $t \to \infty$ for (almost) any initial conditions (see, for example, [31–34]). Thus in the limit $\alpha \ll 1$ the problem of synchronization of the CFM system is reduced to the synchronization of the low-frequency chaotic oscillators $CH_{tr,re}$. However, if parameter α is of the order of 1, the dynamics of the full system (3.12) is more complicated and synchronization may not occur.

Frequencies of the $VCO_{tr,re}$ change within the range $\Delta\Omega = m_1\omega_0$ around the central frequency ω_0. However, in the neighborhood of the synchronized state, the frequency of the VCO_{re} is close to that of VCO_{tr}, and therefore the bandwidth of the LPF can be made small compared to the bandwidth of the chaotically modulated carrier $\Delta\Omega$ (or, equivalently, $T \gg (m_1\omega_0)^{-1}$). In fact, the signal which should pass through the filter, u, has the bandwidth of the low-frequency chaotic oscillators $CH_{tr,re}$, and so the bandwidth of LPF should be determined by the bandwidth of the chaotic oscillator itself ($T < \tau_{ch}$). This provides selectivity of the proposed scheme with respect to in-band interference, because signals with frequencies not close to the current transmitter frequency are effectively filtered out by the low-pass filter in PLL. Only during short intervals when frequencies of the carrier and the interference are sufficiently close, the PLL is disturbed by the latter, and some deviations from the synchronized state occur. Unlike the simple FM modulation scheme (Figure 3.6), the gain of the phase lock loop m_2 can be made much smaller than m_1, and therefore even large in-band interference is significantly attenuated in the PLL and does not destroy synchronization.

3.3.3 Numerical Simulations of the Synchronization Scheme

We performed numerical simulations of the synchronization scheme shown in Figure 3.7 for two third-order chaotic systems described earlier [33], with $\boldsymbol{x} = y, z$, and

$$f(x, y, z) = y,$$
$$\boldsymbol{F}(x, y, z) = \left\{ \begin{matrix} -x - \delta y + z, \\ \gamma(\mu G(x) - z) - \sigma y \end{matrix} \right\} \qquad (3.14)$$

The parameters of the systems are $\delta = 0.43$, $\mu = 24.7$, $\sigma = 0.72$, $\gamma = 0.1$, and

$$G(x) = \begin{cases} x(1 - x^2), & \text{at } -1.2 < x < 1.2, \\ 0.528, & \text{at } x < -1.2, \\ -0.528, & \text{at } x > 1.2. \end{cases} \qquad (3.15)$$

The phase discriminator function was chosen $\phi(\varphi) \equiv \sin(\varphi)$. We indeed found a large parameter region for a stable synchronized operation of the CFM system. Figure 3.9a shows the region of lock-in (convergence from any arbitrary initial conditions to the synchronous state) on the parameter plane (λ, ϵ) for $\beta = 5$ and two different values of α. For small $\alpha = 0.01$, the lock-in occurs for $\epsilon > 0.15$ except for very small λ. For larger $\alpha = 0.1$, the lock-in region shrinks. It is important to emphasize that this synchronization scheme is robust against

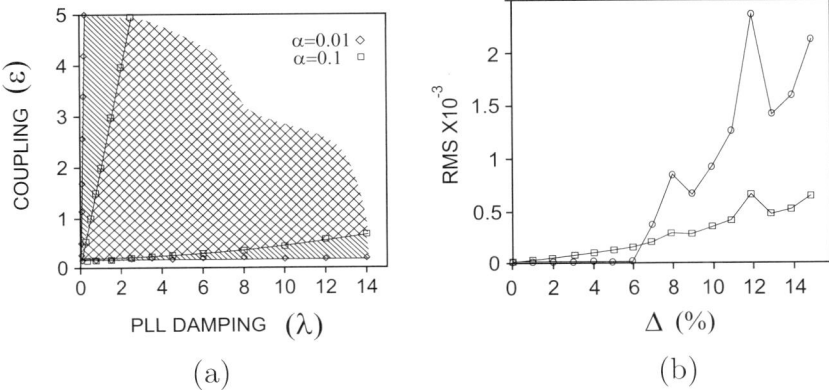

Fig. 3.9. (a) Region of stable CFM synchronization in the parameter plane (λ, ϵ) for system (3.12), (3.14), (3.15) for $\beta = 5$; (b) RMS values of $x_{tr}(t) - x_{re}(t)$ (squares) and $\omega_{tr}(t) - \omega_{re}(t)$ (circles) as functions of the parameter mismatch $\Delta = (\gamma_{re} - \gamma_{tr})/\gamma_{tr}$. (Reprinted from [30], ©1999 John Wiley & Sons, Ltd.)

small detunings of parameters. We investigated the influence of the parameter mismatch between transmitter and receiver chaos generators on the quality of synchronization. We varied parameter γ_{re} in (3.14) of the receiver while keeping $\gamma_{tr} = 0.1$ at the transmitter. Figure 3.9b shows RMS values of the difference $x_{tr}(t) - x_{re}(t)$ and the VCO frequency difference $\omega_{re}(t) - \omega_{tr}(t)$ as functions of the parameter mismatch $\Delta = (\gamma_{re} - \gamma_{tr})/\gamma_{tr}$. For small values of the mismatch, the difference $x_{tr}(t) - x_{re}(t)$ is small, and the PLL is able to adjust the frequency of VCO_{re} to keep exact synchronization ($\omega_{re} = \omega_{tr}$). At larger values of $\Delta > 6\%$, the PLL is unable to maintain synchronization, because the open-loop PLL frequency detuning $m_1\omega_0|x_{tr}(t) - x_{re}(t)|$ occasionally exceeds the PLL hold range $m_2\omega_0$.

The robustness of the proposed method against the periodic in-band interference signal has been studied by adding a sinusoidal component at the frequency ω_1 within the CFM range to the transmitted signal. The amplitude of the interference component was 20% of the amplitude of the transmitted signal. The results of the simulations are shown in Figure 3.10. Small splashes of the PLL output signal u (Figure 3.10a) are produced when the frequency

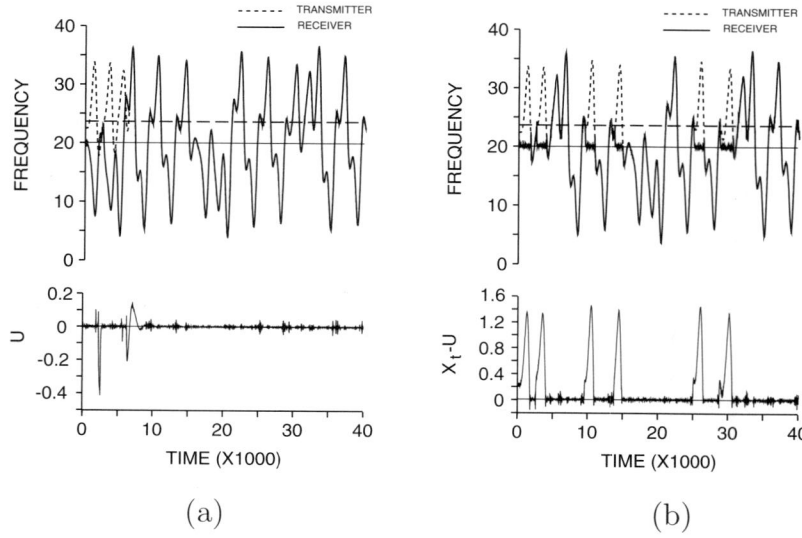

Fig. 3.10. (a) CFM synchronization in the presence of in-band interference. Dashed line in the top panel indicates the frequency of the interference signal $\omega_1 = 23$; (b) Synchronization of chaotic oscillators via standard FM link in the presence of in-band interference at the frequency $\omega_1 = 23$. (Reprinted with permission from [30], ©1999 John Wiley & Sons, Ltd.)

of the transmitted signal is close to ω_1. Nevertheless, because $m_2 < m_1$, frequency of VCO_r remains close to VCO_t, and the system does not lose synchronization. In contrast, similar interference in the conventional FM scheme (Figure 3.10b) produced large perturbations of the frequency of VCO_{re} and loss of synchronization. In this scheme, the receiver PLL must have much larger gain m_1 in order to vary the frequency of VCO_{re} within the range of the transmitter operation. Therefore, in the standard FM scheme all in-band interference signals directly affect the chaotic oscillator at the receiver.

3.4 Communication Using CFM Signals

3.4.1 Differential BPSK Modulation

The proposed method of synchronization of CFM oscillators can be readily utilized for information transmission. Indeed, because the PLL provides phase synchronization between the transmitter and the receiver, differential binary phase shift keying modulation at the transmitter can be easily detected at the receiver. In order to maintain synchronization irrespective of phase switching, the phase lock loop must operate at the second harmonic of the carrier signal, so a multiplier and a high-pass filter should be added to the scheme (see Figure 3.11). An example of the binary information transmission using this

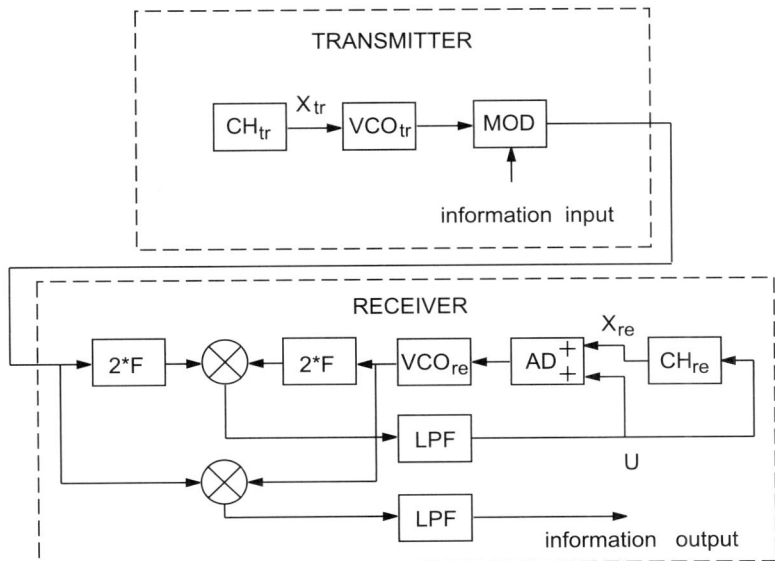

Fig. 3.11. Block diagram CFM DBPSK communication scheme. (Reprinted from [30], ©1999 John Wiley & Sons, Ltd.)

system is shown in Figure 3.12. After initial transient, phase switching is readily detected at the receiver without any loss of synchronization. Because the receiver in this scheme can be synchronized in both in-phase and anti-phase mode, an initialization procedure is required to remove the ambiguity in the decoded information.

3.4.2 Binary Frequency Shift Keying Modulation

In traditional FH systems the frequency of the carrier signal is modulated by a discrete time PN spreading code. In this situation, when the carrier frequency is hopping from one value to another, it is very difficult to keep signals phase-locked in order to use phase modulation and less efficient BFSK modulation is usually used instead. Although a smooth frequency variation in CFM signals allows the coherent detection as shown above, the BFSK can be used with CFM as a simpler and more cost-effective solution.

The block diagram of a CFM BFSK communication scheme is shown in Figure 3.13. The same PLL circuit as shown in Figure 3.7 is used as a receiver. The transmitter is similar to that shown in Figure 3.7, but the analog or binary

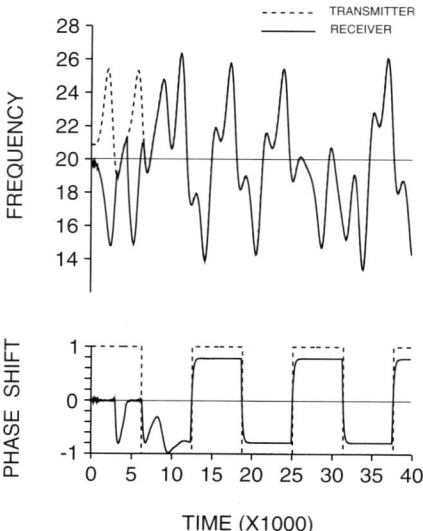

Fig. 3.12. Digital communication using BPSK in conjunction with CFM: upper panel: frequencies of transmitter and receiver VCOs, lower panel: phase variation at the transmitter (dashed line) and the signal detected in the receiver (solid line). (Reprinted with permission from [30], ©1999 John Wiley & Sons, Ltd.)

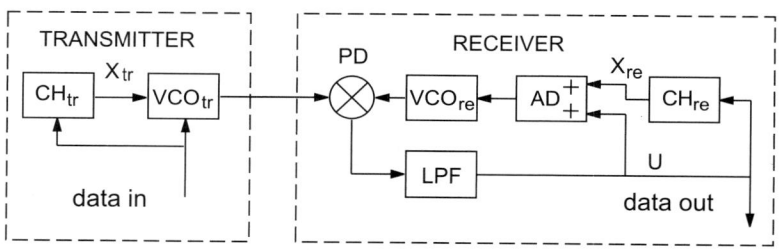

Fig. 3.13. Block diagram of CFM BFSK communication scheme

data signal modulates the VCO_{tr} frequency in addition to $x_{tr}(t)$. At the same time the information signal is applied to the CH_{tr}. The VCO_{tr} generates a CFM signal with the frequency:

$$\omega_{tr} = \dot{\varphi}_{tr} = \omega_0(1 + m_1 x_{tr}(t) + m_2 b_i), \qquad (3.16)$$

where: ω_0 is the VCO_{tr}'s free running frequency; m_1 and m_2 are the modulation parameters; b_i is the information signal.

The information is encoded as a frequency deviation from the original CFM signal. To ensure the regime of identical synchronization between the transmitter and the receiver, the information signal b_i should be applied to the CH_{tr} in the same way as the detected signal $u(t)$ is applied to the CH_{re}. In this case, despite the changes in the CH's dynamics caused by b_i and $u(t)$, both CHs can oscillate synchronously. We assume that the amplitude of the information signal is small enough to keep the CHs in the chaotic regime. In the receiver the phase discriminator (PD) generates a signal, which is a function of the phase difference between received and local CFM signals. The local CFM signal frequency is:

$$\omega_{re} = \dot{\varphi}_{re} = \omega_0(1 + m_1 x_{re}(t) + m_2 u). \tag{3.17}$$

The local VCO, PD, and the low-pass filter (LPF) make a phase locked loop, which tends to synchronize the local VCO to the transmitter. Assuming that the PD is a multiplier and the VCOs generate sinusoidal waveforms, the PD output is $A\sin(\varphi)$, where φ is the phase difference between the PD input signals $\varphi = \varphi_{tr} - \varphi_{re}$. We consider a LPF with the transfer function in the form $K(s) = 1/(1 + T_f s)$. In this case the joint dynamics of the transmitter and the receiver can be described by the following equations

$$\dot{\mathbf{X}}_{tr} = \mu \mathbf{F}(x_{tr} + \epsilon b_i; \bar{\mathbf{x}}_{tr}), \tag{3.18}$$

$$\begin{aligned} \dot{\mathbf{X}}_{re} &= \mu \mathbf{F}(x_{re} + \epsilon u; \bar{\mathbf{x}}_{re}) \\ \dot{\varphi} &= (\beta(x_{tr} - x_{re}) + b_i - u)/\lambda \\ \dot{u} &= \lambda(\sin(\varphi) - u), \end{aligned} \tag{3.19}$$

where $\mathbf{F}$ is the vector function of the state variables $\mathbf{X}_{\mathbf{re,tr}} \equiv \{x_{re,tr}; \bar{\mathbf{x}}_{re,tr}\} \in \mathbb{R}^d$ (d is the dimension of CHs) that represents CH dynamics in receiver and transmitter, $x_{re,tr}$ are scalar components through which the CHs are coupled, $\bar{\mathbf{x}}_{re,tr} \in \mathbb{R}^{d-1}$ represent remaining $d - 1$ components of the state vectors, ϵ is a coupling parameter, and u is the LPF output voltage; μ controls the characteristic time scale of the CHs; $\beta = m_1/m_2$ is the relative modulation depth; and $\lambda = 1/\sqrt{Am_2\omega_0 T_f}$ is the PLL damping parameter. The time in (3.18) and (3.19) is normalized by $\sqrt{T_f/Am_2\omega_0}$ and the phase variables $\{x_{tr,re}, \mathbf{x}_{\mathbf{tr,re}}, u\}$ and b_i by A.

The set of equations (3.18) and (3.19) has a synchronous solution:

$$\mathbf{X}_{\mathbf{re}}(t) \equiv \mathbf{X}_{\mathbf{tr}}(t), \quad u = b_i, \quad \varphi = \arcsin(u), \tag{3.20}$$

which corresponds to the regime with identical synchronization between the CHs and constant phase shift between the VCOs. In this regime the information signal b_i can be decoded by measuring the PLL output voltage u. In order to demonstrate that the synchronous regime is stable we assume that $\mu \ll 1$. In this case the equations (3.18) and (3.19) can be split into fast (the PLL, described by φ and u) and slow (the CHs, described by $\mathbf{X_{tr}}$ and $\mathbf{X_{re}}$) subsystems. The fast subsystem for which $\mathbf{X_{tr}}, \mathbf{X_{re}}, b_i$ can be considered constant, has an equilibrium state:

$$u = u_s = \beta(x_{tr} - x_{re}) + b_i, \varphi = \varphi_s = \arcsin(u_s), \qquad (3.21)$$

This solution exists and is stable if $\beta(x_{tr} - x_{re}) + b_i < 1$. As soon as the fast subsystem converges to steady state (u_s, φ_s), the CHs become dissipatively coupled. It has been shown in many examples that the dissipative coupling can provide stable synchronization of chaotic systems (see, for example, [34]). Assuming that the conditions for the CHs synchronization are satisfied, we conclude that the synchronous regime (3.18), (3.19) exists and is stable. In this regime $u = b_i$ and, therefore, the transmitted information can be detected in the receiver. This communication system can use binary or analog information signals b_i.

The modulation scheme was tested in computer simulations. For the chaotic oscillators we used the Rössler oscillators of the form:

$$\begin{aligned}
\dot{x}_1 &= \mu(-x_2 - x_3 - d) \\
\dot{x}_2 &= \mu(2.25x_1 + 0.15x_2) \\
\dot{x}_3 &= \mu(0.04 + 5x_3(x_1 - 2))
\end{aligned} \qquad (3.22)$$

with $x_{t,r} = x_1$ and $d = b_i$ in the transmitter and $d = u$ in the receiver. The parameters were set as the following: $\mu = 0.05$, $\omega_0 = 40$, $m_1 = 0.01$, $m_2 = 0.2$, and $Bi = \pm 0.4$.

The results of numerical simulations are presented in Figure 3.14, which shows the waveforms of the CH_{re} output, errors of chaos synchronization, and decoded binary signal given by random sequence of bits.

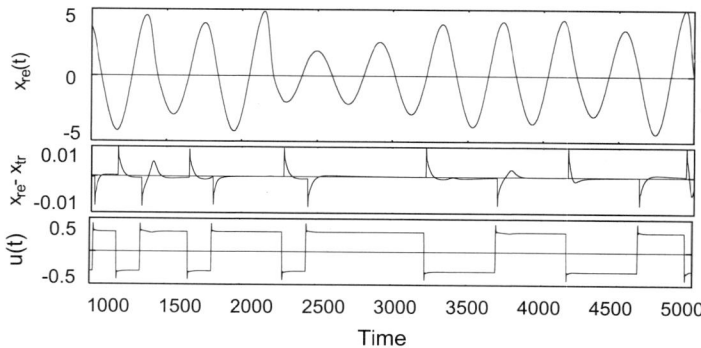

Fig. 3.14. The simulated waveforms for the system (3.18), (3.19) with random binary information signal b_i

3.4.3 Multiuser CFM Communication System

As shown above, the CFM signals with phase or frequency modulation can be used for information transmission. In order to characterize the efficiency of a communication system the BER performance of the scheme in the presence of different kinds of interferences should be evaluated. In this section we present the BER characteristics of CFM systems with additive channel noise and propose a multiple access technique based on the synchronization of chaotic oscillators in different users.

To describe the application of the CFM method for multiuser communications we propose the following general multiuser communication scheme (Figure 3.15). The base station (BS) and all the mobile units (MU) have the chaos generators (CH) with closely matched parameter values. The base station uses its CH to generate a CFM signal which it broadcasts to all the mobile units. Each MU receives this signal, and uses it for synchronization of its CH to the BS chaos generator. Once all chaos generators in the cell are synchronized, unit i can transmit information to unit k or to the BS using the chaotic waveform to generate its own information-bearing CFM signal. In order to minimize user-to-user interference, each transmitter changes the chaotic waveforms, which are synchronized and therefore identical in all users, by applying its own unique transformation f_i to the chaotic waveform before generating its CFM carrier. The channel in which the mobile units are transmitting the CFM signals should be separated from the synchronization channel, in which the base station is transmitting the reference CFM signal, to avoid interference with the synchronization signal. Let us consider first the synchronization channel. In the following, we denote the variables and acronyms corresponding to the base station by subscript b, and the ones corresponding to the mobile units by subscript u. The chaos generator of the base station CH$_b$ is described by the following equation

$$\tau \dot{\boldsymbol{x}}_b = \boldsymbol{F}(\boldsymbol{x}_b), \tag{3.23}$$

where $\boldsymbol{x}$ is the vector of state variables of CH$_b$; $\boldsymbol{F}$ is the nonlinear vector field function; and τ denotes the characteristic time constant of the chaos generator. One of the variables of the chaotic system at the base station, x_b, is used to control the frequency of the VCO$_b$ producing the reference CFM signal. The equation for the instantaneous phase of the reference CFM signal reads:

$$\dot{\varphi}_b \equiv \omega_b = \omega_0(1 - mx_b), \tag{3.24}$$

where ω_0 is the "natural" frequency of VCO$_b$, and m is the modulation gain coefficient. The instantaneous phase of the CFM signal generated by the VCO$_u$, φ_u, is described by a similar equation:

$$\dot{\varphi}_u \equiv \omega_u = \omega_0(1 - mx_u - \tilde{m}v), \tag{3.25}$$

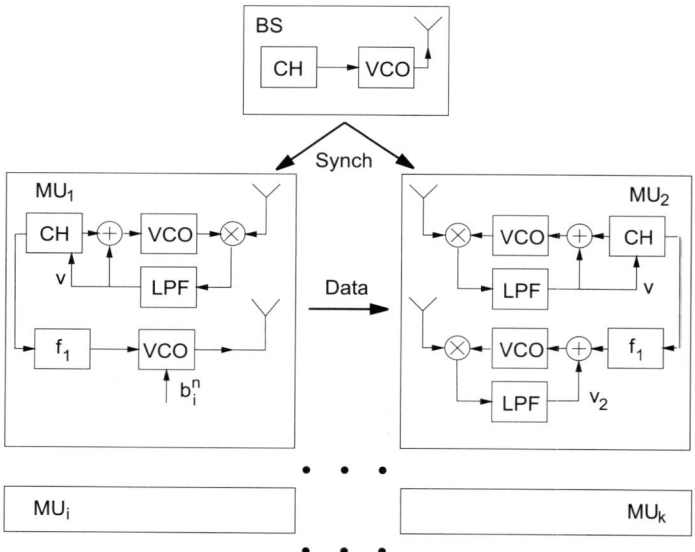

Fig. 3.15. Block diagram of the multiuser CFM.

where v is the phase-lock loop output signal, $\tilde{m}$ is the VCO sensitivity, and $x_u(t)$ is the chaotic signal generated by the CH_u. The latter is described by the equation:

$$\tau \dot{\boldsymbol{x}}_u = \boldsymbol{F}(\boldsymbol{x}_u) + av. \qquad (3.26)$$

This chaos generator is driven by the PLL output signal v; a determines the coupling strength. The PLL is a combination of phase discriminator ($\otimes$), LPF, and VCO (see Figure 3.15). For the first-order LPF $K(s) = 1/(1 + Ts)$ the equation for the PLL can be written as follows,

$$T\dot{v} = \Phi(\varphi_u - \varphi_b) - v, \qquad (3.27)$$

where $\Phi(x) = \sin(x)$ if the phase detector is implemented using a multiplier and the carriers have sinusoidal waveforms.

We assume that the frequency range of the information transmission does not overlap with the frequency range of the synchronization signal, so the information-bearing signals do not interfere with the synchronization signal. In this case the analysis of the synchronization process between the base station and the mobile units is equivalent to the one made in Section 3.3.2. If the characteristic time constant τ of the chaos generators is much greater than $\sqrt{(T/(\tilde{m}\omega_0))}$, then there exists a range of parameters $T, m, \tilde{m}, \omega_0, a$, in which the synchronized solution of Equations (3.23)–(3.27),

$$\boldsymbol{x}_u = \boldsymbol{x}_b,$$
$$\varphi_u = \varphi_b,$$
$$v = 0,$$
(3.28)

is globally stable.

Now let us turn to the information transmission among the mobile units. Once CHs of the mobile units are synchronized to the CH_b of the base station, the mobile units can use the CFM signal for information transmission. If the frequency of the CFM signal transmitted by the ith mobile unit (transmitter) is modulated by the binary information signal (BFSK), it can be described by the following equation,

$$\omega_i \equiv \dot{\varphi}_i = \omega_1(1 + m_1 f_i(\boldsymbol{x}_i) + m_2 b_i^n), i = 1, N.$$
(3.29)

Here N is the total number of transmitting mobile units, $f_i(\boldsymbol{x})$ is a unique function assigned to the ith unit, and $b_i^n = \pm 1$ is the nth bit of its information sequence.

If the information from the ith unit is sent to the kth unit, the latter must use the same function $f_i(\boldsymbol{x})$ to be able to stay tuned to the signal sent by the ith unit. The dynamics of the PLL in the signal channel at the kth unit is controlled by the equations

$$\dot{\varphi}_k \equiv \omega_k = \omega_1(1 + m_1 f_i(\boldsymbol{x}_k) + m_3 v_k),$$
$$T\dot{v}_k = \sum_{i=1, i \neq k}^{N} \Phi(\varphi_k - \varphi_i) - v_k.$$
(3.30)

If only one ith unit is transmitting ($N = 1$), and the channel is noise- and distortion-free and the CHs are synchronized $\boldsymbol{x}_i = \boldsymbol{x}_k$, the set of Equations (3.29)–(3.30) possesses a synchronized solution:

$$v_k = m_3^{-1} m_2 b_n^i,$$
$$\varphi_k - \varphi_i = \Phi^{-1} v_k$$
(3.31)

Therefore, the receiver k can detect the transmitted bit b_i^n without errors by determining the sign of the PLL output v_k integrated over the bit duration, which means that the PLL in the data channel works as a frequency discriminator in respect to the CFM signal.

In a multiuser system, other units besides i are transmitting the information-bearing signal in the same frequency band near ω_1. This causes an interference signal in the PLL of the kth receiver, which along with the channel noise leads to certain errors. In the next section we study the bit error rate performance in the presence of an additive channel noise and user interference.

3.4.4 Performance Evaluation

The BER performance of a CFM communication link in a noisy channel was studied in an experiment with electronic circuits. The block diagram of a

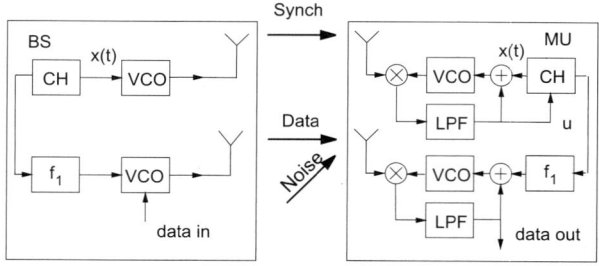

Fig. 3.16. Block diagram of experimental circuit used for measurement of BER characteristics of CFM communication system in noisy channel.

simplified system is shown in Figure 3.16. This simplified system models one-directional communication between the BS and a MU. The synchronization channel only has the relatively small internal noise of the circuit. The additional wideband noise was mixed with the transmitted signal in the data channel. In both experiments and numerical simulations we used the following simple chaotic oscillator (model B from [35]),

$$
\begin{aligned}
\dot{x} &= yz \\
\dot{y} &= x - y \\
\dot{z} &= 1 - xy.
\end{aligned}
\tag{3.32}
$$

The hardware implementation of this system in the electronic circuit is shown in Figure 3.17. This circuit uses one quadruple operational amplifier TL084N and two multipliers AD633. The chaotic attractor of system (3.32) is shown in Figure 3.18a. The VCOs and PLLs of the base station and the mobile unit were implemented using a standard 74HC4046 PLL chip. The transmitter and receiver circuits are shown in Figure 3.19. The nominal frequency f_0 of the VCOs was set to 1.8 MHz. The characteristic time constant of the chaos generators was 56 μmsec. In these experiments the base station and the mobile units were connected by a wire. As a function $f(\mathbf{x})$ controlling the VCOs of the mobile units, we used the second variable y of the chaotic generator (3.32) (see Figure 3.17).

The quality of the synchronization between the base station and the mobile units can be illustrated by the plot x_u versus x_b (Figure 3.18b).

The power spectra of the chaotic signal ("baseband") and the CFM signal are shown in Figure 3.20. We transmitted binary information (pseudo-random sequence of bits) between two mobile units by increasing the instantaneous frequency of the VCO_u by 30 kHz with respect to f_u for bit "1" transmission, and decreasing it by 30 kHz for bit "0". The output signal from the phase lock loop of the receiving mobile unit is shown in Figure 3.21. It was integrated over the bit duration for detecting the information bit. To evaluate the system

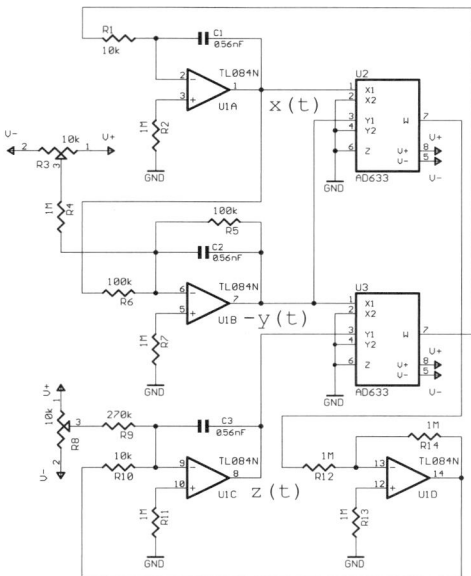

Fig. 3.17. Circuit digram of the chaotic oscillator used in the experiments.

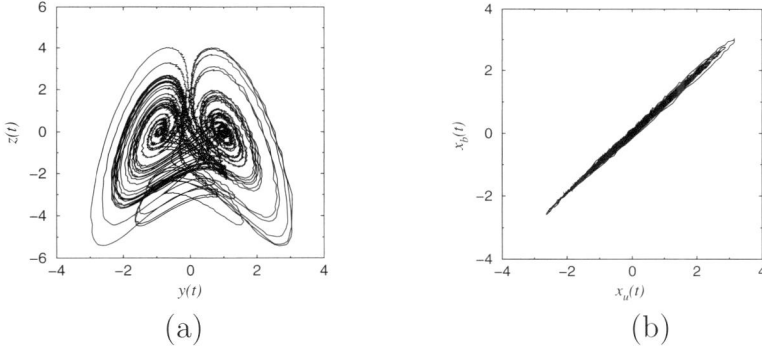

Fig. 3.18. (a) chaotic attractor of system (3.32) obtained experimentally in a circuit shown in Figure 3.17; (b) synchronization between the chaos generators of the base station and a mobile unit in electronic circuit.

performance, we added white Gaussian noise to the transmission channel, and calculated the bit-error rate as a function of the normalized signal-to-noise ratio (more precisely, ratio of the energy per bit to the spectral density of noise E_b/N_0). The obtained performance curve can be compared with the standard noncoherent FSK performance [1, 2] (see Figure 3.22a). As one can see, the performance is slightly worse that optimal noncoherent FSK (by about

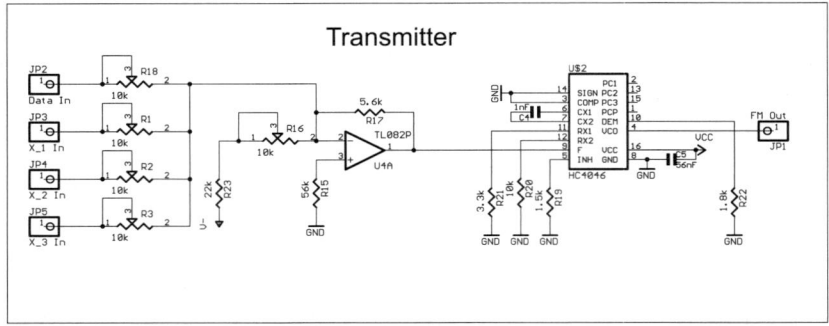

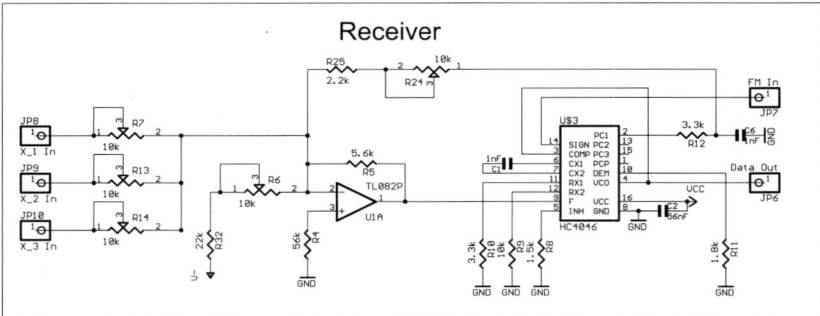

Fig. 3.19. Circuit diagrams of the transmitter (top) and the receiver (bottom).

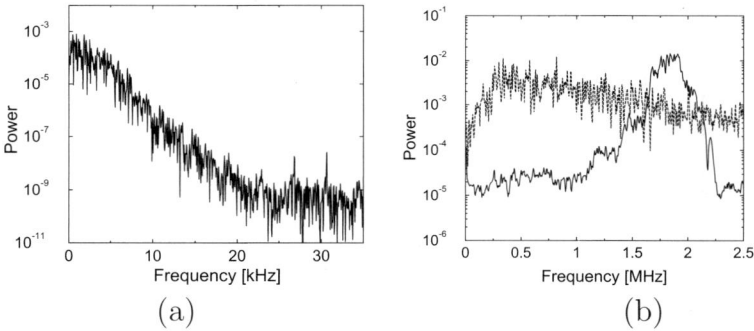

Fig. 3.20. Power spectra of the baseband chaotic signal (a) and the CFM signal (b). In the right plot, the power spectrum of the channel noise is also shown by the dotted line.

4 dB) for BER $> 10^{-3}$, however, the curve flattens out at lower BER. This behavior is a result of occasional events of PLL desynchronization due to the CHs synchronization errors and the internal noise in the receiver. When the CHs were perfectly synchronized the system performed at lower BER, but the

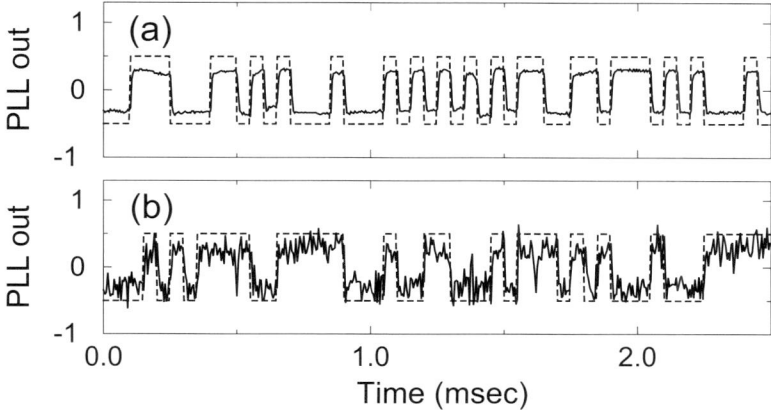

Fig. 3.21. The output voltage v from the PLL of the mobile unit for two different values of the noise level: (a) $E_b/N_0 = 20dB$; (b) $E_b/N_0 = 7dB$. Dotted lines indicate the original bit sequence.

saturation still took place due to the the internal noise in the experimental setup. The performance of the system described in the previous section for

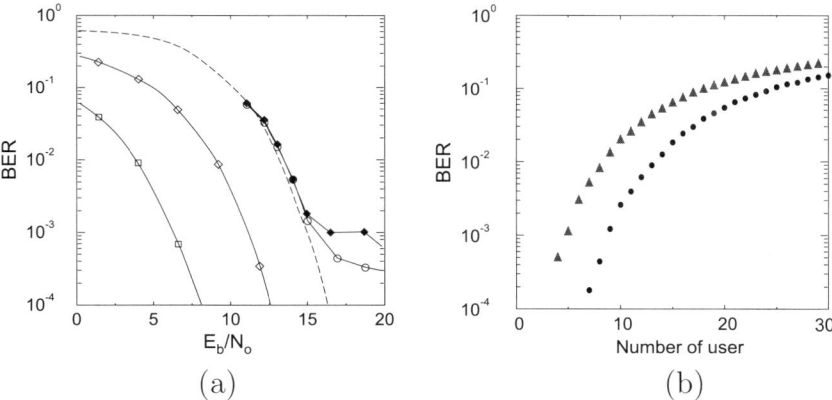

Fig. 3.22. (a) Bit-error rate in the single-user CFM transmission as a function of the normalized signal-to-noise ratio E_b/N_0 (solid squares), same for perfectly synchronized chaotic oscillators at the transmitter and the receiver (open circles), BPSK (open squares), optimal non-coherent FSK (diamonds), and FSK with PLL detector (dashed line). (b) Bit-error rate in the multi-user CFM transmission as a function of the number of users for two values of bit length: 20τ (triangles) and 40τ (circles).

multiple pairs of mobile units was studied in the numerical simulations. We chose the parameters of the simulation to satisfy the stability criteria of the synchronization in a single-user system, namely, $m_u = 0.05, m_1 = 2, T = 20, \tau = 50$. Here T and τ were set in reference to ω_1, which was set to 1. The synchronization plot x_u versus x_b is similar to the experimental one presented in Figure 3.18b. In order to generate statistically independent signals using a single "source" signal $x(t)$ from the base station, transmitters used time-delayed versions of the chaotic signal $x(t - \Delta T_i)$ with different time delays ΔT_i corresponding to different users. Figure 3.22b shows the bit-error rate as a function of the number of users for bit lengths 20τ and 40τ. The BER smoothly increases with the number of users which indicates that the system has good multiple access capabilities.

3.5 Conclusions

We presented two schemes for communication systems based on chaotic signals and frequency modulation. For the case when a chaotic signal is used as the carrier, the frequency modulation is realized by varying the time-scale parameter of the chaotic oscillator. In the receiver this frequency modulated chaotic signal can synchronize the local chaotic oscillator by using two feedback control loops independently operating in the time (τ-loop) and the phase space (x-loop) domains. The $x - \tau$ synchronization scheme is a generalization of the phase locked loop for chaotic systems and the main properties of this scheme are very similar to the ordinary PLL. Therefore the majority of applications of a PLL can be directly transformed to the chaotic systems with an $x - \tau$ synchronization loop. In particular, modulation of the time-scale parameter in a chaotic oscillator by an information signal results in generating a frequency modulated chaotic signal, whose information content can be demodulated in the τ-loop. If the chaotic waveform is multiplied by ± 1 in accordance with a binary information signal, which is equivalent to the BPSK modulation with $\pm \pi/2$ phase shift, a similar synchronization scheme operating with squared signals can be used as a demodulator.

In the second scheme the chaotic signal is used for frequency modulation of a periodic carrier to broaden the spectrum. This system is similar to a system with interpolated frequency hopping where the pseudo-noise spreading sequence is digitally interpolated to ensure smooth carrier frequency variation instead of random hopping. This method differs from a simple modulation/demodulation technique, in which a phase lock loop directly reconstructs the low-frequency chaotic signal in the receiver. In the proposed method the frequencies of the voltage-controlled oscillators in both transmitter and receiver are modulated by the chaotic generators. The phase lock loop detects the error (difference) between the chaotic signals, which is used for synchronization of the chaotic oscillators. This allows a reduced PLL gain and provides good selectivity of this scheme with respect to the in-band interference.

The PLL-based synchronization scheme has a wide region in the parameter space where the synchronous regime is globally stable. In contrast with traditional frequency hopping, in this system the receiver is always synchronized to the transmitted carrier signal and both phase and frequency modulation can be used for the information signal. Our computer simulations show that this scheme is robust against channel noise and interference and has good bit-error-rate characteristics.

Acknowledgments

The authors acknowledge numerous discussions on the subject with H. Abarbanel, L. Larson, and M. Sushchik. This work was supported by the U.S. Army Research Office under MURI grant DAAG55-98-1-0269 and by the U.S. Department of Energy under grant DOE/DE-FG03-95R14516.

References

1. R. A. Scholtz. The Origins of Spread-Spectrum Communications. *IEEE Trans. Commun.*, COM-30(5), pp. 822–854, 1982.
   R. L. Pickholtz, D. L. Schilling, and L. B. Milstein. Theory of Spread-Spectrum Communications–A Tutorial *IEEE Trans. Commun.*, com-30 (5), pp. 855–884, 1982.
2. M. K. Simon, J. K. Omura, R. A. Scholtz, and B. K. Levitt. Spread Spectrum Communications, Computer Science Press, Rockville, MD, 1985.
   M. K. Simon, J. K. Omura, R. A. Scholtz, and B. K. Levitt. Spread Spectrum Communication Handbook (McGraw-Hill, New York, 1994).
3. J. Karaoguz. High-rate wireless personal area networks, *IEEE Commun. Mag* vol. 39 (12), pp. 96–102, 2001. 39:96–102.
4. M. P. Kennedy and G. Kolumban (Eds.) *IEEE Trans. Circuit. Syst. I*, vol. 47(12), pp. 1661–1732, 2000.
5. L. Kocarev, G. M. Maggio, M. Ogarzalek, L. Pecora, and K. Yao (Eds.) *IEEE Trans. Circuit. Syst. I* vol. 48 (12), pp. 1385–1527, 2001.
6. L. Pecora and T. Carroll. Synchronization in chaotic systems, *Phys. Rev. Lett.* vol. 64 (8), pp. 821–824, 1990.
7. M. P. Kennedy, G. Kolumban, G. Kis, and Z. Jako, Performance evaluation of FM-DCSK modulation in multipath environments *IEEE Trans. Circuit Syst. I*, vol. 47 (12), pp. 1702–1711, 2000. M. P. Kennedy, G. Kolumban, G. Kis, Simulation of the multipath performance of FM-DCSK digital communications using chaos, in *Proceedings of the 1999 IEEE International Symposium on Circuits and Systems VLSI, ISCAS'99* (Cat. No.99CH36349), vol. 4, pp. 568–571, 1999.
8. L. Kocarev, K, S. Halle, K. Eckert, L. O. Chua, and U. Parlitz, Experimental Demonstration of Secure Communications via Chaotic Synchronization *Int. J. Bifurc. Chaos* vol. 2 (3), pp. 709–713, 1992.
   L. Kocarev, U. Parlitz, General Approach for Chaotic Synchronization with

Applications to Communication *Phys. Rev. Lett.* vol. 74(25), pp. 5028–5031, 1995.

K. Cuomo and A. V. Oppenheim, Circuit implementation of synchronized chaos with applications to communications, *Phys. Rev. Lett.* vol. 71 (1) pp. 65–68, 1993.

T. Carrol and L. Pecora, Cascading synchronized chaotic systems, *Physica D* vol. 67 (2), pp. 126–140, 1993.

A. R. Volkovskii, N. F. Rul'kov, Synchronous chaotic response of a nonlinear oscillator system as a principal for detection of the information component of chaos, *Sov. Tech. Phys. Lett.*, vol. 19, p. 97, 1993.

9. U. Feldmann, M. Hasler, W. Schwarz, Communication by chaotic signals: The inverse system approach, *Int. J. Circuit Theory and Applications* vol. 24 (5), pp. 551–579, 1996.

10. U. Parlitz and L. Kocarev, Multichannel communication using auto-synchronization, *Int. J. Bifurcation and Chaos*, vol. 6 (3), pp. 581–588, 1996.

11. W. P. Torres, A. V. Oppenheim, and R. R. Rosales, Generalized frequency modulation, *IEEE Trans. Circuit. Syst. I*, vol. 48 (12), pp. 1405–1412, 2001.

12. C. C. Chen and K. Yao, Numerical evaluation of error probabilities of self-synchronizing chaotic communications, *IEEE Comm. Lett.*, vol. 4 (2), pp. 37–39.

C. C. Chen and K. Yao, Stochastic-calculus-based numerical evaluation and performance analysis of chaotic communication systems, *Trans. Circuit. Syst. I*, vol. 47 (12), pp. 1663–1672, 2000.

13. G. M. Maggio and Z. Galias, Applications of symbolic dynamics to differential chaos shift keying, *IEEE Trans. Circuit. Syst. I*, vol. 49 (12),pp. 1729–1735, 2002.

14. N. F. Rulkov and L. S. Tsimring, Synchronization methods for communication with chaos over band-limited channels, *Int. J. Circuit Theory Appl.*, vol. 27 (6), pp. 555–567, 1999.

15. M. M. Sushchik et al., Chaotic pulse position modulation: a robust method of communicating with chaos *IEEE Comm. Lett.*, vol. 4 (4), pp. 128–130, 2000.

16. G. Heidari-Bateni and C. D. McGillem, A chaotic direct-sequence spread-spectrum communication system, *IEEE Trans. Commun.*, vol. 42 (3), pp. 1524–1527, 1994.

G. Heidari-Bateni, C. D. McGillem, and M. F. Tenorio, A novel multiple address digital communication. *In: IEEE Int. Conf. Communications ICC'92* pp. 1232–1236, 1992.

T. Kohda and A. Tsuneda, Statistics of chaotic binary sequences, *IEEE Trans. Information Technology*, vol. 43, (1), pp. 104–112, 1997.

R. Hegger, H. Kantz, L. Matassini, Chaos-based asynchronous DS-CDMA systems and enhanced rake receivers: measuring the improvements, *IEEE Trans. Circuit. Syst. I* vol. 48 (12), pp. 1445–1453, 2001.

17. C. C. Chen, K. Yao K, K. Umeno, and E. Biglieri Design of spread-spectrum sequences using chaotic dynamical systems and ergodic theory, *IEEE Trans. Circuit. Syst. I* vol. 48 (9) pp. 1110–1114, 2001.

18. L. Cong and L. Shaoqian, Chaotic spreading sequences with multiple access performance better than random sequences, *IEEE Trans. Circuit. Syst. I*, vol. 47 (3), pp. 394–397, 2000.

19. L. Cong and S. Songgeng, Chaotic frequency hopping sequences, *IEEE Trans. Commun.* vol. 46 (11), pp. 1433–1437, 1998.

L. Cong; W. Xiaofu, Design and realization of an FPGA-based generator for chaotic frequency hopping sequences, *IEEE Trans. Circuit. Syst. I*, vol. 48 (5), pp. 521–532, 2001.

20. P. A. Bernhardt, Chaotic frequency modulation, *in Proceedings of SPIE - the International Society for Optical Engineering*, vol. 2038 pp. 162–181, 1993.
    P. A. Bernhardt, Communications using chaotic frequency modulation, *Int. J. Bifurc. Chaos in Appl. Sci. Eng.*, vol. 4 (2), pp. 427–40, 1994.

21. A. Volkovskii, Synchronization of chaotic systems using phase control *IEEE Trans. Circuit. Syst. I* vol. 44 (10), pp. 913–917, 1997.

22. F. M. Gardner, Phaselock Techniques (Wiley, New York, 1979).

23. E. Ott, C. Grebogi, and J. A. Yorke, Controlling chaos, *Phys. Rev. Lett*, vol. 64 (11), pp. 1196–1199, 1990.

24. Y. C. Lai and C. Grebogi, Synchronization of chaotic trajectories using control, *Phys. Rev. E*, vol. 47 (4), pp. 2357–2359, 1993.

25. D. Vassiliadis, Parametric adaptive control and parameter identification of low-dimensional chaotic systems, *Physica D*, vol. 71 (3), pp. 319–341, 1994.

26. A. S. Dmitriev, Y V. Kislov, and S O. Starkov, Experimental study of appearance and interaction of strange attractors in the circle type self-oscillator, *Sov. Phys. Tech. Lett.* vol. 30, pp. 1439–1441, 1985.

27. N. F. Rulkov, A. R. Volkovskii, A. Rodriguez-Lozano, E. Del Rio, and M. G. Velarde, Mutual synchronization of chaotic self-oscillators with dissipative coupling, *Int. J. Bif. Chaos* vol. 2, pp. 669–676, 1992.

28. N. M. Filiol, C. Plett, T. Riley, and M. A. Copeland, An interpolated frequency-hopping spread-spectrum transceiver, *IEEE Trans. Circuits Syst* vol. 45, pp. 3–12, 1998.

29. A. J. Viterbi, Principles of Coherent Communication (McGraw-Hill, New York, 1966).

30. A. R. Volkovskii and L. S. Tsimring, Synchronization and communication using chaotic frequency modulation *Int. J. Circ. Theor. Appl.*, vol. 27 pp. 569–576, 1999.

31. H. Fujisaka and T. Yamada, Stability Theory of Synchronized Motion in Coupled-Oscillator Systems, *Progress of Theoretical Physics* (Japan), vol. 69 (1), pp. 32–47, 1983.

32. V.S. Afraimovich, N.N. Verichev, and M.I. Rabinovich, Stochastic Synchronization of Oscillations in Dissipative Systems, *Radio Phys. and Quantum Electron.*, vol. 29, pp. 747–751, 1986.

33. N. F. Rulkov, A. R. Volkovskii, A. Rodriguez-Lozano, E. Del Rio and M. G. Velarde, Mutual synchronization of chaotic self-oscillators with dissipative coupling, *Int. J. Bif. and Chaos*, vol. 2 pp. 669–676, 1992

34. L. Pecora, T. Carrol, G. Johnson, and D. Mar, Fundamentals of synchronization in chaotic systems, concepts and applications, *Chaos*, vol. 7 (4) pp. 520–543, 1997.

35. J. C. Sprott, Some simple chaotic flows, *Phys. Rev. E*, vol. 50 pp. 647–650, 1994.

4

Ultra-Wideband Communications Using Pseudo-Chaotic Time Hopping

David C. Laney and Gian Mario Maggio

Summary. Pseudo-chaotic time hopping (PCTH) is a recently proposed encoding/modulation scheme for UWB (ultra-wide band) impulse radio. PCTH exploits concepts from symbolic dynamics to generate aperiodic spreading sequences, resulting in a noiselike spectrum. In this chapter we present the signal characteristics of single-user PCTH as well as a suitable multiple access technique. In particular, we provide analytical expressions for the BER (bit-error-rate) performance as a function of the number of users and validate it by simulation.

4.1 Background

Merriam-Webster [1] defines radio as, " the wireless transmission and reception of electric impulses or signals by means of electromagnetic waves." For over eighty years, communication via wireless electromagnetic waves has been a cornerstone of the development of modern society [2]. Only the automobile has had as big an impact on society as the radio. The goal of early radio was to achieve wireless communication over greater and greater distances. As understanding and technology improved, the focus shifted to reducing the cost and complexity of radios to ensure their commercial viability. More recently, the focus has been to improve the density of multiuser communications given a fixed available bandwidth. By and large, this has been accomplished by the continued improvement of narrowband (NB) communications facilitated by ever-improving electronic components and the use of information theory to maximize the information density of transmitted signals. Now that almost all of the usable spectrum is dedicated to specific and protected users, how can the multiuser density be improved further? Possibly by the development of ultra-wideband communications where the signal from each user overlays other ultra-wideband users as well as existing narrowband users.

The goal of narrowband systems is to transmit as much information as possible on a small slice of frequency space. Usually, we would like to transmit the information as far as possible for a given amount of power. Other system users must be able to do the same without catastrophic interference. This has driven high data rate narrowband communications to higher and higher center frequencies with small percentage bandwidths expressed as

$$\%BW = \frac{Signal\ Bandwidth}{Center\ Frequency}. \tag{4.1}$$

So, a 10 GHz signal with a 100 MHz bandwidth would have a percentage bandwidth of 1%. Multiple users are enabled by having channels separated in the frequency, time, and/or code domains. In each of these domains, the character of the transmitted signal is concentrated in frequency. Users must be coordinated in time, frequency and/or code domains to keep their data separated from one another.

Another method of supporting multiple users is to spread the energy from each user over a very wide bandwidth. If the percentage bandwidth of the signal,

$$\%BW \geq 20\%, \tag{4.2}$$

then it is called an *ultra-wideband* signal (UWB). Alternatively, a signal that occupies $\geq$500 MHz of bandwidth is also called an ultra-wideband signal.

One possible way of generating UWB signals is through *impulse radio*, that is, generating ultra-short pulses characterized by an extremely broad spectrum. In contrast with narrowband communication systems, this type of signal is generally localized in time rather than frequency. Although impossible to implement, a conceptually simple UWB signal is an impulse function. If the transmitted waveform were a single impulse function, the transmitted power spectral density (PSD) would be flat over all frequencies and the spectrum of one UWB user's transmitted signal would overlay every other UWB and NB user's spectrum. Within any small range of frequencies, the power in the transmitted signal would be very low. This leads to the hope that UWB signals will not interfere very much with NB users (still an open research question). Ideally, the UWB signals would have spectral notches at frequencies occupied by narrowband users to reduce or prevent interference. This is another current topic in the research literature. The next best spectral shape of a UWB signal is flat over a range of frequencies. Interference from one UWB user to another is visualized in the time domain as signals from multiple users being transmitted at the same time (colliding) or being transmitted at different times.

To transmit a stream of information with a UWB system, multiple impulses may be used. Recall that a periodic series of impulses spaced every T seconds has a Fourier transform that is also periodic, with frequency separation $2\pi/T$. On the other hand, an infinite series of impulses randomly spaced in time has a continuous spectrum (*i.e.*, without spectral lines). So, a good modulation format for impulse radio would consist of repeating an impulse function with random interpulse timing. The basic idea of this chapter is to utilize chaos, which still is deterministic, to generate the randomness necessary to produce a feature-free spectrum. In particular, we apply aperiodic time hopping to a UWB impulse radio system. The use of aperiodic modulation preserves the nice spectral features of impulse radio, and the deterministic quality can be used for error correction purposes.

4.1.1 Why Ultra-Wideband Impulse Radio?

The need for wireless infrastructure is growing rapidly. Someday in the near future wireless infrastructure will be as prevalent as telephony and electrical power are today. The explosion in the number of users has been facilitated by the constantly falling cost of portable computers, personal digital assistants, cell phones, and improving information and communications technologies.

As the density of users of digital communications equipment increases, the need to support more users per unit area grows. An emerging network performance metric for personal area, local area, and wide area networks is the spatial information capacity measured in bits per second per unit area reflecting the need to support a large number of users within a confined space each requiring a high data rate. Notice that long-range transmission is not the goal here as is typically the case with communications systems. In order to have a large spatial capacity, it is desirable for the signal strength to decay quickly enough to allow frequency reuse in nearby areas. UWB impulse radio may offer good performance for this application.

The spatial capacity of IEEE 802.11b with 100 meter range and aggregate speed of 33 Mbps is about 1 Kbps per square meter [5], and 802.11a with 50 meter range and an aggregate speed of 648 Mbps results in a spatial capacity of approximately 83 Kbps per square meter. Bluetooth, on the other hand, is targeted for 10 m applications with an aggregate speed of 10 Mbps or a density of 30 Kbps per square meter. The spatial capacity of UWB systems varies from system to system but in [5], it is projected that six users can be supported at 50 Mbps with 10 m range resulting in 1000 Kbps per square meter, an enormous increase in density.

One of the reasons UWB systems have the potential for such a large spatial capacity lies in the channel capacity of wideband signals. In 1948, Shannon formalized Nyquist's work to optimize band-limited channels operating in additive white Gaussian noise (AWGN). Shannon's channel capacity, C, is expressed as

$$C = W log_2 \left(1 + \frac{P}{W N_0} \right), \tag{4.3}$$

where W is the band-limited channel bandwidth, P is the transmit power, and N_0 is the noise power spectral density. If the data rate R is less than C ($R < C$), error-free data transmission is theoretically possible. If $R > C$, errors will occur regardless of what processing is performed at the transmitter and receiver. Notice the channel capacity increases linearly with bandwidth but only logarithmically with signal-to-noise ratio (transmit power). So, the linear expansion of channel capacity with channel bandwidth provides some insight into why UWB signals are natural for high data rate applications. It should be noted, however, that nothing comes for free. High data rates require high pulse repetition frequencies (PRFs) resulting in increased transmit power and interference to existing narrowband users. Recent work by Fontana [6] suggests that PRFs greater than the resolution bandwidth of a narrowband receiver causes interference to the narrowband receiver proportional to the square of the PRF.

UWB impulse radio has the potential to be power efficient by scaling the pulse repetition frequency to the desired data rate. In typical coherent narrowband systems, a sinusoidal carrier must be transmitted continuously to maintain phase lock between the transmitter and receiver. Synchronous UWB impulse radio must also maintain timing lock between the transmitter and receiver but with rapid rise-time and short duration impulses. The average transmitted power can be very low for low data rates and higher for higher data rates whereas the peak power in both cases is very high with a constant energy per bit. Time Domain Corporation claims a prototype impulse radio system that achieves a 39 Kbps transmission rate with 250 microwatts of transmit power at ranges up to 10 miles [7]. UWB impulse radio transceivers are also potentially simple and inexpensive to produce because no voltage-controlled oscillators or mixers are required.

UWB impulse radio has the potential for good narrowband jamming immunity because the signal is spread over an ultra-wide bandwidth. In this case, the spectrum of the narrowband jammer occupies a small portion of the UWB users' spectrum leaving a large portion of the spectrum unjammed. Of course, the specific receiver design is important and if not designed carefully could be intolerant of narrowband jammers. Because the power spectral density of UWB signals is low, the probability of detection and intercept is low using traditional narrowband detection equipment. This, of course, will lead to better detection equipment, possibly operating in the time domain.

The Federal Communications Commission (FCC) has provisionally approved intentional radiation between 3.1–10.6 GHz for UWB products under a strict power-level mask [8]. The levels approved were already acceptable under FCC Part 15 guidelines for unintentional radiators. The fact that the FCC already allowed unintentional interference to existing narrowband users in the band is an indication of how low the permissible power level is. Figure 4.1 indicates the permissible power level over the band for indoor devices. The low permissible power levels encourage the design of systems that produce outputs with a flat transmit spectrum. It is shown in [9] that pseudo-chaotic time hopping of impulse radio produces a flat spectrum without the spectral comb lines produced when periodicity is introduced into the modulation as in [7].

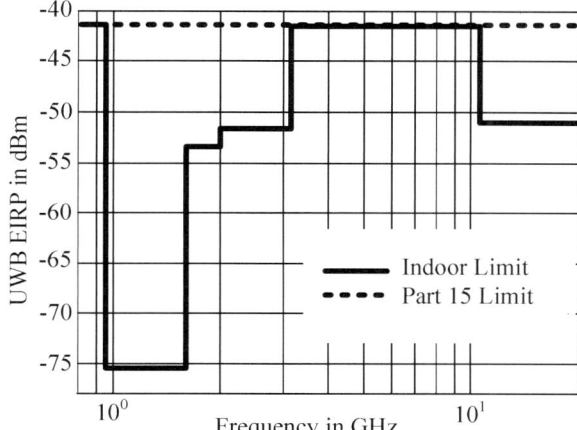

Fig. 4.1. Maximum allowable emission limits provisionally granted by the FCC for indoor ultra-wideband radio. (Reprinted with permission from [8].)

4.1.2 Pseudo-Chaotic Modulation

Over the last decade, there has been great interest in UWB impulse radio communication systems. These systems make use of ultra-short duration ($<$1 ns) pulses that yield ultra-wide bandwidth signals characterized by extremely low power spectral densities [10,11]. UWB systems are particularly promising for short-range wireless communications as they potentially combine reduced complexity with low power consumption, low probability of intercept (LPI), and immunity to multipath fading. The successful deployment of UWB technology depends strongly on the development of efficient multiple access techniques. Existing UWB communication systems employ pseudo-random noise (PN) time hopping for multiple access purposes, combined with pulse-position modulation (PPM) for encoding the digital information. An analysis of the multiuser capabilities of such systems has been presented by Scholtz et al. in [10,12–15].

Recently, it has been suggested to use aperiodic (chaotic) codes in order to enhance the spread-spectrum characteristics of UWB systems by removing the spectral features of the transmitted signal, thus resulting in a low probability of intercept. In addition, the absence of spectral lines may translate into a reduced interference toward other services such as GPS (Global Positioning System) [16]. In [17], the use of aperiodic sequences of pulses in the context of a chaos-based communication system was first proposed. A few schemes with chaotic modulation of the interpulse intervals were then studied in [18, 19]. In [20], a similar scheme was designed for the transmission of binary information and named chaotic pulse-position modulation (CPPM). Also, a scheme introducing a frequency modulation on top of the chaotic time hopping has been reported in [21].

Pseudo-chaotic time hopping is a modulation scheme for UWB impulse radio, was first proposed in [9, 22]. PCTH exploits concepts from symbolic dynamics [23] to generate aperiodic spreading sequences that, in contrast to fixed (periodic) PN sequences, depend on the input data. The PCTH scheme combines pseudo-chaotic encoding with multilevel pulse-position modulation.

Pseudo-chaotic time hopping (PCTH), a modulation scheme for UWB impulse radio that exploits concepts from symbolic dynamics [23] to generate aperiodic spreading sequences that, in contrast to fixed (periodic) PN sequences, depend on the input data. The PCTH scheme combines pseudo-chaotic encoding with multilevel pulse-position modulation. The pseudo-chaotic encoder operates on the input data in a way that resembles a convolutional code [4]. Its output is then used to generate the time hopping sequence resulting in a random distribution of the interpulse intervals, and thus a noise-like spectrum. Significant spreading demands a large number of levels in the transmitter. This, in general, would require at the receiver a convolutional decoder with a large number of states. In [9] it is shown that the PCTH signal can be decoded with a Viterbi detector [4] of reduced complexity, i.e., with a limited number of states. Moreover, detectors of different complexity (and performance) may coexist while decoding the same transmitted PCTH signal. This scalability property, which is not present in conventional convolutional coding, adds flexibility in terms of the receiver design.

Multiaccess PCTH (MA-PCTH) is an extension of PCTH. The basic idea consists of replacing each pulse transmitted by the original PCTH scheme with a pulse train, different for each user. The pulse train is a unique user "signature". Each user is then demodulated with a filter matched to its signature.

4.2 Single-User Pseudo-Chaotic Time Hopping

4.2.1 Introduction

The communication system to be considered in this work comprises a transmitter shown in Figure 4.2 containing a channel encoder and pulse-position modulator, and a receiver containing a demodulator, detector, and channel decoder. Multilevel pulse-position modulation is considered to be driven by a pseudo-chaotic channel encoder based on the Bernoulli shift map.

PCTH [9, 22] exploits concepts from symbolic dynamics [23] to generate aperiodic spreading sequences that depend on the input data. Its output is then used to generate a time-hopping sequence that results in a random distribution of the inter-pulse intervals, and thus a noiselike spectrum. As shown

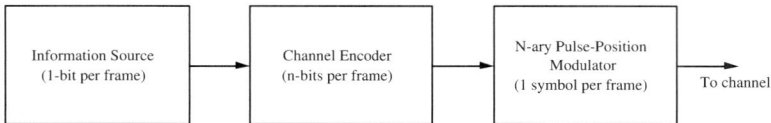

Fig. 4.2. Generic transmitter.

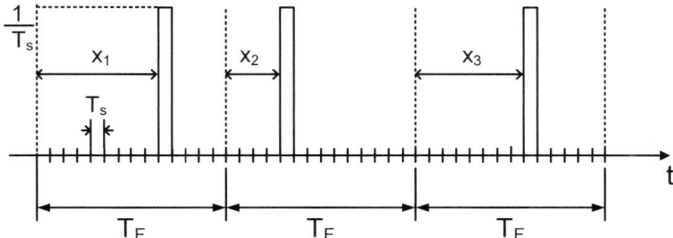

Fig. 4.3. Sketch of the periodic frame. The frame period and the slot period associated with each PPM level are denoted by T_F and T_s, respectively.

in the following, the pseudo-chaotic time hopping code can be considered to be a specific convolutional code.

The next section describes the modulation format used by PCTH. The following section describes the PCTH channel encoder in terms of its characteristic polynomial. The first receiver considered is a matched filter, the optimum frame-by-frame receiver in additive white Gaussian noise (AWGN). It utilizes the maximum likelihood criterion followed by a threshold decoder followed by the second receiver based on maximum likelihood sequence estimation (MLSE) via the Viterbi algorithm. Finally, the bit-error-rate performance for hard decoding is presented.

4.2.2 Signal Description

In this section, the modulation used by PCTH is described. The waveforms are N-ary ($N = 2^n$) orthogonal based on pulse-position modulation (PPM). As shown in Figure 4.3, within the period of each frame, T_F, one of the N signals $\{s_n(t), n = 0, 1, 2, \ldots, (N-1)\}$ is transmitted according to the encoded data presented to the modulator. For simplicity, a rectangular pulse shape has been assumed where

$$s_j(t) = w(t - jT_s), \quad j \in \{0, 1, \ldots, (N-1)\} \text{ and } T_F = NT_s \qquad (4.4)$$

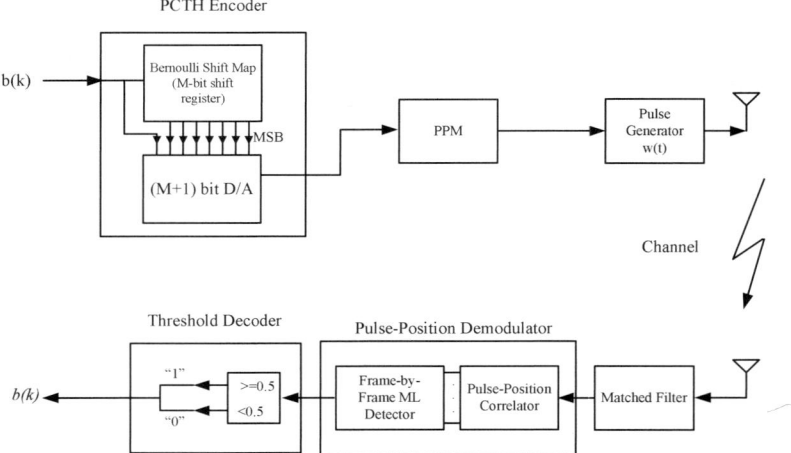

Fig. 4.4. Simplest block diagram of the single-user PCTH receiver utilizing the maximum likelihood criterion on a frame-by-frame basis.

$$w(t) = \begin{cases} \frac{1}{T_s}, & 0 \le t < T_s \\ 0, & \text{otherwise.} \end{cases} \tag{4.5}$$

The functions $\{s_j(t)\}$ form a set of N orthonormal basis functions where

$$\int_{-\infty}^{\infty} s_i(t)s_j(t) \, dt = \delta_{ij}. \tag{4.6}$$

For the kth frame transmitting the kth encoded information bit, the transmitted signal is

$$s^k(t) = w(t - x_k T_s - kT_F), \quad \text{where } x_k \in \{0, 1, 2, \dots, (N-1)\}. \tag{4.7}$$

4.2.3 Modulation

In this section, the basics of the single-user PCTH scheme as shown in Figure 4.4 are presented [9,27]. Let us start by recalling some useful concepts from the symbolic dynamics associated with the shift map. Symbolic dynamics may be defined as a "coarse-grained" description of the evolution of a dynamical system [23]. The idea is to partition the state space and to associate a symbol with each partition. Consequently, a trajectory of the dynamical system can be analyzed as a symbolic sequence. The Bernoulli shift [24] is a simple example of a chaotic map and is defined as

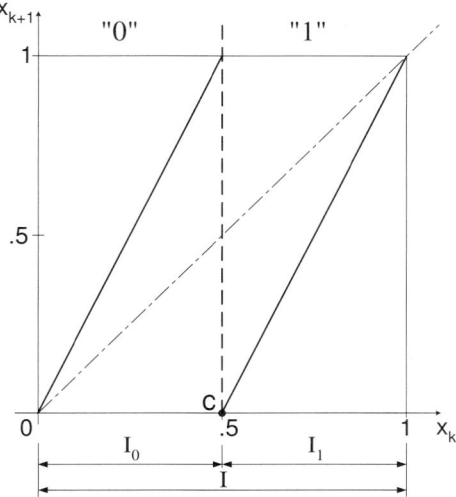

Fig. 4.5. The symbolic dynamics of PCTH.

$$x_{k+1} = 2x_k \bmod 1, \tag{4.8}$$

and whose graph is shown in Figure 4.5. The initial state, x_1, can be expressed as a binary expansion

$$x_1 = 0.b_1 b_2 b_3 \ldots = \sum_{j=1}^{\infty} 2^{-j} b_j, \tag{4.9}$$

with b_j equal to either "0" or "1", and $x_1 \in I = [0,1)$. The next state of the map, x_2, is just

$$x_2 = 0.b_2 b_3 b_4 \ldots \tag{4.10}$$

For this map, a Markov partition [23] can be selected by splitting the interval $I = [0,1)$ into two subintervals: $I_0 = [0,0.5)$ and $I_1 = [0.5,1)$. Then, in order to obtain a symbolic description of the dynamics, the binary symbols "0" and "1" are associated with the subintervals I_0 and I_1, respectively. Figure 4.5 shows the symbolic dynamics used by the PCTH scheme. It should be noted that the most significant bit of the state x_k determines which subinterval the current state occupies.

If the initial condition $x_1 \in \Re$ is chosen randomly with a uniform probability density in the interval $[0,1)$, the points that lie on a periodic orbit are *dense*. This means that for any x_1, and any ϵ, no matter how small, there is at least one point in the interval $[x - \epsilon, x + \epsilon]$ that lies on a periodic orbit. The number of points on a periodic orbit are countably infinite whereas the number of points that do not lie on a periodic orbit are uncountably infinite [24]. As a result, randomly chosen initial conditions (with uniform probability density) lie on a periodic orbit with probability zero. This alleviates any worry that randomly chosen initial conditions will result in periodicity in the signal made up of the initial condition and successive iterates of the Bernoulli shift map.

In PCTH, the initial condition x_1 is considered to be a long sequence of binary information expressed using Equation (4.9) rather than a real number. This long stream of i.i.d. data can be considered to be equivalent to a randomly chosen initial condition, $x_1 \in [0, 1)$. Next, the Bernoulli shift is approximated by the first $(M + 1)$ bits of the sequence. An M-bit shift register R holds the last M information bits. Together with the most recent information bit (LSB in the register), which does not need to be stored until the next state, an $(M + 1)$ bit word is generated that approximates the entire long sequence with $(M + 1)$ significant binary digits. The state space of x has 2^M states. Multiplication by 2 in Equation (4.8) corresponds to a left shift (b_2 goes to b_1, etc.), and the modulo one operation is realized by discarding the most significant bit (MSB). At each clock impulse the most recent bit of information is assigned the least significant bit (LSB) position in the shift register.

¿From the viewpoint of information theory, the shift register implementing the Bernoulli shift may be seen as a particular convolutional encoder [23, 25]. The memory of the structure is given by the shift register which stores the last M input bits. Each input bit causes an output of $(M + 1)$ bits; thus the overall rate is $1/(M + 1)$. In general, the shift register may be followed by a transformation unit for generating more complex chaotic maps. For example, the simple transformation achieved using a Gray/binary converter on the Bernoulli shift leads to the tent map, described by Equation (4.11),

$$x_{k+1} = 1 - 2\,|x_k - 0.5|. \tag{4.11}$$

Figure 4.6 shows the system block diagram including the transformation block reported in [9]. In this work, no transformation unit is used and the basic PCTH scheme of Figure 4.6 is considered. In PCTH, the output of the pseudo-chaotic encoder is used to drive an N-ary pulse position modulator ($N = 2^{(M+1)}$). Each pulse is allocated, according to the pseudo-chaotic modulation, within a periodic frame of duration T_F as shown schematically in Figure 4.3. Only one pulse is transmitted within each symbol period, T_F. If the pulse occurs in the first half of the frame a "0" is being transmitted, otherwise a "1". Each pulse can occur at any of $N = 2^{(M+1)}$ discrete time instants.

The simplest PCTH receiver in a memoryless AWGN channel comprises a pulse-position correlator, or matched filter, followed by an optimum detector utilizing the maximum likelihood criterion as the decision rule. This receiver is optimum in that it minimizes the probability of error on a frame-by-frame basis but does not take advantage of the dependence of each symbol (code word)

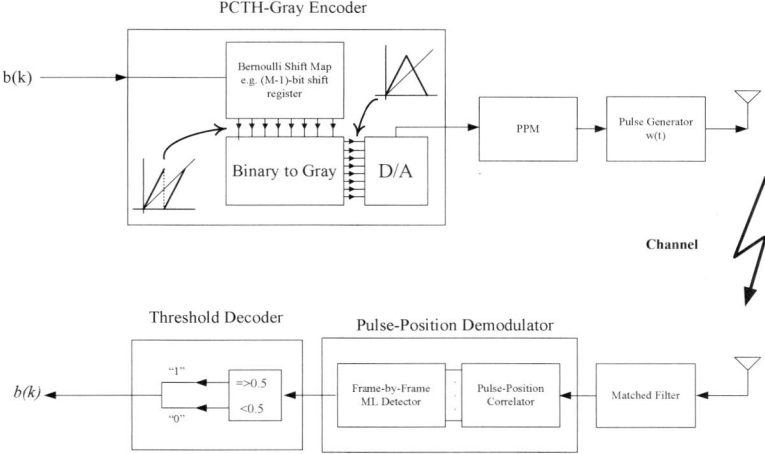

Fig. 4.6. Block diagram of the single-user PCTH transceiver with a map transformation from the Bernoulli shift to the tent map. Frame-by-frame maximum likelihood detection is shown.

on the previous M symbols as would an optimum decoder using maximum-likelihood sequence estimation. In this case, the binary information may be retrieved by means of a threshold decoder at the output of the detector. Because the threshold of the decoder is set to 0.5, the effect of the decoder is to strip off the most significant bit of the detector output. In terms of the symbolic dynamics, the MSB of the detector output specifies in which interval the detected symbol resides, I_0 or I_1. Although not required and often difficult to implement, ideally a matched filter to the transmit pulse-shape is the first block in the receive chain and is shown in Figure 4.6.

4.2.4 Bernoulli Shift as a Convolutional Encoder

In PCTH, the Bernoulli shift takes the current information bit in addition to M past information bits to generate a code word. Thus, the convolutional encoder [23] has memory of length M and a rate of $1/(M+1)$. In this work two cases are considered: 3-bit ($M = 2$) and 8-bit ($M = 7$) implementations of the Bernoulli shift map. Correspondingly, the encoders may be seen as convolutional encoders with constraint lengths $K = (M + 1) = 3$ with $s = 2^2 = 4$ states and $K = (M + 1) = 8$ with $s = 2^7 = 128$ states. Figure 4.7 shows an implementation of the 8-bit Bernoulli shift encoder. By inspection, one can also write down the generator matrix in the standard octal form for the 128-state and 4-state Bernoulli shift maps as

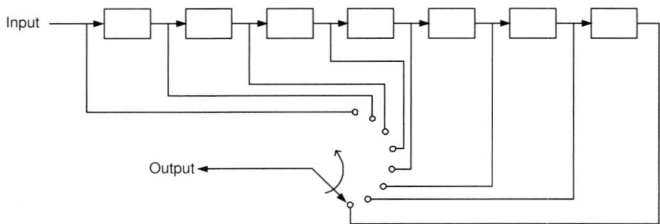

Fig. 4.7. Equivalent convolutional encoder to the 7-bit Bernoulli shift map ($M = 7$). Note that the values contained in the 7-bit shift register correspond to the next 7 information bits.

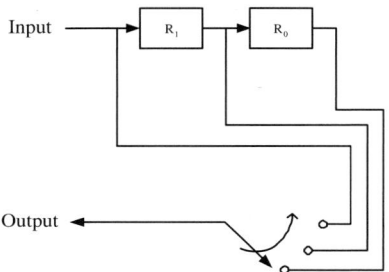

Fig. 4.8. Four-state Bernoulli shift convolutional encoder ($M = 2$).

$$G_{B8} = [200\ 100\ 040\ 020\ 010\ 004\ 002\ 001]^T , \qquad (4.12)$$

and

$$G_{B3} = [4\ 2\ 1]^T . \qquad (4.13)$$

State Diagram and Polynomial Description

Every convolutional encoder is a finite state machine (FSM). The current state of the machine was determined by the previous state and the most recent information bit (least significant bit). There are a limited number of paths that lead to any state. The decoder takes advantage of this and as a result provides coding gain to the receiver.

The 8-bit Bernoulli shift has a state diagram with 128 states. Because this is an unwieldy number, a smaller 4-state Bernoulli shift ($M = 2$) is considered for the following discussion. This reduced-state encoder is shown in Figure 4.8. Assigning symbols to each state in Table 4.1, and noting the transitions between states and their accompanying output symbols in the state table shown in Table 4.2, the state diagram in Figure 4.9 can be generated.

State	R_1	R_0
S_0	0	0
S_1	0	1
S_2	1	0
S_3	1	1

Table 4.1. State Assignments for a Four-State Bernoulli Shift

State	$In = 0$	$In = 1$
0 0	0 0 0 / 0 0	0 0 1 / 1 0
0 1	1 0 0 / 0 0	1 0 1 / 1 0
1 0	0 1 0 / 0 1	1 1 0 / 1 1
1 1	0 1 1 / 0 1	1 1 1 / 1 1

Table 4.2. State Table with Output/Next State for a Four-State Bernoulli Shift

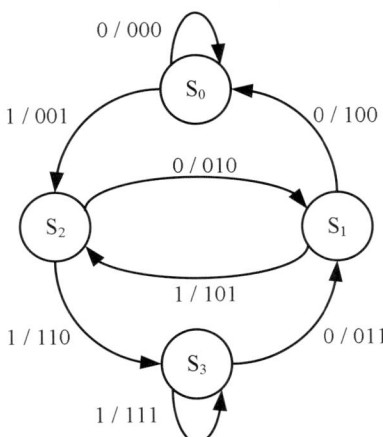

Fig. 4.9. State diagram for the four-state Bernoulli shift convolutional encoder ($M = 2$).

The performance of convolutional codes in terms of coding gain depends on the decoding algorithm and the distance properties of the code itself [28]. The

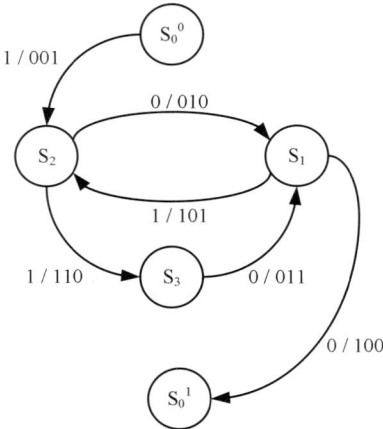

Fig. 4.10. State diagram used to calculate the minimum Hamming weight sequence that differs from the all-zero sequence of the four-state Bernoulli shift.

Input	1	0	0
State	$S_0^0 \to S_2$	$\to S_1$	$\to S_0^1$
Output	001	010	100

Table 4.3. Shortest Path in the Four-State PCTH Encoder Starting in the All-Zero State S_0^0 and Ending in the All-Zero State S_0^1

most important distance property of a convolutional code is the free distance, d_{free}. It characterizes the minimum Hamming weight of all the paths in the state diagram that diverge and remerge with the all-zero state of the code. To determine the free distance of the code consider Figure 4.10. The free distance is the smallest Hamming weight of the output sequence generated along the path from S_0^0 to S_0^1. For the four-state Bernoulli shift, this path is shown in Table 4.3. This path from S_0^0 to S_0^1 has a Hamming weight of three (also a Hamming distance of three to the all-zero path). The second shortest path from S_0^0 to S_0^1 is shown in Table 4.4. This path has a Hamming weight of six. Both of these are possible error paths that the decoder could take and each one contributes to the bit-error rate of the communication system. There are paths of greater distance as well. In fact, the number of paths and their lengths depend on the length of the input sequence making a precise BER determination impossible. As a result, error bounds must be used. Not surprisingly, the minimum distance error path dominates the BER performance of the system under high signal-to-noise ratio conditions [4].

In	1	1	0	0
State	$S_0^0 \rightarrow S_2$	$\rightarrow S_3$	$\rightarrow S_1$	$\rightarrow S_0^1$
Output	001	010	011	001

Table 4.4. Second Shortest Path in the Four-State PCTH Encoder Starting in the All-Zero State S_0^0 and Ending in the All-Zero State S_0^1

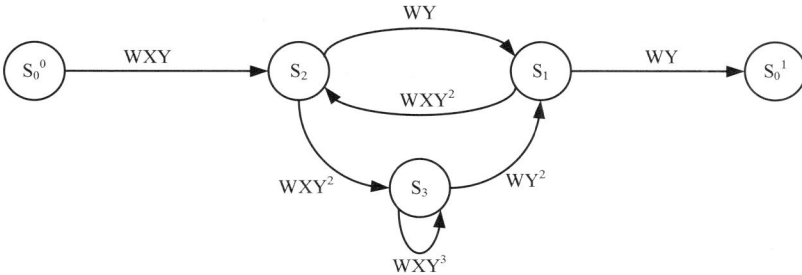

Fig. 4.11. Signal-flow graph used to generate the polynomial description of the four-state Bernoulli shift.

The complete distance properties can be determined from the transfer function of the code derived from the state diagram. The transfer function is determined as a ratio of polynomials representing the number of branches traversed expressed as the exponent of W, the Hamming weight of the input expressed as the exponent of X, and the Hamming weight of the output expressed as the exponent of Y. Figure 4.11 shows the state diagram with the branches relabeled with the appropriate polynomials.

The transfer function can be determined from the system of equations evident in the signal-flow graph of Figure 4.11:

$$E_1 = WYE_2 + WY^2E_3 \qquad (4.14)$$
$$E_2 = WXY^2E_1 + WXYE_0^0 \qquad (4.15)$$
$$E_3 = WXY^2E_2 + WXY^3E_3 \qquad (4.16)$$
$$E_0^1 = WYE_1. \qquad (4.17)$$

Solving Equation (4.16) for E_3, and substituting into Equation (4.14), we see

$$E_1 = \left(WY + \frac{W^2XY^4}{1 - WXY^3} \right) E_2. \qquad (4.18)$$

Substituting for E_1 in Equation (4.17) leads to

$$E_0^1 = WY \left(WY + \frac{W^2 XY^4}{1 - WXY^3} \right) E_2. \tag{4.19}$$

Rewriting Equation (4.15) for E_0^0 and substituting Equation (4.18) for E_1,

$$E_0^0 = \frac{E_2}{WXY} \left[1 - WXY^2 \left(WY + \frac{W^2 XY^4}{1 - WXY^3} \right) \right]. \tag{4.20}$$

Finally, dividing Equation (4.19) by Equation (4.20), the transfer function is

$$T(W, X, Y) \equiv \frac{E_0^1}{E_0^0} = \frac{WY \left(WY + \frac{W^2 XY^4}{1-WXY^3} \right)}{\frac{1}{WXY} \left[1 - WXY^2 \left(WY + \frac{W^2 XY^4}{1-WXY^3} \right) \right]}$$

$$= \frac{W^3 XY^3}{1 - WXY^3 - W^2 XY^3}. \tag{4.21}$$

Using the transfer function, the Hamming weights of all the output sequences of code words (powers of Y) and input sequences (powers of X) that start in S_0 and end in S_0 can be found by setting $W = 1$ resulting in

$$T(X, Y) \equiv T(1, X, Y) = \frac{XY^3}{1 - 2XY^3}, \tag{4.22}$$

or, by also setting $X = 1$, just the output sequences

$$T(Y) \equiv T(1, 1, Y) = \frac{Y^3}{1 - 2Y^3}. \tag{4.23}$$

Expanding $T(Y)$ in a power series for $(1 - x)^{-1} = 1 + x + x^2 + x^3 + \cdots$

$$T(Y) = Y^3 (1 + 2Y^3 + 4Y^6 + 8Y^9 + \cdots) = Y^3 + 2Y^6 + 8Y^{12} + \cdots. \tag{4.24}$$

The interpretation of this result is that starting from the all-zero state and returning to it there is one path around the state diagram with a Hamming weight of three, two with a Hamming weight of six, four with a weight of nine, and so on. So, 4-state PCTH has a free distance, $d_{free} = 3$.

Observe from Table 4.2 that the number of states visited between the first deviation from the all-zero path and the remergence with the all-zero path is equal to the constraint length, $K = 3$. Because the Bernoulli shift corresponds to a single bit-error propagating through the shift register, the Hamming weight of the output sequence is also equal to the constraint length. Thus, for the Bernoulli shift, the minimum Hamming distance between the all-zero sequence and minimum distance error path through the trellis is equal to the constraint length. Accordingly, the free distance of the 8-bit Bernoulli shift is $d_{free} = 8$.

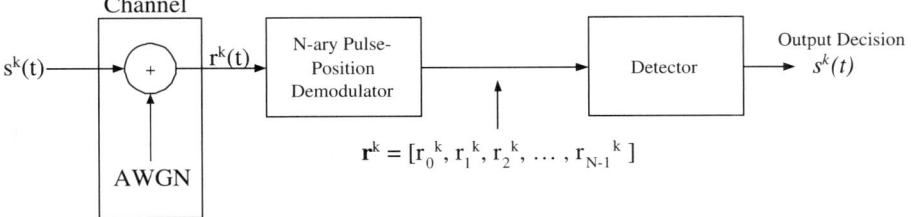

Fig. 4.12. Basic N-ary receiver configuration.

4.2.5 Optimum Frame-by-Frame Receiver

Let us assume the N-ary orthogonal signal described in Section 4.2.2 is transmitted over a channel that is memoryless and corrupted by additive white Gaussian noise (AWGN) as shown in Figure 4.12. The received signal for the kth frame is

$$r^{(k)}(t) = s^{(k)}(t) + n(t), \qquad (4.25)$$

where $n(t)$ is AWGN with power spectral density $\sigma^2 = N_0/2$ W/Hz. The job of the receiver is to classify correctly $r^{(k)}(t)$ into the correct slot x_k for each k. The two approaches to classification are the Neyman-Pearson (NP) approach and the Bayesian approach. The NP approach maximizes the probability of detection subject to a fixed false alarm rate. This approach is often used in sonar and radar applications. The Bayesian approach minimizes the cost of the decision made and leads to the minimization of the probability of classification error [29] for each frame.

Dropping the frame index k, and breaking down $r(t)$ into its components in terms of the orthonormal basis functions s_j, we have

$$\begin{aligned}
r_j &= \int_{-\infty}^{\infty} r(t)s_j(t)\,dt \\
&= \int_{-\infty}^{\infty} s(t)s_j(t)\,dt \ + \ \int_{-\infty}^{\infty} n(t)s_j(t)\,dt \\
&= \rho_j + n_j, \qquad (4.26)
\end{aligned}$$

where

$$\rho_j = \int_{-\infty}^{\infty} s(t)s_j(t)\,dt.$$

The multiple hypotheses for each frame from which the receiver must choose are

$$H_i : r_i = s_i + n_i, \tag{4.27}$$
$$r_k = n_k,$$
$$\forall k \neq i.$$

If the cost of making an error from each $s_j(t)$, $\forall j$ to any of the other $(N-1)$ possible signals is equal, then the maximum likelihood decision rule minimizes the cost function by choosing the hypothesis H_i, if

$$Pr(\mathbf{s_i} \mid \mathbf{r}) > Pr(\mathbf{s_k} \mid \mathbf{r}), \quad \forall i \neq k. \tag{4.28}$$

This selection criterion is the maximum a posteriori probability (MAP) criterion. Because these probabilities are unknown, Bayes' rule is used to express

$$Pr(\mathbf{s_i} \mid \mathbf{r}) = \frac{Pr(\mathbf{r} \mid \mathbf{s_i}) Pr(\mathbf{s_i})}{Pr(\mathbf{r})}.$$

If each of the signals $\mathbf{s_i}$ is equally probable and $Pr(\mathbf{r})$ is independent of $\mathbf{s_i}$, choosing the largest $Pr(\mathbf{r} \mid \mathbf{s_i})$ is equivalent to choosing the largest $Pr(\mathbf{s_i} \mid \mathbf{r})$. Given the noise statistics, these probabilities can be computed for each $\mathbf{s_i}$.

Because the noise components n_j are uncorrelated Gaussian random variables, they are also independent allowing the probability density function of $Pr(\mathbf{r} \mid \mathbf{s_i})$ to be written as the product [30]

$$p(\mathbf{r} \mid \mathbf{s_i}) = \prod_{k=0}^{(N-1)} p(r_k \mid s_{i,k}),$$
$$k = 0, 1, \ldots, (N-1),$$

leading to

$$p(\mathbf{r} \mid \mathbf{s_i}) = \frac{1}{\sqrt{\pi N_0}} \exp\left[-\frac{\sum_{k=0}^{(N-1)}(r_k - s_{i,k})^2}{N_0} \right], \tag{4.29}$$
$$i = 0, 1, \ldots, (N-1).$$

The term in the exponential is the distance between the received vector $\mathbf{r}$ and $\mathbf{s_i}$ where

$$\mathbf{r} = [r_0, r_1, r_2, \ldots, r_{N-1}], \tag{4.30}$$

and

$$\mathbf{s_i} = [s_{i,0}\ s_{i,1},\ s_{i,2},\ \ldots\ s_{i,(N-1)}], \tag{4.31}$$

where, for example,

$$s_3 = [0_{3,0}\ 0_{3,1},\ s_{3,2},\ 1_{3,3}, \ldots,\ 0_{3,(N-1)}].$$

The terms from the product are [4, 30],

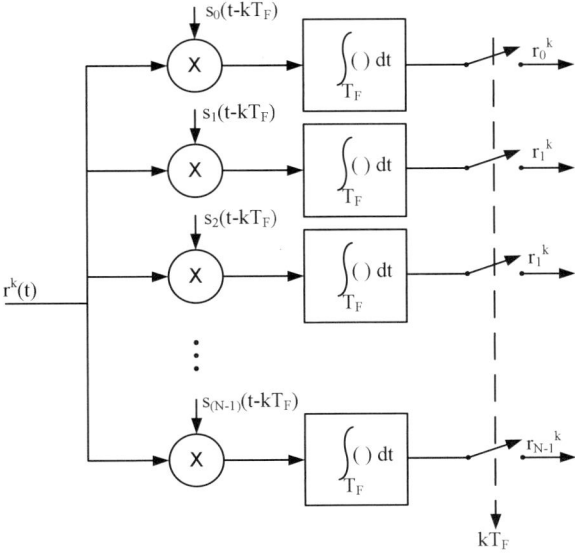

Fig. 4.13. Correlation demodulator for the kth frame.

$$\sum_{k=0}^{(N-1)} (r_k - s_{i,k})^2 = \|\mathbf{r}\|^2 - 2\mathbf{r} \cdot \mathbf{s_i} + \|\mathbf{s_i}\|^2.$$

The first term is the received energy and the third term is the energy of the transmitted signal $\mathbf{s_i}$. Assuming equal energy $\mathbf{s_i}$, $\forall i$, these are constants as well. The only signal-dependent term is the middle term, the projection of $\mathbf{r}$ on $\mathbf{s_i}$. The demodulator diagrammatically shown in Figure 4.13 performs the cross-correlation.

Without loss in generality for any particular frame [4], assume $\mathbf{s_0}$ was sent by the transmitter. Then the outputs of the correlator will be

$$\mathbf{r} = [a^2 + an_0, an_1, an_2, \ldots, an_{N-1}],$$

where the energy of the transmitted signal is a^2. Dividing each amplitude by a, an equivalent vector is

$$\mathbf{r} = [a + n_0, n_1, n_2, \ldots, n_{N-1}]. \tag{4.32}$$

The probability density functions for the $(N-1)$ slots where $j \neq 0$ are [4]

$$p(x_j) = \frac{1}{\sqrt{\pi N_0}} \exp\left[-\frac{(x_j)^2}{N_0}\right],$$
$$j = 1, \ldots, (N-1),$$

and for the first slot, r_0,

$$p(x_0) = \frac{1}{\sqrt{\pi N_0}} \exp\left[-\frac{(x_0 - a)^2}{N_0}\right].$$

The probability of making a correct decision is then [4]

$$P_c = \int_{-\infty}^{\infty} Pr(n_1 < r_0 \mid r_0) Pr(n_2 < r_0 \mid r_0) \cdots Pr(n_{N-1} < r_0 \mid r_0)\, p(r_0)\, dr_0.$$

(4.33)

Each of the terms involving n_i for $i \neq 0$, has the form

$$Pr(n_i < r_0 \mid r_0) = \int_{-\infty}^{r_0} p(x_i)\, dx_i$$

$$i \neq 0,$$

and so

$$P_c = \int_{-\infty}^{\infty} \left[\int_{-\infty}^{r_0} \frac{1}{\sqrt{\pi N_0}} \exp\left(-\frac{x^2}{N_0}\right) dx\right]^{(N-1)} p(r_0)\, dr_0. \qquad (4.34)$$

Finally, because the probability of symbol error [4, 30], $P_{es} = 1 - P_c$,

$$P_{es} = 1 - \frac{1}{(\pi N_0)^{N/2}} \int_{-\infty}^{\infty} \left[\int_{-\infty}^{z} \exp\left(-\frac{x^2}{N_0}\right) dx\right]^{(N-1)} \exp\left[-\frac{(z-a)^2}{N_0}\right] dz$$

$$= 1 - \frac{1}{(\pi N_0)^{N/2}} \int_{-\infty}^{\infty} [1 - Q(z)]^{(N-1)} \exp\left[-\frac{(z-a)^2}{N_0}\right] dz$$

$$= \frac{1}{\sqrt{2\pi}} \int_{-\infty}^{\infty} [1 - \Phi(z)^{N-1}] \exp\left[-\frac{(z-a)^2}{N_0}\right] dz, \qquad (4.35)$$

where

$$\Phi(x) = \frac{1}{\sqrt{2\pi}} \int_{-\infty}^{x} e^{-\frac{t^2}{2}} dt.$$

This is the probability of error for any equal energy N-ary orthogonal signal set. Figure 4.14 compares this probability of error to that obtained by Monte Carlo simulation of the PCTH system in AWGN.

4.2.6 Maximum Likelihood Sequence Estimation

The Viterbi algorithm is used to implement both hard and soft maximum likelihood sequence estimation. Hard decoding utilizes the Hamming distance (the difference in the number of ones) between candidate and received code words to select the most likely transmitted code sequence. The Hamming distance between two paths through the trellis i and k can be expressed as

$$H(i, k) = l_i \oplus l_k,$$

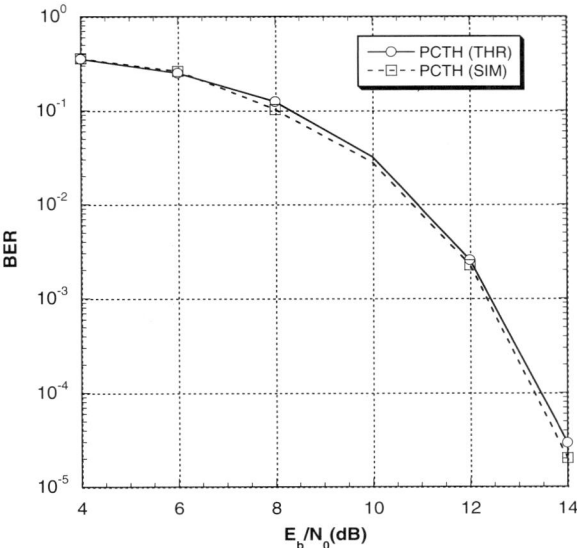

Fig. 4.14. Simulated versus analytical BER performance of single-user PCTH in the presence of AWGN. Note that in this case the error probability coincides with orthogonal 256-PPM. (Reprinted with permission from [27], ©2002 IEEE.)

where l_i and l_k are the sequences of output code words along paths i and k and $\oplus$ indicates modulo 2 addition.

Without loss in generality, due to the linearity of convolutional codes (evidenced by the fact that the operation of the code can be expressed using linear algebra), a bound on the probability of error can assume that the all-zero information sequence is transmitted. This is true because the set of Hamming distances per code word between the all-zero sequence l_0, and any code sequence l_a, is the same as the set of distances between any other sequence l_b and l_a.

The Viterbi algorithm compares the received sequence to each surviving path through the trellis and chooses the path with the smallest distance. So, the decoder can be considered to perform a pairwise comparison between the received sequence l_R and each sequence through the trellis l_i. Assume the all-zero sequence is transmitted, say the decoder is comparing the distance between l_R and l_0 with the distance between l_R and some other path, l_D. Let the Hamming distance between l_0 and l_D be $H(0, D) = d$. If the Hamming distance between the received sequence and the all-zero sequence $H(0, R) < \frac{1}{2}(d+1)$, then the received path is closer to the all-zero path than l_D resulting in the correct path being chosen. Because the all-zero path is being transmitted, the Hamming weight of the received path is also the number of errors experienced over the channel. If $H(0, R) \geq \frac{1}{2}(d+1)$, then the wrong path will be selected. The probability that the wrong path is selected for an odd distance d in this pairwise comparison is then [4, 32, 33]

$$P_2(d) = \sum_{k=\frac{d+1}{2}}^{d} \binom{d}{k} p^k (1-p)^{d-k}, \qquad (4.36)$$

where p is the probability of bit error over the channel. This is simply the binomially distributed probability that between $\frac{1}{2}(d+1)$ and d errors occur over the channel. If d is even, the incorrect path is chosen when $H(0, R) > \frac{1}{2}d$. When $H(0, R) = \frac{1}{2}d$, there is a tie between the distance to the all-zero path and the competing path l_D. So, an average of $\frac{1}{2}$ of these cause an error. Then the probability of error for an even distance d is

$$P_2(d) = \sum_{k=\frac{d}{2}+1}^{d} \binom{d}{k} p^k (1-p)^{d-k} + \frac{1}{2} \binom{d}{\frac{d}{2}} p^{d/2} (1-p)^{d/2}. \qquad (4.37)$$

A long information sequence results in many paths that diverge from the all-zero path and remerge at any particular node. The number of such paths depends on the length of the information sequence causing the probability of error to be dependent on the length of the information sequence. An upper bound can be expressed as the union bound of the pairwise error probabilities,

$$P_e = \sum_{d=d_{free}}^{\infty} a_d P_2(d), \qquad (4.38)$$

where a_d is the number of paths Hamming distance d from the all-zero path and d_{free} is the free distance of the code. The values of a_d are just the coefficients in the expansion of $T(Y)$. Note that some values of $a_d = 0$ where there is no term in the expansion of $T(Y) \propto Y^d$. The probability of information bit error can be found by observing that the exponent of X in each term of $T(X, Y)$ is equal to the number of ones in the information sequence (and the number of errors from the all-zero information sequence). This is the number of bit errors experienced along the path. By taking the derivative and observing its form,

$$\frac{\partial T(X, Y)}{\partial X}\bigg|_{X=1} = \beta_{d_{free}} Y^{d_{free}} + \beta_{d_a} Y^{d_a} + \beta_{d_b} Y^{d_b}, \qquad (4.39)$$

we find the probability of bit error to be

$$P_b = \sum_{d=d_{free}}^{\infty} \beta_d P_2(d), \qquad (4.40)$$

where again some values of β_d are 0 due to the absence of terms (exponents of X with value d in the expansion of $T(X, Y)$). For M-ary orthogonal signals ($M = 2^k$),

$$p = \frac{2^{k-1}}{2^k - 1} \cdot \frac{1}{\sqrt{2\pi}} \int_{-\infty}^{\infty} \left[1 - (1 - Q(y))^{M-1} \right] \exp \left[-\frac{\left(y - \sqrt{2S_0} \right)^2}{2} \right] dy.$$

(4.41)

PCTH

For the three-bit Bernoulli shift, evaluating Equation (4.21) at W=1 results in Equation (4.22),

$$T(1, X, Y) = T(X, Y) = \frac{XY^3}{1 - 2XY^3}.$$

(4.42)

Applying Equation (4.39) we find

$$\left. \frac{\partial T_{B3}(X, Y)}{\partial X} \right|_{X=1} = Y^3 + 4Y^6 + 12Y^9 + \cdots,$$

(4.43)

or

$$[d_{free}, d_a, d_b] = [3, 6, 9]$$

$$[\beta_{dfree}, \beta_a, \beta_b] = [1, 4, 12],$$

which results in a probability of bit error bounded by

$$P_b \leq 1 \cdot P_2(3) + 4 \cdot P_2(6) + 12 \cdot P_2(9).$$

(4.44)

Figure 4.15 shows this bound as a function of E_b/N_0, as well as the lower bound taken by considering only the first term in Equation (4.43), the free distance term. Also shown are Monte Carlo simulation results with 100 errors counted per point.

4.3 Multiple Access for Pseudo-Chaotic Time Hopping

4.3.1 Introduction

Most communications systems have a requirement to support multiple users simultaneously. With a fixed portion of electromagnetic spectrum, this could be accomplished by breaking up the spectrum into channels and assigning a channel to each user (FDMA), or assigning the entire spectrum to one user at a time by assigning a timeslot to each user (TDMA), or the same spectrum can be shared by multiple users at the same time by assigning codes to each user (CDMA).

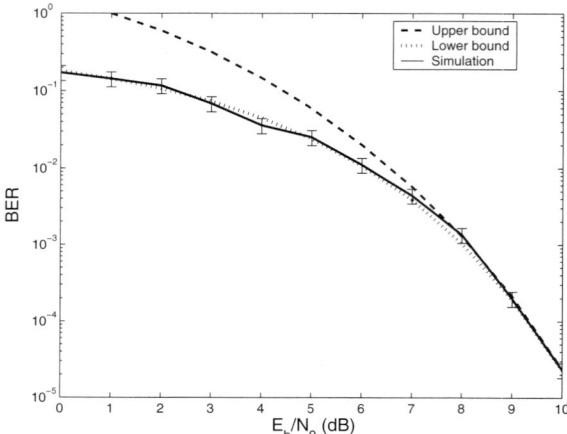

Fig. 4.15. Simulated versus analytical BER performance of three-bit single-user PCTH using hard Viterbi decoding in the presence of AWGN. (Reprinted with permission from [25], ©2003 IEEE.)

In this work the single-user PCTH described in Section 4.2 is extended to a multiaccess system by assigning unique signatures to each user. Here, randomly selected 32-bit binary codes were chosen to each have 16 ones and 16 zeros. Slotted multilevel pulse-position modulation is considered where the entire signature for each user is pulse-position modulated according to each user's channel encoder. A slotted system assumes synchronism between users and modulation that permits user signatures to overlap uniquely so that only periodic cross-correlations need to be considered.

4.3.2 System Description

Figure 4.16 shows a block diagram for the proposed multiple-access scheme based on pseudo-chaotic time hopping, denoted by MA-PCTH. The transceiver architecture shown in Figure 4.16 refers to the generic jth user and includes a map transformation unit for generality [22]. For the remainder of this work, the pseudo-chaotic encoder considered has no transformation unit but rather is the simplest PCTH encoder where the shift register drives the symbolic digital-to-analog converter directly (as shown in Figure 4.4). The input to the system is an independent and identically distributed source of binary data $b_k^{(j)}$, where the lower index denotes the kth information bit. The input sequence feeds the pseudo-chaotic encoder, whose operation has been described in Section 4.2. In single-user PCTH, the output of the pseudo-chaotic encoder

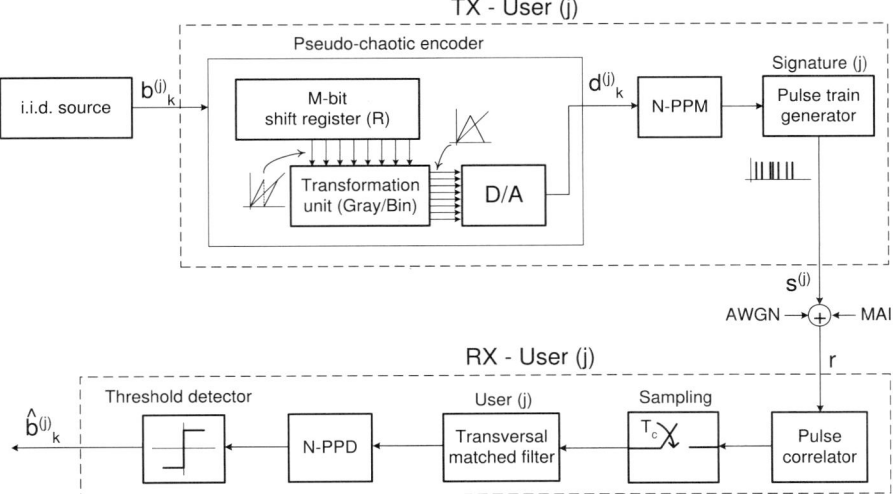

Fig. 4.16. Simplified block diagram of the MA-PCTH scheme using the optimum frame-by-frame maximum likelihood receiver. (Reprinted with permission from [27], ©2002 IEEE.)

$d_k^{(j)}$ drives the N-PPM modulator producing the time hopping. In MA-PCTH, though, the output of the modulator is used to trigger a pulse-train generator corresponding to the specific signature $c^{(j)}$ associated with the jth user. In this work, for simplicity, we consider a slotted system where the periodic frames of all users are synchronized. Each frame is subdivided into N slots of duration $T_s = T_F/N$. In turn, each slot contains N_c chips; correspondingly, the chip duration is given by $T_c = T_s/N_c$. In this analysis it is assumed that the signature for each user is confined within one slot duration T_s; that is, the user signatures do not invade adjacent slots. This also implies that, within a given frame period T_F, two generic users (j) and (k) will either transmit in different slots or collide in the same slot. The situation for a single frame period is illustrated in Figure 4.17. The transmitted signal $s^{(j)}(t)$ for the jth user can be expressed, for each frame, as

$$s^{(j)}(t) = \sum_{l=0}^{N_c-1} c_l^{(j)} w_p(t - lT_c - d_k^{(j)} T_s), \qquad t \in [0, T_F), \quad k = 0, 1, 2, \dots, \quad (4.45)$$

where $c_l^{(j)} \in \{0, 1\}$ is the binary sequence representing the jth user's signature for $l \in \{0, \dots, N_c - 1\}$. On the other hand, $w_p(t)$ is the pulse waveform that in this work is assumed to be rectangular,

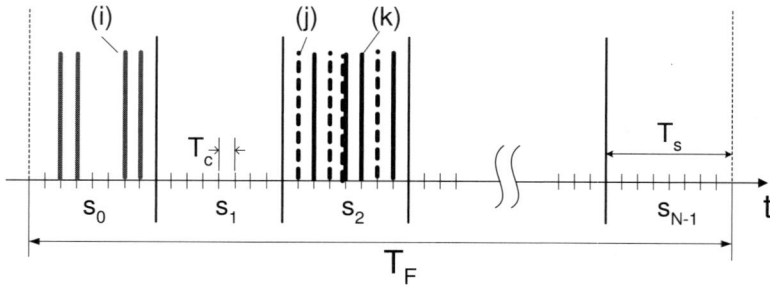

Fig. 4.17. Sketch of the periodic frame for the MA-PCTH scheme with three users. The frame period, T_F, is divided into N slots, each of duration $T_s = T_F/N$. Note the different "signatures" associated with the different users. Users (j) and (k) exhibit a collision in the third slot, s_2. (Reprinted with permission from [27], ©2002 IEEE.)

$$w_p(t) = \begin{cases} \frac{1}{T_p}, & 0 < t < T_p, \\ 0, & \text{otherwise}, \end{cases} \qquad (4.46)$$

where T_p is the pulse duration, and $T_p \leq T_c$. So, for each information bit $b_k^{(j)}$, a pseudo-chaotic iterate $d_k^{(j)} \in \{0, \dots, N-1\}$ is generated and the pulse train for the jth user is transmitted in the corresponding slot within the frame.

In the case of a single user, the signal received is simply

$$r(t) = s(t) + n(t), \qquad (4.47)$$

where $n(t)$ is additive white Gaussian noise. In general, with N_u users transmitting simultaneously, the input to the jth receiver will be

$$r^{(j)}(t) = s^{(j)}(t) + n_u^{(j)}(t) + n(t), \qquad (4.48)$$

where the term $n_u^{(j)}(t)$ takes into account the multiple-access interference (MAI) caused by the remaining $(N_u - 1)$ users sharing the channel.

Referring to Figure 4.16, the jth receiver comprises a pulse correlator for the pulse waveform $w_p(t)$. The output of the correlator for the ith chip of the sth slot is given by

$$\phi_{si} = \int_{iT_c+sT_s}^{(i+1)T_c+sT_s} w_p(\tau) r(\tau) d\tau, \qquad i = 0, \dots, N_c - 1, \qquad (4.49)$$

which is sampled at each chip, T_c. The samples ϕ_{si} are then fed into a digital transversal matched filter [34]. In the case under consideration, the weights a_i should coincide with the user signature; that is,

$$a_i \equiv c_i^{(j)}, \tag{4.50}$$
$$i = 0, \ldots, N_c - 1.$$

Thus, the output of the matched filter for slot s is

$$y_s^{(j)} = \sum_{i=0}^{N_c-1} c_i^{(j)} \phi_{si}, \quad s = 0, \ldots, N-1, \tag{4.51}$$

where the subscript s runs over the number of slots per frame. The pulse-position demodulation is carried out by applying the maximum-likelihood criterion, that is, selecting the largest sample at the output of the matched filter for each frame period, T_F. Namely, the most likely slot $\hat{s}^{(j)}$ is

$$\hat{s}^{(j)} = \arg\max_s \{ y_s^{(j)}, s = 0, \ldots, N-1 \}. \tag{4.52}$$

Finally, the estimate $\hat{b}_k^{(j)}$ of the transmitted bit for the jth user can be obtained by means of a threshold detector or decoder.

4.3.3 Theoretical Bit-Error-Rate

In this section, we analyze the BER performance of the MA-PCTH scheme in the presence of AWGN. The SNR (signal-to-noise ratio) is defined by E_b/N_0, where E_b is the energy per user bit, and N_0 is the AWGN single-sided power spectral density with $\sigma_n^2 = N_0/2$. Without loss in generality, system performance is considered to be from the perspective of *user* 1 in the presence of multiple access interference introduced by the $(N_u - 1)$ other users. The cross-correlation value with each user is normalized to the auto-correlation value of *user* 1. A detailed analysis for the two- and three-user cases follows as well as a general BER expression for an arbitrary number of users. The baseline behavior is represented by the single-user PCTH scheme. As discussed in Section 4.2, the maximum likelihood receiver coincides with orthogonal N-ary PPM ($N = 2^{M+1}$).

Two-User Case

For each frame, two users can either transmit in different slots or in the same slot. By denoting these mutually exclusive events with A and B, respectively, then the probability of detecting *user* 1 in the wrong slot P_e is given by

$$P_e = P(error|A)P(A) + P(error|B)P(B), \tag{4.53}$$

where if each user's transmitted symbol is equally likely to be in each of the N slots,

$$P(A) = \frac{N-1}{N}$$

$$P(B) = \frac{1}{N}. \tag{4.54}$$

$P(A)$ and $P(B)$ are the probabilities that the two users transmit in different slots (event A) and the same slot (event B), respectively. $P(error|A)$ and $P(error|B)$ are the probabilities of error given each of these events. $P(error|A)$ is obtained by modifying the symbol error probability of N-ary orthogonal signaling given in Equation (4.35) by considering that $user$ 1 and $user$ 2 do not transmit in the same slot,

$$P(error|A) = \frac{1}{\sqrt{2\pi}} \int_{-\infty}^{\infty} \left[1 - \Phi(y - \sqrt{2S_1}\gamma_2)\Phi(y)^{N-2} \right] e^{-\left((y-\sqrt{2S_1})^2/2\right)} dy, \tag{4.55}$$

where

$$\Phi(x) = \frac{1}{\sqrt{2\pi}} \int_{-\infty}^{x} e^{-(t^2/2)} dt.$$

$S_1 = E_1/N_0$ is the SNR of $user$ 1, with transmitted energy E_1, and

$$\gamma_2 = \sum_{i=0}^{N_c-1} c_i^{(1)} c_i^{(2)} \tag{4.56}$$

is the periodic cross-correlation between $user$ 2 and $user$ 1, the user of interest. Similarly, $P(error|B)$ is obtained by considering that $user$ 1 and $user$ 2 transmit in the same slot:

$$P(error|B) = \frac{1}{\sqrt{2\pi}} \int_{-\infty}^{\infty} \left[1 - \Phi(y)^{N-1} \right] e^{-\left((y-\sqrt{2S_1}(1+\gamma_2))^2/2\right)} dy. \tag{4.57}$$

Combining Equation 4.53 to 4.55, and 4.57, the average probability of symbol error for two users is,

$$P_e = \frac{1}{\sqrt{2\pi}} \frac{N-1}{N} \int_{-\infty}^{\infty} \left[1 - \Phi(y - \sqrt{2S_1}\gamma_2)\Phi(y)^{N-2} \right] e^{-\left((y-\sqrt{2S_1})^2/2\right)} dy$$

$$+ \frac{1}{\sqrt{2\pi}} \frac{1}{N} \int_{-\infty}^{\infty} \left[1 - \Phi(y)^{N-1} \right] e^{-\left((y-\sqrt{2S_1}(1+\gamma_2))^2/2\right)} dy. \tag{4.58}$$

Three-User Case

If three users are present, all three can transmit in different slots (event denoted by A), all three can transmit in the same slot (event B), or each of the three possible pairs of users can transmit in the same slot (events C_{12}, C_{23}, C_{13}). Specifically, C_{ij} corresponds to users i and j transmitting in the same slot and the remaining user transmitting in a different slot. The average error probability P_e of detecting $user$ 1 in the wrong slot is given by

$$P_e = P(error|A)P(A) + P(error|B)P(B) + P(error|C_{12})P(C_{12})$$
$$+ P(error|C_{23})P(C_{23}) + P(error|C_{13})P(C_{13}). \tag{4.59}$$

Assuming all users transmit independent and identically distributed binary data,

$$P(A) = \frac{(N-1)(N-2)}{N^2},$$

$$P(B) = \frac{1}{N^2},$$

$$P(C_{12}) = P(C_{23}) = P(C_{13}) = \frac{N-1}{N^2}. \tag{4.60}$$

$P(error|A)$ is obtained by modifying the symbol error probability of N-ary orthogonal signaling in Equation (4.35), by considering that users 1, 2, and 3 each transmit in different slots.

$$P(err|A) = \frac{1}{\sqrt{2\pi}} \int_{-\infty}^{\infty} \left[1 - \Phi\left(y - \sqrt{2S_1}\gamma_2\right) \Phi\left(y - \sqrt{2S_1}\gamma_3\right) \Phi(y)^{N-3} \right]$$
$$e^{-\left((y-\sqrt{2S_1})^2/2\right)} dy, \tag{4.61}$$

where

$$\gamma_j = \sum_{i=0}^{N_c-1} c_i^{(1)} c_i^{(j)}, \tag{4.62}$$

denotes the periodic cross-correlation between *user* 1 and *user* j. Note that increasing γ_j decreases the value of Φ in the integrand, thus increasing the error probability. On the other hand, $P(error|B)$ is obtained by considering that the interference due to *user* 2 and *user* 3 appears in the same slot occupied by *user* 1.

$$P(error|B) = \frac{1}{\sqrt{2\pi}} \int_{-\infty}^{\infty} \left[1 - \Phi(y)^{N-1} \right] e^{-\left((y-\sqrt{2S_1}(1+\gamma_2+\gamma_3))^2/2\right)} dy \tag{4.63}$$

The effect of *users* 2 and 3 transmitting in the same slot as *user* 1 can be readily seen from Equation (4.63) as effectively improving the SNR and decreasing the error probability.

The probability of error for the remaining events C_{12}, C_{23}, and C_{13} can be calculated using

$$P(error|C_{ij}) = \frac{1}{\sqrt{2\pi}} \int_{-\infty}^{\infty} \left[1 - \Phi\left(y - \sqrt{2S_1}\gamma_d\right) \Phi(y)^{N-2} \right] e^{-\left((y-\sqrt{2S_1}(1+\gamma_s))^2/2\right)} dy, \tag{4.64}$$

where γ_d denotes the total cross-correlation of users transmitting in the same slot, but different from the slot occupied by *user* 1, and γ_s indicates the total cross-correlation of users occupying the same slot as *user* 1. Then, $P(error|C_{12})$ is found by setting $\{\gamma_d = \gamma_3, \gamma_s = \gamma_2\}$, $P(error|C_{13})$ is found by setting $\{\gamma_d = \gamma_2, \gamma_s = \gamma_3\}$, and $P(error|C_{23})$ can be obtained by setting $\{\gamma_d = \gamma_2 + \gamma_3, \gamma_s = 0\}$.

General Case

For N_u users let us generalize the previous considerations to the following interference event denoted by C. There are n slots indexed by $i = 1, \ldots, n$, different from the slot used by *user* 1, and slot i contains α_i interfering (constructively) users. The slot occupied by *user* 1 receives contributions from

$$N_I = N_u - 1 - \sum_{i=1}^{n} \alpha_i \tag{4.65}$$

interferers and all the other slots are not used. The probability that *user* 1 is detected in the wrong slot is given by [35],

$$P(error|C) = \frac{1}{\sqrt{2\pi}} \int_{-\infty}^{\infty} \left[1 - \prod_{i=1}^{n} \Phi\left(y - \sqrt{2S_1} \sum_{k=1}^{\alpha_i} \gamma^{(i,k)} \right) \Phi(y)^{N-1-n} \right. $$
$$\left. \cdot \exp\left[-\frac{1}{2}\left(y - \sqrt{2S_1}\left(1 + \sum_{k=1}^{N_I} \gamma^{(k)} \right) \right)^2 \right] \right] dy, \tag{4.66}$$

where $\gamma^{(i,k)}$ represents the cross-correlation between *user* 1 and the interferer indexed by k, in the slot indexed by i, whereas $\gamma^{(k)}$ is the cross-correlation between the interfering user indexed by k, and *user* 1.

In order to calculate the average probability of error in the general case, we need an expression for the probability of each of the possible interference events. The average probability of error is

$$P_e = P(error|A)P(A) + P(error|B)P(B) + P(C'). \tag{4.67}$$

Again, A denotes the event where all users transmit in the same slot, B is the event where all users transmit in different slots, and C' denotes the collection of all other interference events. It follows that

$$P(A) = \frac{(N-1)(N-2)\cdots(N-(N_u-1))}{N^{N_u-1}}$$
$$P(B) = \frac{1}{N^{N_u-1}}. \tag{4.68}$$

Moreover,

$$P(error|A) = \frac{1}{\sqrt{2\pi}} \int_{-\infty}^{\infty} \left[1 - \Phi(y)^{N-N_u} \prod_{i=2}^{N_u} \Phi\left(y - \sqrt{2S_1}\gamma_j \right) \right] e^{-\left((y-\sqrt{2S_1})^2/2 \right)} dy, \tag{4.69}$$

$$P(error|B) = \frac{1}{\sqrt{2\pi}} \int_{-\infty}^{\infty} \left[1 - \Phi(y)^{N-1} \right] e^{-\left((y-\sqrt{2S_1}(1+\sum_{j=2}^{N_u}\gamma_j))^2/2 \right)} dy. \tag{4.70}$$

where again γ_j denotes the periodic cross-correlation between *users* 1 and j. In addition,

$$P(C') = \sum_{C \in \mathcal{C}} P(error|C)P(C), \tag{4.71}$$

where $\mathcal{C}$ indicates the set of all interference events, except A and B. If the users are equicorrelated, $P(C')$ can be calculated as the weighted average [27] of the probability of error of all events in $\mathcal{C}$,

$$P(C') = \frac{1}{N^{N_u - 1}} \sum_{\lambda=1}^{\lambda_{max}} \sum_{a_\lambda=2}^{a_{\lambda-1}} \cdots \sum_{a_2=2}^{a_1} \sum_{a_1=2}^{a_0} \binom{N_u}{a_1} \binom{N_u - a_1}{a_2} \cdots$$

$$\cdot \binom{N_u - a_1 - \cdots - a_{\lambda-1}}{a_\lambda} \cdot (N-1)(N-2) \cdots (N - (N_u - \sum_{i=1}^{\lambda-1} a_i) - (\lambda - 1))$$

$$\cdot \left(\frac{1}{\beta_2! \beta_3! \ldots}\right) \cdot \frac{1}{N_u} \cdot [a_1 Pr_e(A_1^1, A_2, \ldots, \alpha_0) + a_2 Pr_e(A_1, A_2^1, \ldots, \beta_0) + \cdots$$

$$+ \alpha_0 Pr_e(A_1, A_2, \ldots, \alpha_0^1)] \quad (4.72)$$

where

$$a_0 = min(N_u - 1, N_u - \sum_{i=2}^{\lambda} a_i). \tag{4.73}$$

β_2 is the number of slots in which two users transmitted, and β_3 is the number of slots in which three users transmitted, and so on. From the perspective of *user* 1, the user of interest, equicorrelated interferers can be accommodated by assigning the same code sequence with the desired cross-correlation to each interferer. This is done for simplicity rather than finding a series of sequences that are equicorrelated. In Equation (4.72),

$$\lambda_{\max} = \lfloor (\frac{N_u}{2}) \rfloor \tag{4.74}$$

is the maximum number of different possible interference events within a single frame. This occurs when $N_u/2$ pairs of users interfere. The number of users who transmit in slots with no interfering users is

$$\alpha_0 = N_u - \sum_{i=1}^{\lambda} a_i. \tag{4.75}$$

In Equation (4.72), the events $A_1, \ldots, A_\lambda$ correspond to a_1 users transmitting in the same slot, a_2 users transmitting in the same slot but different from the A_1 users, and so on. The superscript 1 indicates that *user* 1, the user of interest, is included in that set.

For all cases, in order to calculate the BER, we need to convert symbol errors to bit errors. Namely, we convert the probability of detecting *user* 1 in a wrong slot, P_e, to the bit error probability P_b. The errors that consist of confusing the slot used by *user* 1 with any of the other $N-1$ slots are equiprobable and occur with probability

$$\frac{P_e}{N-1} = \frac{P_e}{2^M - 1},\qquad (4.76)$$

where $\log_2 N = M$ coded bits per symbol. Let's assume without loss of generality that the binary information digit transmitted by *user* 1 is zero; then the probability that the receiver makes a bit error is the probability of confusing the slot where *user* 1 is transmitting with any of the last $N/2$ slots in the frame [4]. Thus,

$$P_b = \frac{2^{M-1}}{2^M - 1} P_e \approx \frac{P_e}{2},\quad M \gg 1. \qquad (4.77)$$

Example: Four-User Case

The probability of detecting *user* 1 in the wrong slot can by subdivided into the following cases.

1. The four users transmit in the same slot, with probability $1/N^3$.
2. All users transmit in different slots, with probability $(N-1)(N-2)(N-3)/N^3$.
3. Two users transmit in the same slot and the two remaining users transmit in independently different slots, with probability $(N-1)(N-2)/N^3$.
4. Three users transmit in the same slot and the remaining user transmits in a different slot, with probability $(N-1)/N^3$.
5. Two users transmit in the same slot and the two remaining users transmit in the same slot different from the previous two, with probability $(N-1)/N^3$.

If the cross-correlation between *user* 1 and all other users is not equal, then each of the above events must be subdivided further. For instance, the probability that two users transmit in the same slot and the two remaining users transmit in the same slot, different from the previous two (case 5), is the sum of the probabilities that:

5a. *Users* 1 and 2 transmit in the same slot and *users* 3 and 4 transmit in the same slot different from the other users, with probability $\frac{1}{3}(N-1)/N^3$.

5b. *Users* 1 and 3 transmit in the same slot and *users* 2 and 4 transmit in the same slot different from the other users, with probability $\frac{1}{3}(N-1)/N^3$.

5c. *Users* 1 and 4 transmit in the same slot and *users* 2 and 3 transmit in the same slot different from the other users, with probability $\frac{1}{3}(N-1)/N^3$.

The probability of error for each of these events can be evaluated by applying Equation (4.66). For example, the probability of case 5a is obtained by setting $n = 1$, $\alpha_1 = 2$, $\gamma^{(1,1)}$ (resp., $\gamma^{(1,2)}$) being the cross-correlation between *users* 1 and 3 (resp., the cross-correlation between *users* 1 and 4) and $\gamma^{(1)}$ the cross-correlation between *users* 1 and 2.

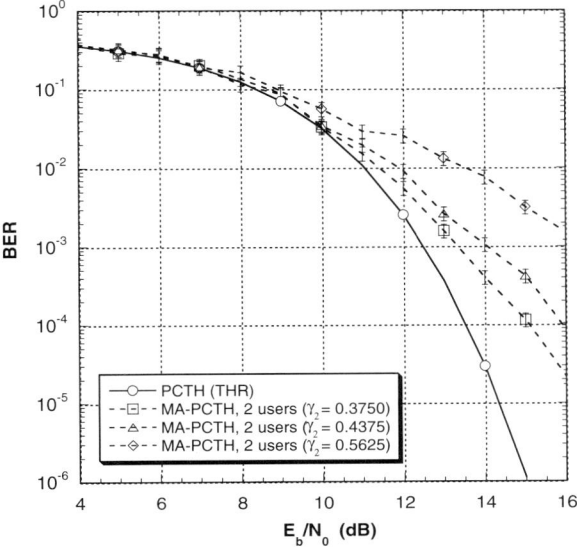

Fig. 4.18. Simulated BER performance of MA-PCTH with two users for various values of cross-correlation. Note that the performance improves with decreasing cross-correlation. (Reprinted with permission from [27], ©2002 IEEE)

4.3.4 Simulation Results

This section reports the simulation results for the MA-PCTH scheme and compares them with the theoretical predictions. The results of the analysis are presented in terms of BER probability versus the signal-to-noise ratio at the receiver, expressed in dB.

Figure 4.14 shows the simulated and analytical BER for single-user PCTH. The analytical calculation uses the method outlined in Section 4.3.3. We used $M = 8$ bits corresponding to $N = 256$ PPM levels, with $N_c = 32$ chips/slot. As mentioned in Section 4.3.3, the BER performance coincides with orthogonal 256-PPM. This is the baseline from which to compare the multiuser cases because it represents the best performance that could be possibly achieved using the optimum frame-by-frame receiver. For multiple users this would correspond to orthogonal signature sequences or zero cross-correlation between signatures.

In each of the multiuser cases a unique 32-bit signature sequence was assigned to each user. The binary sequences that were chosen were randomly selected. The only constraint imposed on the sequence selection process was that each sequence contain an equal number of ones (specifically 16 ones and 16 zeros). This maintains a constant energy across all users. The randomly selected sequences have periodic cross-correlation values to *user* 1, the user of interest, of 0.3750, 0.4375, and 0.5625. The family of curves in Figure 4.18 shows the two-user simulated performance with each of these cross-correlation values. Note how the BER performance improves with decreasing cross-correlation (see Equation (4.58)). As the cross-correlation of the second user's sequence increases, the probability of choosing the slot in which *user* 2 transmitted increases, and so does the intended user's probability of error. This is consistent with the fact that orthogonal signaling results in the best possible BER performance for this receiver. Figure 4.19 shows a comparison of the simulated performance versus the theoretical predictions for each cross-correlation value.

Figure 4.20 shows the analytical BER performance using Equation (4.66), and simulated performance of four users. The three interfering users have cross-correlation values to *user* 1 of 0.3750, 0.4375, and 0.5625. Note that, as pointed out in [36], depending on the value of the cross-correlation and/or with enough users an error floor in the BER can develop. Figure 4.21 shows the simulated performance of the same four-user case compared to each of the two-user cases previously discussed. The four-user system performance is dominated by the user with the highest cross-correlation.

Finally, Figure 4.22 shows the BER performance of the simulated system as a function of the number of users N_u. In the simulated two-user data, a sequence with a cross-correlation of 0.5625 is used. For the four-user case the cross-correlation values are 0.3750, 0.4375, and 0.5625, with respect to *user* 1. In the eight-user case, two users have a cross-correlation value of 0.3750, three other users have a value of 0.4375, and the remaining two interfering users a value of 0.5625. As the number of users increases, the BER performance is degraded. This is true provided that the cross-correlation increases with an increasing number of users, as in this case.

4.3.5 Bit-Error-Rate Floor

If only two users are present, for each frame, they transmit either in the same slot or in different slots. The maximum interference, due to the total cross-correlation γ_T has just a single term, that due to *user* 2. If, as required to discriminate among different users, the cross-correlation between users 1 and 2 is less than unity, no error floor is present. For more than two users, the condition for the existence of an error floor is

$$\gamma_T = \sum_{j=2}^{N_u} \gamma_j \geq 1. \tag{4.78}$$

If the users are equicorrelated with $\gamma_j = \gamma = 0.5625$, three users is the minimum number for which an error floor develops, because $\gamma_T = 1.125$. The event causing the error floor is C_{23} in Section 4.3.3. The error floor occurs because

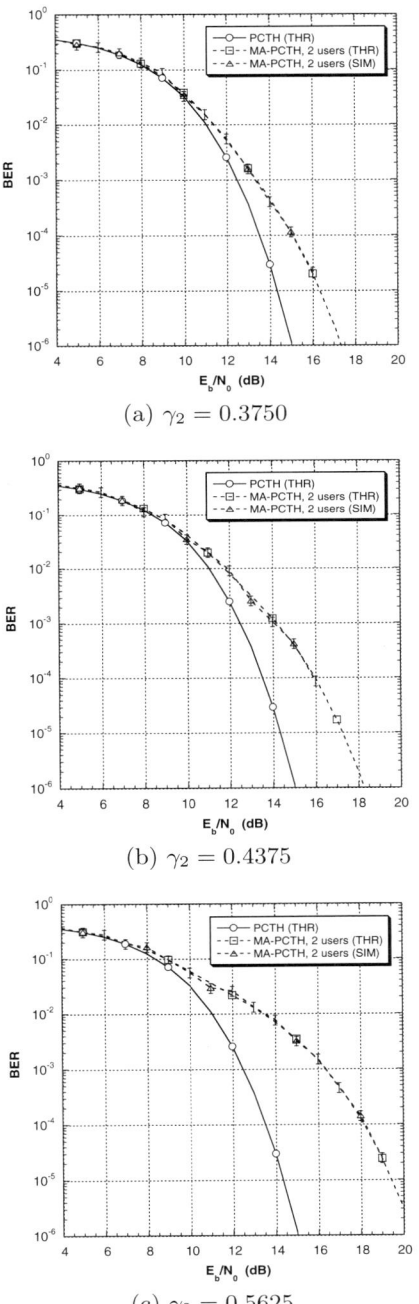

(a) $\gamma_2 = 0.3750$

(b) $\gamma_2 = 0.4375$

(c) $\gamma_2 = 0.5625$

Fig. 4.19. Simulated versus analytical BER performance of MA-PCTH with two users for various values of cross-correlation compared to single-user PCTH. (Reprinted with permission from [27], ©2002 IEEE)

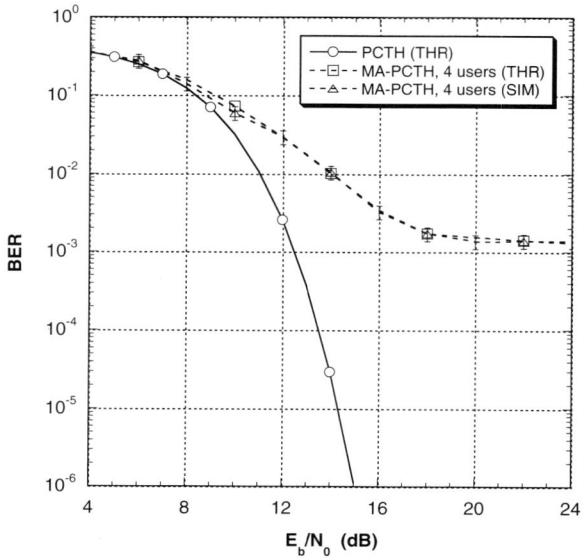

Fig. 4.20. Simulated versus analytical BER performance of MA-PCTH with four users. *Users* 2, 3, 4 have cross-correlation values to *user* 1 of 0.3750, 0.4375, and 0.5625, respectively. Note that an error floor in the BER develops. (Reprinted with permission from [27], ©2002 IEEE)

$$P(error|C_{23}) \to 1 \quad as \quad SNR \to \infty. \tag{4.79}$$

This can be seen from Equation (4.64) which for C_{23} has $\gamma_d = \gamma_T = 1.125$, and $\gamma_s = 0$. If three or more users are present, the value of the error floor is the sum of the event probabilities $P(C')$ where the condition of Equation (4.78) is satisfied. For the equicorrelated three-user case with $\gamma_j = \gamma = 0.5625$, the value of the error floor is the corresponding coefficient $P(C_{23}) = (N-1)/N^2$. Converting from symbol error rate to BER, the error floor has the value

$$P_{e,floor}^{3-user} = P(C_{23}) = \frac{(N/2)(N-1)}{(N-1)(N^2)} = 1.953E - 03, \tag{4.80}$$

where $N = 2^{M+1}$. Figure 4.23a shows the simulated and calculated performance for this three-user case.

Equation (4.72) can be used to calculate the expected performance of the four-user equicorrelated case. After some simplification, the error floor can be found to be equal to

$$P_{e,floor}^{4-user} = P(C_{1,23,4}) + P(C_{1,2,34}) + P(C_{1,24,3}) + P(C_{1,234})$$

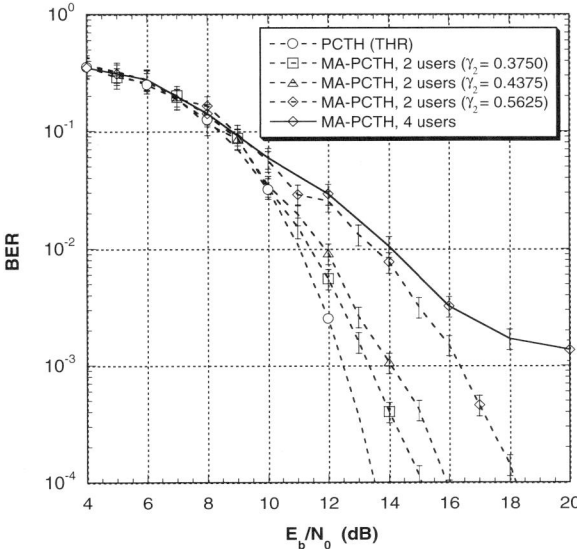

Fig. 4.21. Simulated BER dependence on the cross-correlation. For the four-user case the cross-correlation values are the same as in Figure 4.20. (Reprinted with permission from [27], ©2002 IEEE)

$$= \left(\frac{(N/2)}{N-1}\right)\left(3\frac{(N-1)(N-2)}{N^3} + \frac{(N-2)}{N^3}\right)$$
$$= 5.82E - 03, \tag{4.81}$$

where the subscript notation indicates which users are transmitting in the same slots and which are in different slots. For example, $P(C_{1,2,34})$ indicates that *users* 3 and 4 transmit in the same slot and *users* 1 and 2 transmit in different slots from each other and different from *users* 3 and 4. In Figure 4.23b the corresponding simulated and calculated BER performance is shown.

It follows that the error floor can be expressed formally as

$$P_{e,floor} = P_e(C')\bigg|_{SNR\to\infty}. \tag{4.82}$$

4.4 Conclusions

The success of MA-PCTH as a communication system depends on how many users can be supported at a sufficiently low error rate and a sufficiently high

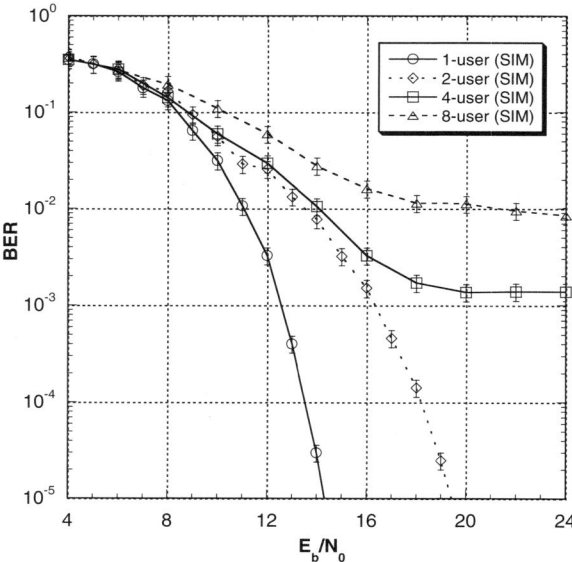

Fig. 4.22. Simulated BER dependence on the number N_u of users. In the two-user case the cross-correlation equals 0.5625. In the four-user case, the cross-correlation values are 0.3750, 0.4375, and 0.5625, with respect to *user* 1. In the eight-user case, two users have cross-correlation 0.3750, three other users 0.4375, and the remaining two interfering users 0.5625. (Reprinted with permission from [27], ©2002 IEEE)

data rate. In this chapter we have shown that the BER performance of single-user PCTH is the same as N-ary orthogonal signaling (e.g. N-PPM). For multiuser communications, there are many parameters that need to be optimized in order to produce the best overall system performance. These include the length of the signature sequences, the cross-correlation between sequences, the dimensionality of the pulse-position-modulator and the system data rate capacity. In this work, we have investigated the influence of the periodic cross-correlation between the generic user and each of the other users on the BER, in a synchronous system. Our analysis indicates that the highest cross-correlation among interferers tends to dominate the BER performance. For future work, in order to improve the BER performance, one needs to find sets of signature sequences exhibiting constant cross-correlation between any two sequences in the set, or that are bounded by an acceptable level. An interesting variation of the proposed scheme is the nonslotted MA-PCTH case, where $NcTc > Ts$. In this case, the length of each user's sequence extends through one slot into an adjacent slot causing intrasymbol interference. When users interfere with each other, they no longer do it in a unique way because there is more than one

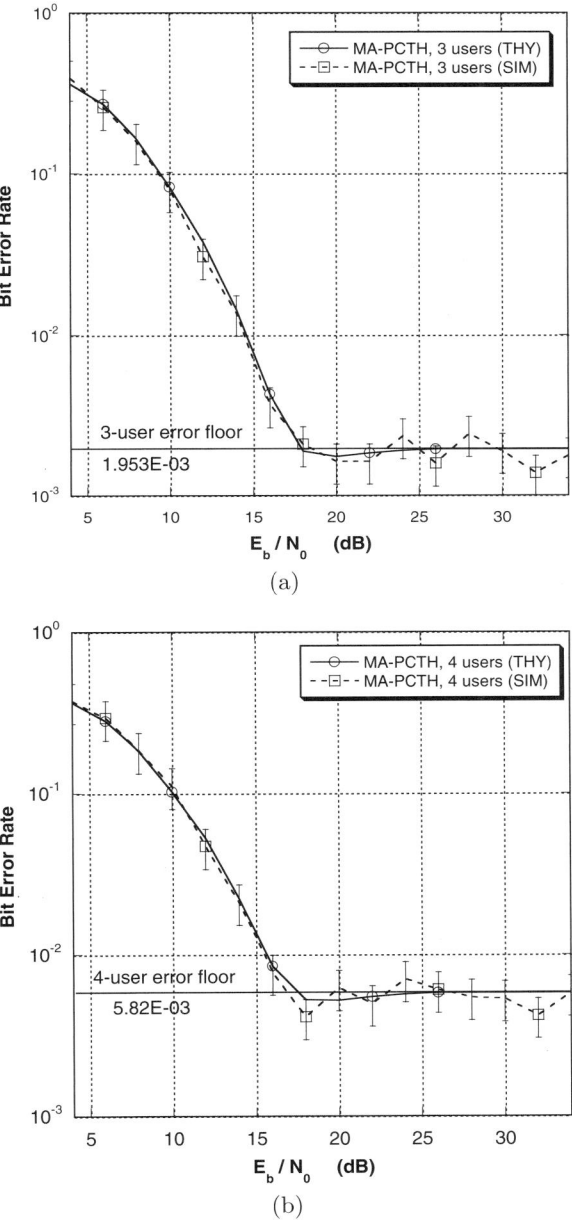

Fig. 4.23. Simulated and analytical BER performance of the three-, and four-user cases of MA-PCTH with $\gamma_j = 0.5625$. (Reprinted with permission from [36], ©2002 IEEE)

way that the sequences can overlap. Aperiodic cross-correlations must now be considered. The possible benefits are a potentially improved data rate, simpler implementation, and the potential to support more users albeit with a possibly higher BER.

Acknowledgments

G.M. Maggio gratefully acknowledges financial support in part by the Army Research Office (John Lavery), grant No. DAAG55-98-1-0269. D. Laney acknowledges financial support under a fellowship from HRL Laboratories, Malibu (CA). Finally, the authors would like to thank Prof. K. Yao, Prof. L. Milstein, Dr. L. Reggiani and Dr. P. Shamain as well as our co-authors for stimulating discussions.

References

1. *Merriam-Webster Collegiate Dictionary.* Merriam-Webster, Inc. online, 2002. http://www.m-w.com.
2. P. J. Nahin, *The Science of Radio.* Woodbury, NY: American Institute of Physics Press, 1996.
3. Nobel Foundation, *Nobel e-Museum.* online, 2002. http://www.nobel.se/physics/laureates.
4. J. G. Proakis, *Digital Communications.* New York, NY: Mc Graw Hill, 3rd ed., 1995.
5. D. G. Leeper, A long-term view of short-range wireless, *IEEE Computer Magazine*, vol. 34, pp. 39–44, 2001.
6. R. J. Fontana, An insight into UWB interference from a shot noise perspective, in *Proceedings of Ultra Wideband Systems and Technologies Conference*, pp. 309–313, 2002.
7. P. Withington, Impulse radio overview. online, 1998. http://www.timedomain.com.
8. Federal Communications Commission, Revision of Part 15 of the Commission's rules regarding ultra-wideband transmission systems. First Report and Order FCC 02-48, ET docket 98-153, April 2002.
9. G. M. Maggio, N. Rulkov, and L. Reggiani, Pseudo-chaotic time hopping for UWB impulse radio, *IEEE Transactions on Circuits and Systems—I*, vol. 48, 2001.
10. M. Z. Win and R. A. Scholtz, Impulse radio: How it works, *IEEE Communications Letters*, vol. 2, pp. 36–38, 1998.
11. S. S. Kolenchery, J. K. Townsend, and J. Freebersyer, A novel impulse radio network for tactical military wireless communications, in *Proceedings of MILCOM*, pp. 59–65, 1998.
12. R. A. Scholtz, Multiple access with time hopping impulse modulation, in *Proceedings of MILCOM*, pp. 447–450, 1993.

13. F. Ramirez-Mireles and R. A. Scholtz, N-orthogonal time-shift-modulated signals for ultra-wide bandwidth impulse radio modulation, in *IEEE Miniconference Proceedings on Commmunication Theory*, 1997.
14. F. Ramirez-Mireles and R. A. Scholtz, Multiple-access performance limits with time hopping and pulse position modulation, in *Proceedings of MILCOM*, pp. 529–533, 1998.
15. M. Z. Win and R. A. Scholtz, Ultra-wide bandwidth time-hopping spread-spectrum impulse radio for wireless multiple access communications, *IEEE Transactions on Communications*, vol. 48, pp. 679–689, 2000.
16. R. G. Aiello, G. D. Rogerson, and P. Enge, Preliminary assessment of interference between ultra-wideband transmitters and the global positioning system: A cooperative study, in *Proceedings of the National Technical Meeting of the Institute of Navigation*, 2000.
17. P. A. Bernhardt, Chaotic frequency modulation, in *Proc. SPIE*, vol. 2038, pp. 162–81, 1993.
18. N. F. Rulkov and A. R. Volkovskii, Threshold synchronization of chaotic relaxation oscillations, *Phys. Lett. A*, vol. 179, pp. 332–336, 1993.
19. H. Torikai, T. Saito, and W. Schwarz, Synchronization via multiplex pulse trains, *IEEE Trans. Circuits and Systems—I*, vol. 46, pp. 1072–1085, 1999.
20. M. Sushchick, N. Rulkov, L. Larson, L. Tsimring, H. Abarbanel, K. Yao, and A. Volkovskii, Chaotic pulse position modulation: A robust method of communicating with chaos, *IEEE Communications Letters*, vol. 4, pp. 128–130, 2000.
21. T. Yang and L. Chua, Chaotic impulse radio: A novel chaotic secure communication system, *Int. J. Bif. and Chaos*, vol. 10, pp. 345–357, 2000.
22. G. M. Maggio, N. Rulkov, M. Sushchik, L. Tsimring, A. Volkovskii, H. Abarbanel, L. Larson, and K. Yao, Chaotic pulse-position modulation for ultrawide-band communication systems, in *Proc. of UWB Conference*, Washington D.C., 1999.
23. D. Lind and B. Marcus, *An Introduction to Symbolic Dynamics and Coding* (Cambridge University Press, 1995).
24. E. Ott, *Chaos in Dynamical Systems* (Cambridge University Press, 1993).
25. D. C. Laney, G. M. Maggio, F. Lehmann, and L. E. Larson, BER and spectral properties of interleaved convolutional time hopping for UWB impulse radio, in *Proc. of Globecom 2003*, San Francisco, CA, December 1-5, 2003.
26. C. Caire, G. Taricco, and E. Biglieri, Bit-interleaved coded modulation, *IEEE Trans. on Inf. Theory*, vol. 44, pp. 932–946, 1998.
27. D. C. Laney, G. M. Maggio, F. Lehmann, and L. E. Larson, Multiple access for UWB impulse radio with pseudo-chaotic time hopping, *IEEE J. on Selected Areas in Comm.*, vol. 20, pp. 1692–1700, 2002.
28. S. Lin and D. J. Costello, *Error Control Coding* (Prentice-Hall, 1983).
29. S. M. Kay, *Fundamentals of Statistical Signal Processing: Detection Theory* (Prentice Hall, 1998).
30. J. M. Wozencraft and I. M. Jacobs, *Principles of Communication Engineering* (Waveland Press, 1965).
31. E. Zehavi, 8-psk trellis code for a rayleigh channel, *IEEE Trans. Comm.*, vol. 40, pp. 873–884, 1992.
32. A. J. Viterbi and J. K. Omura, *Principles of Digital Communication and Coding* (McGraw-Hill, N.Y., 1979).
33. R. D. Gitlin, J. F. Hayes, and S. B. Weinstein, *Data Communication Principles* (Plenum,, N.Y., 1992).

34. L. W. Couch, *Digital and Analog Communication Systems*, 5th ed.. (Prentice-Hall, 1997).
35. G. M. Maggio, D. Laney, F. Lehmann, and L. Larson, A multi-access scheme for UWB radio using pseudo-chaotic time hopping, in *Proceedings of Ultra Wideband Systems and Technologies Conference*, pp. 225–229, 2002.
36. G. Maggio, D. Laney, and L. Larson, BER for synchronous multi-access UWB radio using pseudo-chaotic time hopping, in *Proceedings of IEEE Global Telecommunications Conference (Globecom)*, pp. 1324–1328, 2002.

5

Optimum Spreading Sequences for Asynchronous CDMA System Based on Nonlinear Dynamical and Ergodic Theories

Kung Yao and Chi-Chung Chen

Summary. This chapter presents a tutorial and overview of the interplay among nonlinear dynamical system theory, ergodic theory, and the design and analysis of spreading sequences for CDMA communication systems. We first address some motivational factors in information theory, communication theory, and communication systems to chaotic communication systems. Then we consider some basic issues in CDMA communication system. Next we summarize some properties of nonlinear dynamical system and ergodic theories needed for this study. Some history and details on the design and analysis of optimum chaotic asynchronous and chip-synchronous spreading sequences for CDMA systems are given. These optimum spreading sequences allow about 15% more users than random white sequences/Gold codes in an asynchronous system and 73% more users in a chip-synchronous system. Comparisons of performance of these system under ideal and practical conditions are also made. Finally, some brief conclusions are given.

5.1 Introduction

In the 19th century and first half of the 20th century, practical communication systems included: digital transmission in telegraphy over wires and radio over free-space propagation; analog transmission telephony over wires; and am/fm radio broadcasts over free-space propagation. These system designs were ad hoc and governed mainly by available hardware technology. As new hardware devices came along (e.g.,vacuum tube amplifier, transistor, microelectronic chip/microprocessor, microwave amplifier, laser, etc.), new communication systems came into existence. As communication systems became more complex, there were greater needs for systematic treatments. Since the 1950s, information theory and statistical decision/estimation theory have provided the analytical tools for the successful analysis and design of advanced communication systems. Today, communication theory and systems are mature. Any proposed new communication concepts and technologies have to compete and be justified with the existing information and communication theories and systems.

The basic purpose of communication is to provide reliable transmission of information utilizing minimum bandwidth. Information theory provides the mathematical theory of information processing and transmission using efficient modulation/demodulation in conjunction with coding/decoding. Communication theory provides the concepts and models for analysis and design of communications. Communication system deals with implementation at the hardware system and subsystem levels of communications. There is a close interplay among these three well-developed but different disciplines of information theory, communication theory, and communication system.

Pecora and Carroll [15] in 1991 showed that two chaotic systems can be self-synchronized. This rather remarkable discovery has caused much interest and created the research field of "chaotic communication." Many papers motivated by this work have been published over the last 10 years. Chaos theory is a branch of nonlinear mathematics with a history of over 100 years. Chaos and fractals are intellectually challenging with many not immediately obvious properties and have been popularized by Lorenz, Mandelbrot, and others in the last 30 years.

Many workers in chaotic communication (particularly in the early years) were fundamentally more interested in the synchronization issue in nonlinear circuit and mathematical aspects of the chaotic system and not really in the communication system aspects. A simple justification of this claim is that most publications in "chaotic communication" appeared in physics/mathematics/circuit/system journals and not in information theory/communication theory/communication system journals. A meaningful question is how many chaotic communication concepts have been (or are potentially able to be) translated to practical communication systems?

Perhaps one can categorize "chaotic communication" into two classes. In one class, some chaotic properties are not necessarily implementable nor competitive with conventional communication systems (e.g., systems that are essentially only baseband systems that have no equivalent carrier frequency translatable version, systems that utilize some artificially imposed nonlinearity with interesting chaotic properties, systems that are not competitive with conventional communication systems in terms of SNR, data rate, complexity, interference rejection, etc.) In another class, certain chaotic communication systems only use some chaotic properties to replace some functional parts of conventional communication systems; these systems may still be implemented, analyzed, and shown to have some advantages with respect to their conventional equivalent counterparts (e.g, a chaotic pulse-position-modulation system (PPM) is a PPM system; a FM-DCSK system is a FM digital comm system; a laser communication system that exploits some intrinsic chaotic property is an optical communication system; a chaotic generated sequence CDMA system is a CDMA system; etc.)

We do not deal further with the above-mentioned issues of chaotic communication. In recent years, there have been various special issues and overview papers attempting to address these issues. The purpose of this chapter is to consider chaotic generated sequences for CDMA system applications. Section 2 provides a summary of the basic operations of a CDMA digital communication system. Section 3 considers some basic properties of chaotic spread-spectrum sequences. Section 4 considers a CDMA system model. Section 5 constitutes the heart of this chapter and deals with the design and analysis of chaotic spread-spectrum sequences. Section 6 discusses some aspects of ergodic theory and nonlinear dynamical systems relevant to chaotic spread-spectrum sequences. Section 7 presents results on the design of optimum chaotic spread-spectrum sequences. Section 8 provides a comparison of the performance of some chaotic spread-spectrum CDMA systems under ideal and nonideal environments. Section 9 considers some implementational issues and the construction of optimum chaotic spread-spectrum sequences from the well known Gold codes. Section 10 considers the acquisition time associated with these optimum chaotic spread-spectrum sequences. Section 11 provides some brief conclusions.

5.2 Introduction to CDMA Communication System

There are three well-known multiplexing methodologies used in modern digital communication systems. In a frequency-division-multiplexing-access (FDMA) system, the entire allocated frequency band is divided into various frequency subbands, and an individual user is given a dedicated subband. This is the oldest multiplexing method, and it is conceptually simple and has been used in telephony, digital microwave systems, and so on. Unfortunately, FDMA is not very flexible and does not make use of the total available bandwidth efficiently relative to the total data transmission rate. In time-division-multiplexing-access (TDMA), each user is assigned a time slot with the usage of the full bandwidth. The advantage of the TDMA system is that it allows possible dynamic user assignment and has been used in many satellite communication systems as well as cellular telephony (e.g., GSM-2G system). A TDMA system is more efficient in bandwidth efficiency than a FDMA system. In a code-division-multiplexing-access (CDMA) spread-spectrum (SS) system, the information data of bandwidth R = 1/T Hz is modulated (spread) by a pseudo-random (PN) spreading sequence code with a smaller chip duration of $T_C = T/PG$ to a larger bandwidth of $B = R \times PG$, where T denotes the data symbol duration and PG denotes the processing gain. As we discuss in more detail, a PG much greater than one allows significant rejection of multiple access interferences from other users in the same full frequency band. CDMA digital systems were first used for military communications, but now are used in various cellular telephony systems (e.g., IS-95, WCDMA-3G, etc.) A CDMA system has been shown to be even more efficient in the transmitted data bits per transmission Hertz bandwidth than a TDMA system.

Consider the CDMA methodology implemented on a binary-phase-shift-keyed (BPSK) digital communication system. Figiure 5.1 shows the block

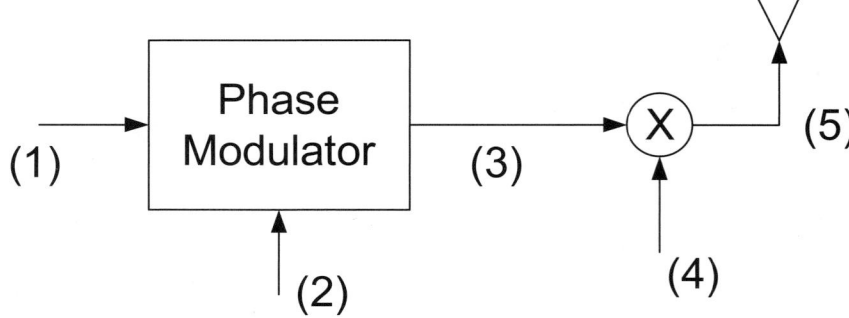

Fig. 5.1. Block diagrams of a BPSK-CDMA spread-spectrum transmitter.

diagrams of a BPSK - CDMA transmitter system marked with (1) - (5) corresponding to Equations (5.1) - (5.5) respectively. Waveforms also marked with (1) - (5) corresponding to these equations are shown in Figure 5.2. Let the binary data be denoted by

$$m(t) = b_k = \pm 1, \quad kT \le t < (k+1)T, \tag{5.1}$$

and the carrier waveform be denoted by

$$q(t) = \sqrt{(2P)} \cos(\omega_0 t), \tag{5.2}$$

where P is the carrier power and $f_0 = \omega_0/(2\pi)$ is the carrier frequency. Denote the modulated waveform by

$$s(t) = q(t)m(t) = \sqrt{(2P)}m(t)cos(\omega_0 t). \tag{5.3}$$

Let the SS sequence be denoted by

$$a(t) = a_k = \pm 1, \quad kT_c \le t < (k+1)T_c, \tag{5.4}$$

and the SS coded waveform given by

$$r(t) = s(t)a(t) = \sqrt{(2P)}m(t)a(t) \cos(\omega_0 t). \tag{5.5}$$

The waveform $r(t)$ is transmitted and is received by the BPSK - CDMA receiver system with the associated waveforms marked by (5') - (9) corresponding to Equations (5.5') - (5.9), respectively, as shown in Figure 5.3. Corresponding received waveforms are shown in Figure 5.4. Denote the received waveform by

$$r'(t) = r(t) + I(t) + n(t), \tag{5.5'}$$

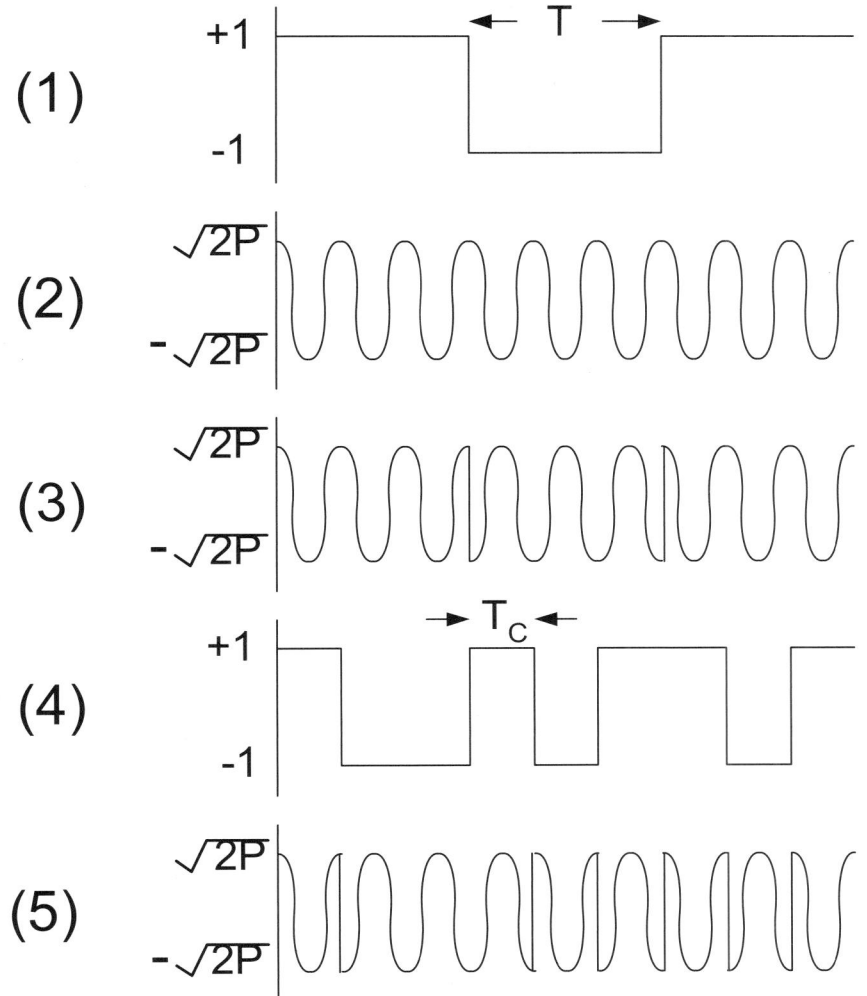

Fig. 5.2. Transmitter waveforms of a BPSK - CDMA spread-spectrum system.

where $r(t)$ is the transmitted waveform, $I(t)$ denotes the other SS - coded waveforms in the CDMA system, and $n(t)$ denotes the channel noise. Denote the phase-locked locally generated SS code sequence by

$$a'(t) \approx a(t) = a_k = \pm 1, \quad kT_c \leq t < (k+1)T_c. \qquad (5.6)$$

Denote the despread SS - decoded waveform by

$$u(t) = r'(t)a'(t) \approx \sqrt{(2P)}m(t)\cos(\omega_0 t). \qquad (5.7)$$

Denote the phase-locked locally generated carrier waveform by

$$q'(t) \approx c_0 \cos(\omega_0 t). \tag{5.8}$$

Finally, the demodulated binary data using a matched filter is given by

$$m'(t) = \int_0^T u(t)q'(t)dt \approx m(t) = b_k = \pm 1, \quad kT \leq t < (k+1)T. \tag{5.9}$$

We note the waveforms on the r.h.s. of (5.6) - (5.9) are valid in the high signal-interference-ratio (SIR) and high signal-to-noise ratio (SNR) scenarios. Indeed, in the absence of other users' mutual interferences and additive channel noise, the $\approx$ sign becomes the $=$ sign in these equations. We also note, the operations in (5.9) - (5.5') in the CDMA receiver are inverse operations of (5.1) - (5.5) in the CDMA transmitter. Of course, in the absence of other users' mutual interferences, there is no need to deploy the CDMA methodology on top of a conventional BPSK digital communication system.

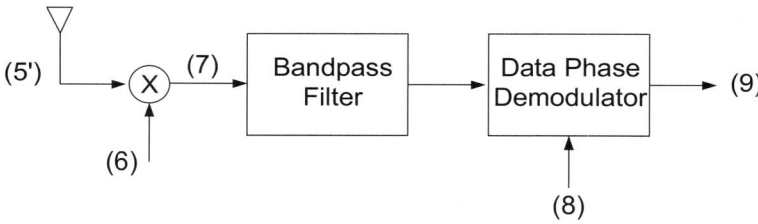

Fig. 5.3. Block diagrams of a BPSK - CDMA spread-spectrum receiver.

5.3 Chaotic CDMA Communication System

Many investigations on spread-spectrum communications have addressed direct sequence code division multiple access (DS-CDMA) systems, where all users transmit on the same band at the same time and are distinguished only by means of a code signature of a spreading sequence. System performance of a DS-CDMA communication system using the single-user matched filter receiver critically depends on the auto-correlation and cross-correlation of the spreading sequences. As is well known, the orthogonal sequences are optimal for the downlink of wireless communication systems, where all users are synchronous, because the multiple-access interferences (MAI) from other users are absent. It is inevitable, however, that the cross-correlations among the spreading sequences are nonzero for the uplink channel because of asynchronism. In this chapter, we consider the design of the spreading sequences for

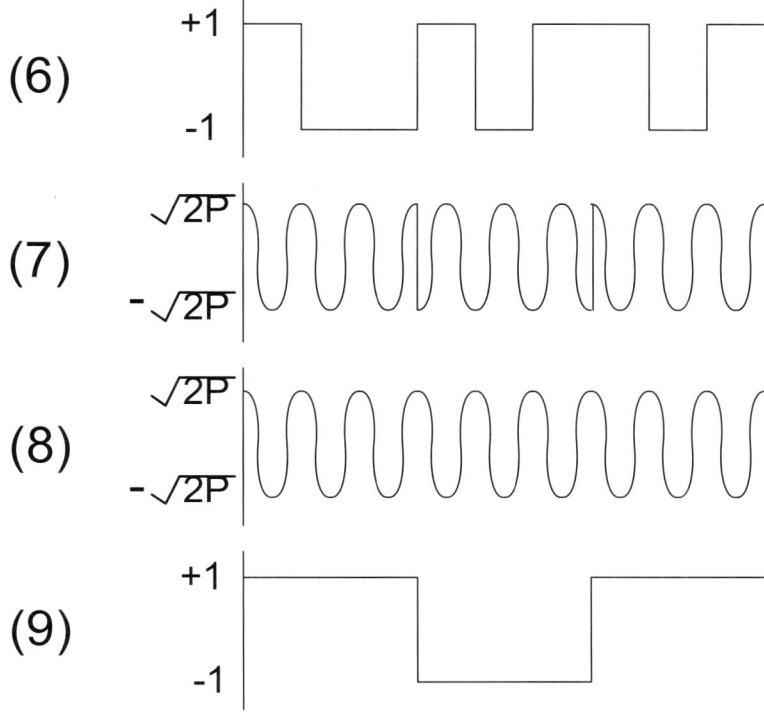

Fig. 5.4. Receiver waveforms of a BPSK - CDMA spread-spectrum system.

two DS-CDMA system models of chip-synchronous CDMA (CS-CDMA) and asynchronous CDMA (A-CDMA) systems for uplink operations.

Earlier efforts in using chaotic dynamical systems/signals for CDMA applications have been studied [1, 12, 19]. By treating the spreading sequences for an A-CDMA system as random processes and assuming they are independent and stationary, Mazzini et al. [14] found the ensemble-averaged auto-correlation function that minimizes the expected interference-to-signal ratio under the standard Gaussian approximation (SGA) and proposed a nearly optimal binary sequence generator using a piece-wise affine map (PWAM). However, this class of chaotic map generators may have practical implementational difficulties due to the need for high slope in the map as well as a finite-precision computational problem if the slope of the map is some power of 2. Chen et al. [6] derived general results on the partial auto-correlation function of the optimal spreading sequences for CS-CDMA and A-CDMA systems to minimize the average error probability under the SGA condition without the assumption of independence and stationarity on the spreading sequences, and also provided a practical implementation of the optimal real-valued spreading sequence from a chaotic dynamical system, particularly a chaotic (ergodic) dynamical system with Lebesgue spectrum.

5.4 CDMA System Models

We consider a CDMA system with K users and spreading factor N. The received signal is

$$r(t) = \sum_{k=1}^{K} s^{(k)}(t - \tau^{(k)}) + n(t), \tag{5.10}$$

where $s^{(k)}(t) = Re\{b^{(k)}(t)a^{(k)}(t)\exp(\omega_c t + \phi_0^{(k)})\}$ is the transmitted signal from the kth user, $n(t)$ is the white Gaussian noise with two-sided spectral density $N_0/2$, $b^{(k)}(t)$ and $a^{(k)}(t)$ are the data and spreading signal, respectively, w_c is the carrier frequency, and $\phi_0^{(k)}$ is the phase. The data signal $b^{(k)}(t)$ is a sequence of ± 1 rectangular pulses with a duration of T_b. The spreading sequence has a period N, and is composed of rectangular pulses with duration T_c and chip amplitude $a_j^{(k)}$ such that $\sum_{j=0}^{N-1} a_j^{(k)^2} = 1$. We assume that $N = T_b/T_c$, and $\tau^{(k)}$ is the time delay.

Because we are concerned with relative phase shifts modulo 2π and relative time delays modulo T_b, there is no loss of generality in assuming $\phi^{(i)} = \phi_0^{(i)} - \omega_c \tau^{(i)} = 0$, $\tau^{(i)} = 0$ for the desired receiver i, and $T_c = 1$ and considering only $0 \le \phi^{(k)} < 2\pi$ (assumed to be uniformly distributed in $[0, 2\pi)$), and $0 \le \tau^{(k)} < N$ for $k \ne i$.

The output of the single-user matched filter at the ith receiver is

$$Z_i = 2 \int_0^{T_b} r(t)a_i(t)\cos(\omega_c t)dt$$

$$= b_0^{(i)} + \sum_{k \ne i}^{K} I^{(k)}(b^{(k)}, \phi^{(k)}, \tau^{(k)}) + \sum_{l=0}^{N-1} a_l^{(i)}\eta_l, \tag{5.11}$$

where η_l is the equivalent Gaussian noise in the lth chip with $E[\eta_l] = 0$ and $E[\eta_i\eta_j] = N_0\delta(i - j)$, and $\sum_{k \ne i}^{K} I^{(k)}(b^{(k)}, \phi^{(k)}, \tau^{(k)})$ is the MAI. The channel SNR is given by SNR $= 1/N_0$.

The interference term $I^{(k)}(b^{(k)}, \phi^{(k)}, \tau^{(k)})$ due to the k-user has been found [16] to be

$$I^{(k)}(b^{(k)}, \phi^{(k)}, \tau^{(k)}) = (b_{-1}^{(k)}R_{k,i}(\tau^{(k)}) + b_0^{(k)}\hat{R}_{k,i}(\tau^{(k)}))\cos(\phi^{(k)}), \tag{5.12}$$

where $R_{k,i}(\tau)$ and $\hat{R}_{k,i}(\tau)$ for $0 \le l \le \tau < l + 1 \le N$, are given by

$$R_{k,i}(\tau) = C_{k,i}(l - N) + [C_{k,i}(l + 1 - N) - C_{k,i}(l - N)](\tau - l),$$
$$\hat{R}_{k,i}(\tau) = C_{k,i}(l) + [C_{k,i}(l + 1) - C_{k,i}(l)](\tau - l), \tag{5.13}$$

where $C_{k,i}(l)$ is the partial cross-correlation between the kth and the ith user and is defined by

$$
C_{k,i}(l) \equiv
\begin{cases}
\displaystyle\sum_{j=0}^{N-l-1} a_j^{(k)} a_{j+l}^{(i)}, & 0 \le l \le N-1 \\[2ex]
\displaystyle\sum_{j=0}^{N+l-1} a_{j-l}^{(k)} a_j^{(i)}, & 1-N \le l \le -1 \\[2ex]
0, & |l| \ge N.
\end{cases}
\tag{5.14}
$$

If $\tau^{(k)}$ for $k \ne i$ is a discrete random variable uniformly distributed in $\{0, 1, 2, \ldots, N-1\}$, the system is called the chip-synchronous CDMA (CS-CDMA) system whereas the system is called the asynchronous CDMA (A-CDMA) system if $\tau^{(k)}$ is a continuous random variable uniformly distributed in $[0, N)$.

5.5 Derivation of Optimal Sequences

5.5.1 Asynchronous CDMA

The overall interference variance for the ith user from all other users can be computed [16] as

$$
\sigma_A^2(i) \equiv E_{\phi^{(k)}, b_{-1}^{(k)}, b_0^{(k)}, \tau^{(k)}} \Big\{ \Big[\sum_{k \ne i}^{K} I^{(k)}(b^{(k)}, \phi^{(k)}, \tau^{(k)}) \Big]^2 \Big\}
$$

$$
= \frac{1}{6N} \sum_{k \ne i}^{K} \sum_{l=1-N}^{N-1} [2C_{k,i}^2(l) + C_{k,i}(l) C_{k,i}(l+1)].
\tag{5.15}
$$

With the SGA assumption, the error probability of the ith user is given by

$$
P_e(i) = Q \left(\sqrt{\frac{1}{\sigma_A^2(i) + N_0}} \right),
\tag{5.16}
$$

where $Q(.)$ is the complementary standard Gaussian probability distribution. The evaluation of the error probability of an A-CDMA system based on the moment space bounding technique and its relationship to the SGA assumption were discussed in [22].

Using the following identity

$$
\sum_{l=1-N}^{N-1} C_{x,y}(l) C_{x,y}(l+n) = \sum_{l=1-N}^{N-1} C_{x,x}(l) C_{y,y}(l+n)
\tag{5.17}
$$

given in [17] and the trivial identity $C_x(l) \equiv C_{x,x}(l) = C_{x,x}(-l)$, (5.15) can be simplified to

$$\sigma_A^2(i) = \frac{1}{6N} \sum_{\substack{k \neq i}}^{K} [2C_k(0)C_i(0) + 4 \sum_{l=1}^{N-1} C_k(l)C_i(l)$$

$$+ \sum_{l=1-N}^{N-1} C_k(l)C_i(l+1)]$$

$$= \frac{1}{6N} \sum_{\substack{k \neq i}}^{K} [2C_k(0)C_i(0) + 4 \sum_{l=1}^{N-1} C_k(l)C_i(l)$$

$$+ \sum_{l=0}^{N-1} C_k(l)C_i(l+1) + C_k(l+1)C_i(l)]. \qquad (5.18)$$

The importance of (5.18) lies in the fact that $\sigma_A^2(i)$ can be computed from the auto-correlation functions alone; the cross-correlation functions are not needed. Moreover, the bit error probability is independent of the distribution of the spreading sequences due to the SGA assumption.

Because the Q-function is convex for positive arguments, the lower bound of average error probability of all users in the system can be attained by assigning the same interference variance to every user. This can be done with $C_i(l) = C_k(l) = C(l)$ for all i, k, l. With the normalization $C_i(0) = 1$, we minimize the average error probability by minimizing the interference power

$$\sigma_A^2 = \frac{K-1}{6N} [2 + 4 \sum_{l=1}^{N-1} C^2(l) + 2 \sum_{l=0}^{N-1} C(l)C(l+1)]. \qquad (5.19)$$

This is a positive quadratic form in $C(l)$ whose unique minimum is achieved when $\partial \sigma^2 / \partial C(l) = 0$ for $l = 1, 2, \ldots, N-1$, that is, when

$$4C(l) + C(l+1) + C(l-1) = 0, \qquad \forall l = 1, 2, \ldots, N-1. \qquad (5.20)$$

The solution to (5.20) is given by

$$C_k(l) = (-1)^l \frac{r^{l-N} - r^{N-l}}{r^{-N} - r^N}, \quad l = 0, 1, 2, \ldots, N-1, \forall k, \qquad (5.21)$$

where $r = 2 - \sqrt{3}$. Substituting (5.21) into (5.19), we obtain the minimum interference power [6] as

$$\sigma_{A-opt}^2 = \frac{\sqrt{3}(K-1)}{6N} \frac{r^{-2N} - r^{2N}}{r^{-2N} + r^{2N} - 2}. \qquad (5.22)$$

The ensemble-averaged partial auto-correlation function obtained in [14] when the spreading sequences are assumed to be stationary and independent random processes is identical to the deterministic constant partial auto-correlation vector derived here. Note that when $l \ll N$, $C_k(l) \approx (-r)^l$ which decays exponentially with alternative sign. Moreover, the minimum interference variance is given by $\sigma^2_{A-opt} = \sqrt{3}(K-1)/6N$ as N is large, which increases by 15% the number of users achieved with white sequences, that is, $(K-1)/3N$.

5.5.2 Chip-Synchronous CDMA

The variance of MAI for a CS-CDMA system with $\tau^{(k)}$ uniformly distributed in $\{0, 1, 2, \ldots, N-1\}$ for $k \neq i$ is computed in a similar way to (5.15) and given by

$$
\begin{aligned}
\sigma^2_C(i) &= \frac{1}{2N} \sum_{k \neq i}^{K} \sum_{l=0}^{N-1} [C^2_{k,i}(l-N) + C^2_{k,i}(l)] \\
&= \frac{1}{2N} \sum_{k \neq i}^{K} \sum_{l=1-N}^{N-1} C^2_{k,i}(l) \\
&= \frac{1}{2N} \sum_{k \neq i}^{K} [C_k(0)C_i(0) + 2 \sum_{l=1-N}^{N-1} C_k(l)C_i(l)],
\end{aligned}
$$

$$(5.23)$$

where the last equality is due to the identity in (5.17). Similarly, the solution that minimizes (5.23) is given by

$$
C_k(l) = \delta(l), \quad \forall k. \tag{5.24}
$$

That is, the optimal sequences for the CS-CDMA system are the random white sequences and the corresponding minimum MAI power is

$$
\sigma^2_{C-opt} = \frac{K-1}{2N}, \tag{5.25}
$$

which decreases by 73% and 50% the number of users achieved with optimal sequences and white sequences for an A-CDMA system, respectively.

5.6 Ergodic Dynamical Systems

5.6.1 Ergodic Theory

The second-order time-averaged statistic of spreading sequences is needed for sequence design and performance analysis. For a spreading sequence generated by a deterministic dynamical system, the performance can be computed analytically or numerically by using the Birkhoff individual ergodic theory which is restated as follows.

Theorem 1. *(4.2.4 in [13]) Let $(X, \mathcal{A}, \mu)$ be a finite measure space and $S : X \mapsto X$ be a measure-preserving and ergodic transformation. Then, for any integrable f, the average of f along the sequence generated by S, that is, $\{S^{(k)}(x)\}_{k=0}^{\infty}$ for any given "initial" $x \in X$, is equal almost everywhere to the average of f over the space X; that is,*

$$\lim_{n \to \infty} \frac{1}{n} \sum_{k=0}^{n-1} f(S^{(k)}(x)) = \frac{1}{\mu(X)} \int_X f(x)\mu(dx) \qquad a.e. \qquad (5.26)$$

The use of this theorem allows us to evaluate the auto-correlation function of a sequence generated by any measure-preserving ergodic transformation. As an example of measure-preserving and ergodic transformation, consider the tent map $S(x)$ defined by

$$S(x) \equiv 1 - 2|x|, \qquad |x| \leq 1. \qquad (5.27)$$

The uniform measure is invariant for this transformation; the auto-correlation function of the sequence generated by the tent map $S(x)$ can be evaluated by

$$< C(l) > \equiv < \frac{1}{n} \sum_{j=1}^{n} S(x_j)S(x_{j+l}) >$$

$$= \frac{1}{2} \int_{-1}^{1} x S^{(l)}(x) dx$$

$$= \frac{1}{3}\delta(l), \qquad (5.28)$$

where $<>$ denotes the ensemble average with respect to the initial condition x_0 under the corresponding invariant measure.

Another ergodic transformation, the nth degree Chebyshev polynomials defined by $T_n(x) \equiv \cos(n \arccos(x))$ over the interval $[-1, 1]$, have been considered as a sequence generator for the synchronous CDMA system [21]. Examples of Chebyshev polynominals are given by

$$T_0(x) = 1, \quad T_1(x) = x, \quad T_2(x) = 2x^2 - 1, \quad T_3(x) = 4x^3 - 3x, \ldots \quad (5.29)$$

Adler and Rivlin [2] have shown Chebyshev polynomials of degree $n \geq 2$ are mixing and thus ergodic, and their invariant measure is given by $\rho(x)dx = dx/(\pi\sqrt{1-x^2})$. Furthermore, by investigating the asymptotical stability of the Frobenius - Perron operator corresponding to Chebyshev transformations, Chebyshev polynomials are shown to be exact and thus mixing and ergodic transformations [13].

The Chebyshev polynominals have the orthogonality

$$\int_{-1}^{1} T_i(x)T_j(x)\rho(x)dx = \delta_{i,j}\frac{1 + \delta_{i,0}}{2}, \qquad (5.30)$$

where $\delta_{i,j}$ is the Kronecker delta function such that

$$\delta_{i,j} = \begin{cases} 1, & i = j, \\ 0, & i \neq j. \end{cases} \tag{5.31}$$

The auto-correlation functions for sequences generated by these Chebyshev polynomials are given by

$$< C(l) > \equiv < \frac{1}{n} \sum_{j=1}^{n} T_p(x_j) T_p(x_{j+l}) >$$

$$= \int_{-1}^{1} T_p(x) T_{p^{l+1}}(x) \rho(x) dx$$

$$= \frac{1}{2} \delta(l). \tag{5.32}$$

5.6.2 Dynamical Systems with Lebesgue Spectrum

One class of ergodic dynamical systems with special property are those with Lebesgue spectrum [3]. These systems, denoted as $\phi(x)$, not only have an ergodic invariant measure, but are also associated with a special set of orthonormal basis functions $\{f_{\lambda,j}(x)\}$ for Hilbert space L_2. This orthonormal basis can be split up into classes and written as $\{f_{\lambda,j}(x) : \lambda \in \Lambda, j \in F\}$, where λ labels the classes and j labels the functions within each class. The cardinality of Λ can be proven to be uniquely determined and is called the *multiplicity of the Lebesgue spectrum*. If Λ is (countably) infinite, we speak of the (countably) infinite Lebesgue spectrum. If Λ has only one element, the Lebesgue spectrum is called simple. The important property that these particular basis functions $f_{\lambda,j}$ have is

$$f_{\lambda,j} \circ \phi = f_{\lambda,j+1}, \qquad \forall \lambda \in \Lambda, j \in F. \tag{5.33}$$

That is, all the other basis functions in the same class can be generated from one of the basis functions by using compositions with powers of the dynamical system $\phi(x)$. Furthermore, because the basis functions are orthogonal, every function is orthogonal both to every other function in the same class, and to every function in other classes.

An example of chaotic dynamical systems with a Lebesgue spectrum is the Bernoulli shift map $\phi(x)$ with invariant measure density $\rho(x) = 1$, as considered in [3], [4], and defined by

$$\phi(x) = \begin{cases} 2x, & 0 \leq x \leq 1/2, \\ 2x - 1, & 1/2 < x \leq 1. \end{cases} \tag{5.34}$$

The associated basis functions for L_2 space are Walsh functions and are defined by

$$w_1(x) = 1,$$

$$w_{k+1}(x) = \prod_{i=0}^{r-1} \operatorname{sgn}\{\sin^{k_i}(2^{i+1}\pi x)\}, \qquad k = 1, 2, \ldots, \qquad (5.35)$$

where the values of k_i, either 0 or 1, are the binary digits of k, that is, $k = \sum_{i=0}^{r-1} k_i 2^i$. Thus, this generator can produce random white binary sequences. A two-dimensional dynamical system with a Lebesgue spectrum can also be exhibited citeArnold, [4].

Another class of chaotic (ergodic) dynamical systems with a Lebesgue spectrum are the Chebyshev polynomial maps as considered above. In particular, we consider the pth degree Chebyshev polynomial map, that is, $\phi(x) = T_p(x)$, where $p \geq 2$ is prime. The associated basis functions for $L_2([-1,1])$ are also Chebyshev polynomials $\{T_i(x)\}_{i=0}^{\infty}$. Then, the $f_{\lambda,j}(x)$ can be defined by

$$f_{\lambda,j}(x) = T_{\lambda \cdot p^j}(x), \qquad \forall \lambda \in \Lambda, j \in F, \qquad (5.36)$$

where $\Lambda = \{n | n$ is nonnegative integer and relative prime to $p\}$, and F is the set of nonnegative integers. To see this, we consider the composition of ϕ with one of the basis functions:

$$\begin{aligned} f_{\lambda,j} \circ \phi(x) &= T_{\lambda \cdot p^j} \circ T_p(x) \\ &= T_{\lambda \cdot p^{j+1}}(x) \\ &= f_{\lambda,j+1}(x). \end{aligned} \qquad (5.37)$$

Note that the basis function $f_{0,j}(x) = T_0(x) = 1$ constitutes its own class and the basis function we used in (5.32) is the particular case when $\lambda = 1$.

5.7 Chaotic Optimal Spreading Sequences Design

5.7.1 Sequences Construction for CS-CDMA Systems

The previous section shows some chaotic dynamical systems that can produce independently, identically distributed (i.i.d.) sequences. Therefore, the system performance of CS-CDMA and A-CDMA systems using these sequences is identical to the random white sequences with the interference power $\sigma^2 = (K-1)/2N$ and $\sigma^2 = (K-1)/3N$, respectively.

5.7.2 Sequences Construction for A-CDMA Systems

Let us consider a polynomial function $G(x)$ in the Hilbert space $L_2([-1,1])$ with the form

$$G(x) \equiv \sum_{j=1}^{N} d_j T_{p^j}(x) = \sum_{j=1}^{N} (-r)^j T_{p^j}(x), \qquad x \in [-1,1], \qquad (5.38)$$

where $p \geq 2$. Thus, the coefficients of the Chebyshev expansion of $G(x)$ are given by

$$d_0 \equiv 0, \quad d_j = (-r)^j \quad \text{for} \quad 1 \leq j \leq N. \tag{5.39}$$

By using ergodic theory, the average of G^2 along the sequence generated by the Chebyshev transformation $T_p(.)$ is given by

$$< C(0) > \equiv < \frac{1}{n} \sum_{i=1}^{n} G^2(x_i) >$$

$$= \int_{-1}^{1} G^2(x)\rho(x)dx$$

$$= \frac{1}{2} \frac{r^2(1 - r^{2N})}{1 - r^2}$$

$$\equiv A, \tag{5.40}$$

and the normalized auto-correlation function of such sequence can be evaluated by

$$< C(l) > /A \equiv \frac{1}{A} < \frac{1}{n} \sum_{i=1}^{n} G(x_i)G(x_{i+l}) >$$

$$= \frac{1}{A} \int_{-1}^{1} G(x)G(T_{p^l}(x))\rho(x)dx$$

$$= \frac{1}{2A} \sum_{m=l}^{N} d_m d_{m+l}$$

$$= \frac{r^2}{2A} \frac{(-r)^l(1 - r^{2(N-l)})}{1 - r^2}$$

$$= (-1)^l \frac{r^{l-N} - r^{N-l}}{r^{-N} - r^N}. \tag{5.41}$$

Thus, with the condition $r = 2 - \sqrt{3}$, the output sequences $\{y_1, y_2, \ldots, y_N\}$ generated by

$$y_j = \frac{1}{\sqrt{A}}G(x_j),$$

$$x_{j+1} = T_p(x_j), \tag{5.42}$$

are the optimal spreading sequences for A-CDMA systems. Because of the property $T_{p^{j+1}}(x) = T_p \circ T_{p^j}(x)$, the function $G(x)$ in (5.38) actually is a noncausal finite impulse response (FIR) filter fed by the input sequence generated by Chebyshev polynomial map $T_p(x)$. This FIR filter can be easily implemented to produce the output sequence $\{y_j\}$ [6].

When the spreading factor N is large, an alternative practical design is given as follows. Because the auto-correlation function of a Chebyshev sequence is a Kronecker delta function, we can design the optimal spreading sequences by passing these Chebyshev sequences through an infinite impulse response (IIR) low-pass filter with a single pole at $(-r)$. That is,

$$y_{j+1} = -ry_j + \sqrt{2(1-r^2)}\, x_j,$$
$$x_{j+1} = T_p(x_j). \tag{5.43}$$

The output sequence $\{y_1, y_2, \ldots, y_N\}$ of the filter will have an exponential auto-correlation function $(-r)^l$. Then each sequence generated by the same Chebyshev map but starting from different initial condition or generated by a different-degree Chebyshev map is assigned to a different user. The sequences designed in (5.43) are used for simulation of asynchronous CDMA systems considered in the next section.

5.8 Performance Comparisons of CDMA Systems

First, we simulate an A-CDMA communication system using optimal and white spreading sequences generated by Gaussian, uniform, Chebyshev, and binary random-number generators. The simulation results in Figure 5.5 show that the error probability is independent of the distribution of the spreading sequences, and that the optimal sequences are better than random white sequences, which justifies our design. Particularly, the performances using optimal Chebyshev sequences and Gaussian sequences are similar, which is consistent with our design of chaotic spreading sequences using ergodic theory. Moreover, the performance difference between optimal and random white sequences becomes more distinct when the number of users becomes larger.

In order to understand the behavior of the sequences, we also performed simulations for different numbers of users, as shown in Figure 5.6. These simulation results show that the optimal sequences are better than random white sequences by about 15% in terms of allowable number of users, which is consistent with the analytical expression of (5.22). We also observe that when the number of users is smaller, simulation results do not quite match with analytical results obtained under the SGA condition. This confirms the well-known fact that the Gaussian approximation is not valid when the user number is small. The A-CDMA system performances of optimal second- and third-degree Chebyshev sequences are shown in Figure 5.7, which has the same parameters as Figure 5.6, and are also better by about 15% when Gold codes are employed. These results mean Gold codes have similar system performances as random white sequences, which is reasonable because Gold codes are designed to mimic random white sequences excluding some "bad" sequences.

The CS-CDMA system performances using Chebyshev sequences, white Gaussian, and binary sequences are also shown in Figure 5.7. These simulation results show these sequences for the CS-CDMA system have similar performance, and are worse than optimal sequences in asynchronous systems by about 73% in terms of allowable number of users, which corroborates the analytic expression in (5.25).

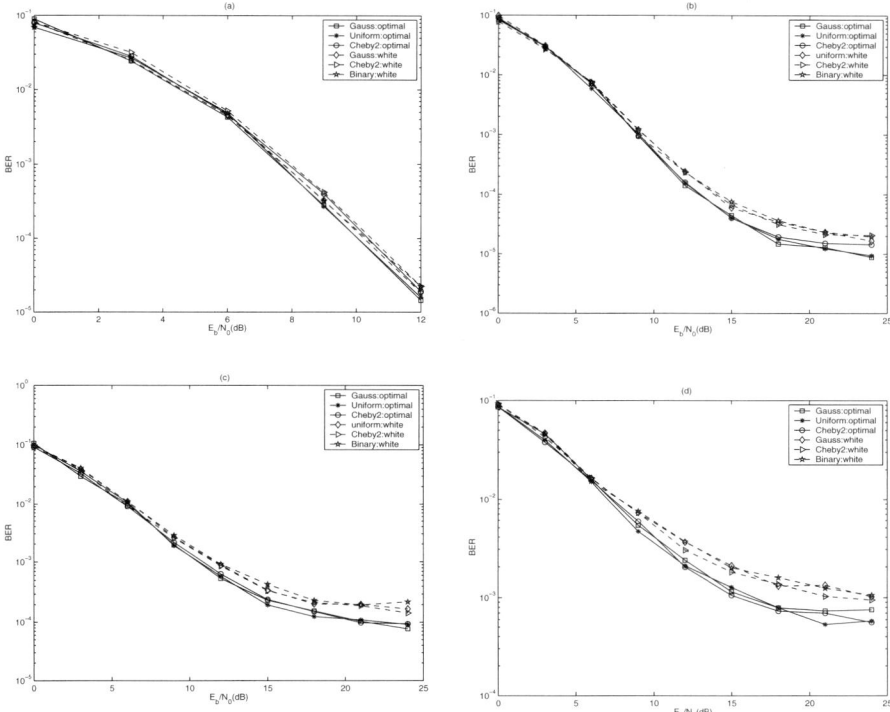

Fig. 5.5. Comparison of error probabilities of an asynchronous CDMA system in AWGN channel using optimal and white sequences generated by Gaussian, uniform, second-degree Chebyshev, and binary random-number generators for sequence length of 31. (a) K = 3, (b) K = 5, (c) K = 7, and (d) K = 10. (Reprinted with permission from [6], ©2001 IEEE.)

As examples of PWAM maps, the tent map and Bernoulli shift map have finite-precision computational difficulty because each iteration of these two maps will shift out one bit of the current value. Unlike PWAM maps, Chebyshev polynomials considered here are expected to be more robust against the finite precision problem. We evaluate the performance loss of A-CDMA systems when a second-degree Chebyshev polynomial generator using finite-precision computation is employed. The simulation results are shown in Figures 5.8 and 5.9. We observe from these simulation results that an A-CDMA system has only slight performance loss by using more than 15-20 bits for sequence generation compared to double precision (52 bits).

Finally, we compare the A-CDMA system performance over two different frequency-nonselective fading channels, Rayleigh and Rician fading channels, when the optimal sequences and Gold codes are employed with the use of the single-user matched filter receiver structure. A Rayleigh nonselective fading

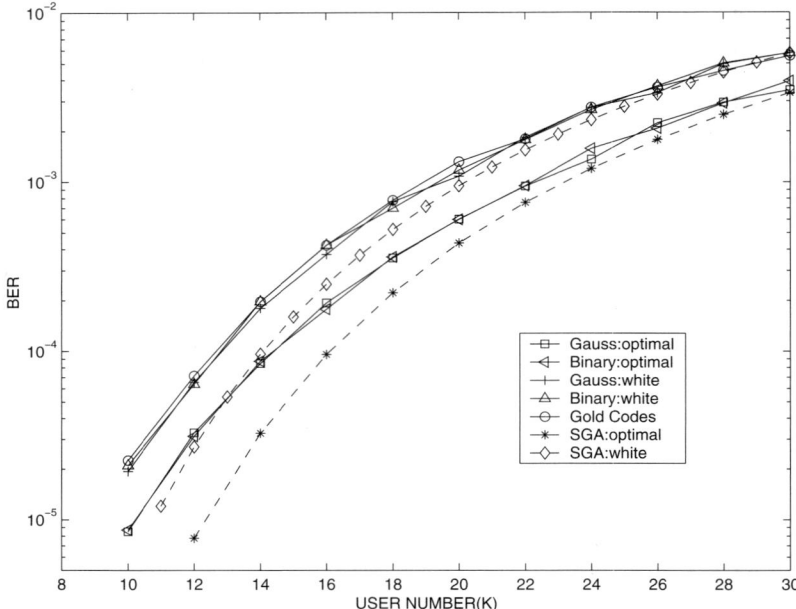

Fig. 5.6. Comparison of error probabilities of an asynchronous CDMA system in AWGN channel using optimal and white sequences generated by Gaussian and binary random-number generators and Gold codes for sequence length of 63 and for different number of users (channel SNR = 25 dB). (Reprinted with permission from [6], ©2001 IEEE.)

channel can be described by the following input - output relationship,

$$y^{(k)}(t) = Re\{A_l^{(k)}s^{(k)}(t)\exp(j\theta_l^{(k)})\}, \quad \text{for} \quad lT_b \le t < (l+1)T_b, \quad (5.44)$$

where $A_l^{(k)}$ is a Rayleigh-distributed random variable with $E\{[A_l^{(k)}]^2\} = 1$ and $\theta_l^{(k)}$ is the phase shift uniformly distributed in $[0, 2\pi)$. All communication links are assumed to fade independently. That is, $A_l^{(k)}$ and $\theta_l^{(k)}$ are independent for all l and k. Assuming the phase shift for the desired ith user is known in the receiver, (i.e., $\phi^{(i)} + \theta_l^{(i)} = 0$,) the bit error probability of the ith user under the SGA assumption is given [9] by

$$P_e(i) = \frac{1}{2}\{1 - [1 + 2(N_0 + \sigma_A^2(i))]^{-1/2}\}$$

$$\approx \frac{1}{2}[N_0 + \sigma_A^2(i)], \quad \text{if} \quad N_0 + \sigma_A^2(i) << 1/2, \quad (5.45)$$

where $\sigma_A^2(i)$ is the (nonfaded) MAI power as defined in (5.15). From (5.45) and the simulation results shown in Figure 5.10, an A-CDMA system using

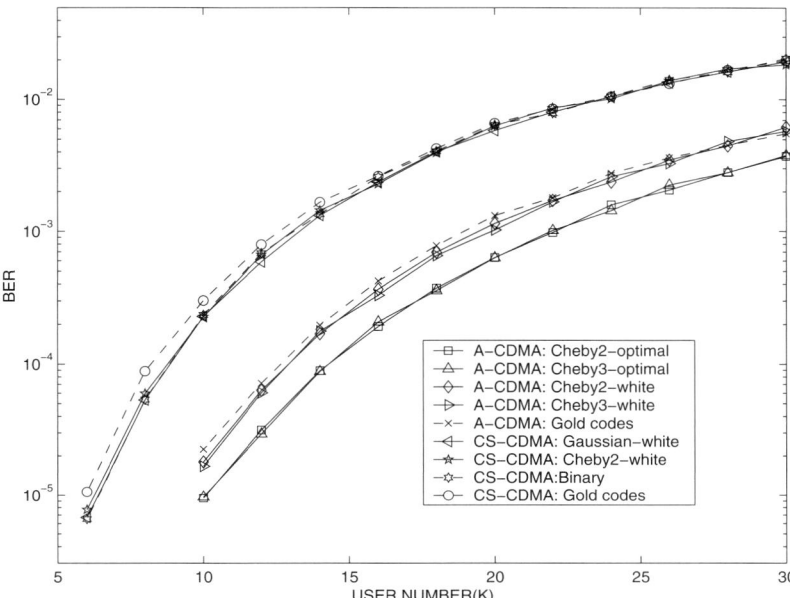

Fig. 5.7. Comparison of error probabilities of an asynchronous CDMA system in AWGN channel using second - and third-degree Chebyshev optimal and white sequences and Gold codes for sequence length of 63 and for different number of users and comparison with chip-synchronous CDMA system using various sequences (channel SNR = 25 dB). (Reprinted with permission from [6], ©2001 IEEE.)

the optimal spreading sequences can allow about 15% more users than Gold codes to have the same performance over the frequency-nonselective Rayleigh fading channel with high channel SNR.

The output signal of a Rician nonselective fading channel is the sum of a nonfaded version of the input signal (specular component) and a nondelayed Rayleigh faded version of the input signal (scatter component). That is, the input - output relationship of a Rician fading channel is given by

$$y^{(k)}(t) = s^{(k)}(t) + Re\{\gamma_k A_l^{(k)} s^{(k)}(t)\exp(j\theta_l^{(k)})\}, \qquad (5.46)$$

for $lT_b \le t < (l+1)T_b$. In (5.46) γ_k is a nonnegative real number and γ_k^2 is the power ratio of a faded component to a specular component. We assume all users have the same faded power ratio γ^2 and only the phase $\phi^{(i)}$ of the specular component is known at the ith user receiver. Under the SGA assumption, the bit error probability over the Rician fading channel is given [9] by

$$P_e(i) = Q\left(\sqrt{\frac{1}{1/2\gamma^2 + (1+\gamma^2)\sigma_A^2(i) + N_0}}\right). \qquad (5.47)$$

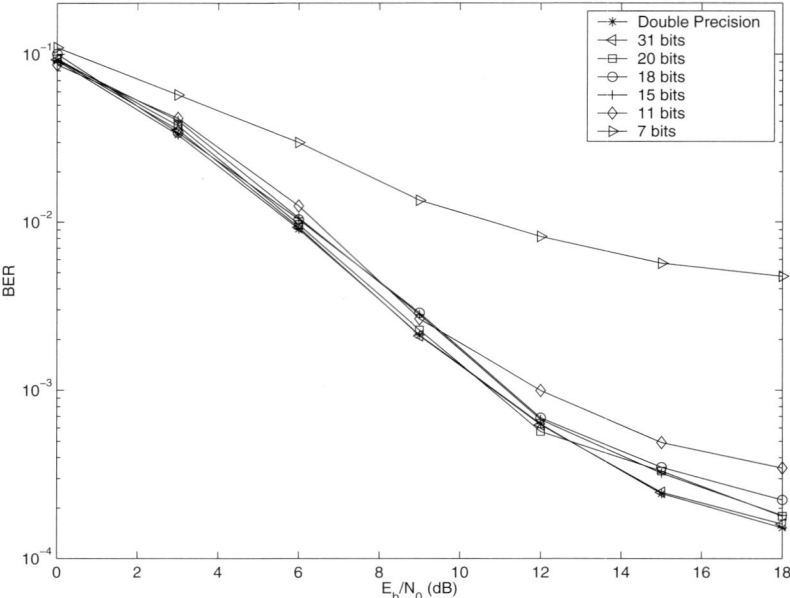

Fig. 5.8. Performance of asynchronous CDMA system in AWGN channel using various finite-precision optimal sequences generated by the second-degree Chebyshev polynomial map with N = 31 and K = 7.

The simulation results of the error probabilities for various γ^2 are shown in Figure 5.11. Clearly, the A-CDMA system performance using the optimal sequences is better than Gold codes because of a smaller $\sigma_A^2(i)$ for the optimal sequences [8].

5.9 Construction of Optimal Spreading Sequences from Gold Codes

As is well known, Gold codes of length $N = 2^n - 1$ are a family of optimal binary sequences that attain the Sidelnikov bound on the maximum $\theta_{\max}$ of periodic nonzero-lag auto-correlation peak θ_a and cross-correlation peak θ_c for any set of N or more binary sequences of period N when n is odd. When n is even, the $\theta_{\max}$ for Gold codes is larger than the Sidelnikov bound by a factor of approximately $\sqrt{2}$ [18]. In other words, Gold codes mimic purely random binary sequences excluding some "bad" sequences, and are expected to have similar performance parameters as purely random binary sequences when employed in an asynchronous CDMA system.

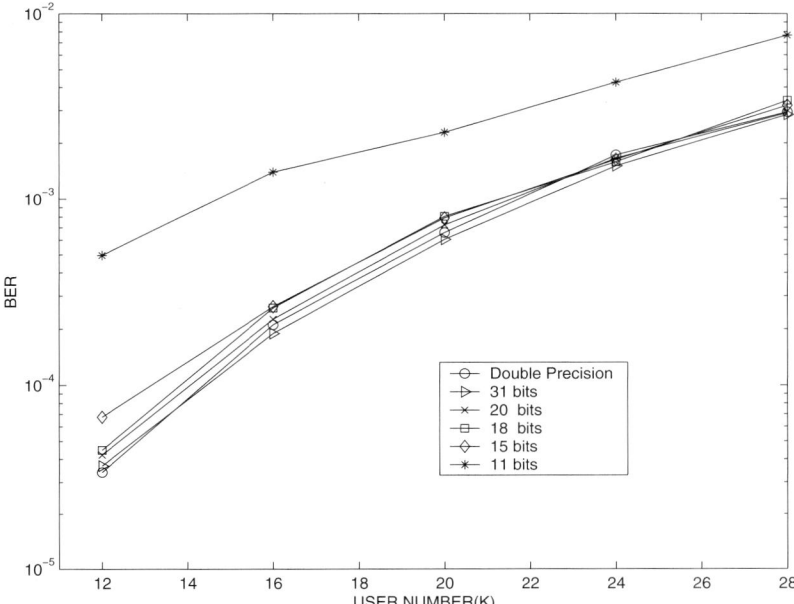

Fig. 5.9. Performance of an asynchronous CDMA system in AWGN channel using various finite-precision optimal sequences generated by the second-degree Chebyshev polynomial map with N = 63 and SNR = 25 dB. (Reprinted with permissions from [6], ©2001 IEEE.)

Neglecting the small values of auto-correlation and cross-correlation on a set of Gold codes, we can design the optimal spreading sequences by passing these Gold codes through an infinite impulse response (IIR) low-pass filter with a single pole at $(-r)$ defined in the previous section. That is,

$$y_0 = x_0,$$
$$y_{j+1} = -ry_j + \sqrt{2(1-r^2)}\,x_{j+1}, \tag{5.48}$$

where $r = 2 - \sqrt{3}$ and $\{x_j\}_{j=0}^{N-1} \in \{-1, 1\}$ is a Gold code. The output sequence $\{y_0, y_1, \ldots, y_{N-1}\}$ of the filter will have an nearly exponential auto-correlation function $(-r)^l$.

First, we simulate an asynchronous CDMA communication system for the spreading factors $N = 31$ using Gold codes and purely random binary sequences. The simulation results, as given in Figure 5.12, are consistently expected. Then we performed the simulation of an asynchronous CDMA system using the proposed optimal spreading sequences, and performance is also shown in Figure 5.12. These simulation results show that the proposed optimal spreading sequences are better than Gold codes and purely random binary sequences by about 15% in terms of allowable number of users, which

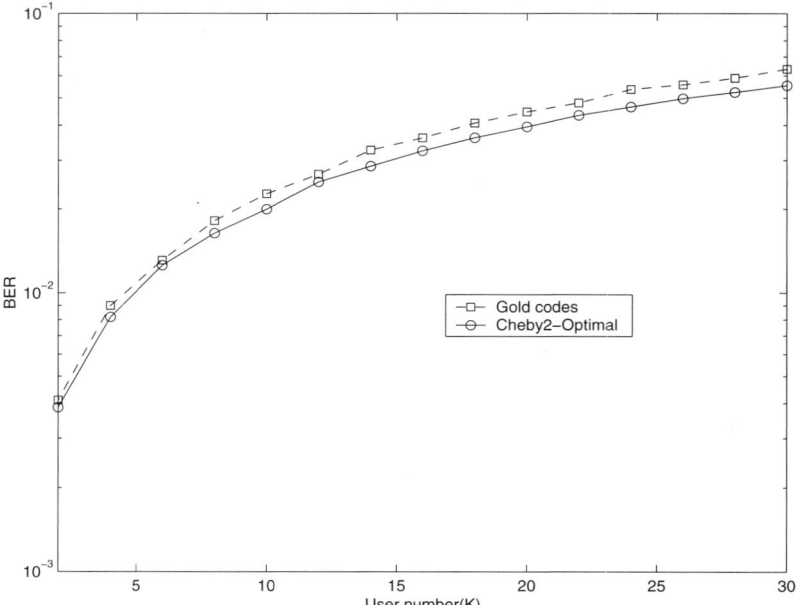

Fig. 5.10. Comparison of error probabilities of an asynchronous CDMA system over frequency-nonselective Rayleigh fading channel using Chebyshev optimal sequences and Gold codes for sequence length of 63 and for different number of users (channel SNR = 25 dB).

is consistent with the analytical expression. We also observe that when the number of users is smaller, simulation results do not quite match with analytical results obtained under the SGA condition. This confirms the well-known fact that the Gaussian approximation is not valid when the user number is small. The performance of asynchronous CDMA systems for the spreading factor $N = 63$ using various sequences is shown in Figure 5.13, which has the same performance characteristics as Figure 5.12.

5.10 Acquisition Time of Optimal Spreading Sequences

In this section [8], we consider the single dwell serial search acquisition model for the spreading sequence [11]. If the correlation error between the desired spreading sequence of the transmitter and local spreading sequence at the receiver is fixed at the update size of the spreading sequence, that is, the probability of detection P_D is constant (time invariant), the mean acquisition time is given in [11] by

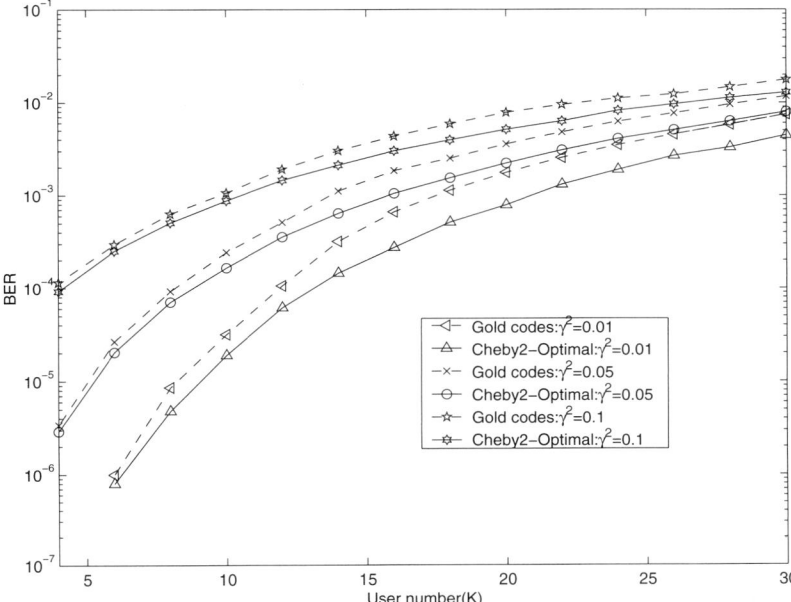

Fig. 5.11. Comparison of error probabilities of an asynchronous CDMA system over frequency-nonselective Rician fading channel with various powers of scatter component using Chebyshev optimal sequences and Gold codes for sequence length of 63 and for different number of users (channel SNR = 25 dB).

$$T_{ac} = \frac{2 + (2 - P_D)(q - 1)(1 + kP_{FA})}{2P_D}\tau_D, \qquad (5.49)$$

where P_{FA} is the probability of false alarm, τ_D seconds is the "dwell time," $k\tau_D$ is the "penalty" time of obtaining a false alarm, and $q = I \times N$ is the total number of cells to be searched, where I is the number of search samples per chip, and N is the period of the spreading sequence.

Firstly, we assume the time delay between the considered transmitter and receiver is some multiple of one-half chip duration to have a constant P_D. We set $k = 4$ for the penalty of a false alarm, $I = 2$, and the detection threshold $\alpha = 0.5$ for comparison with the correlation output at the receiver. The average acquisition time of optimal spreading sequences and Gold codes are shown in Figure 5.14 (as observed in the previous section, since SGA is not appropriate for the MAI when the user number is small, the P_D and P_{FA} are obtained from extensive simulations).

However, the time difference between the transmitter and receiver spreading sequences is not controllable and not necessary to be some multiple of one-half chip duration, and hence is assumed to be uniformly distributed in $[0, N)$. If the phase difference between the transmitted spreading sequence and

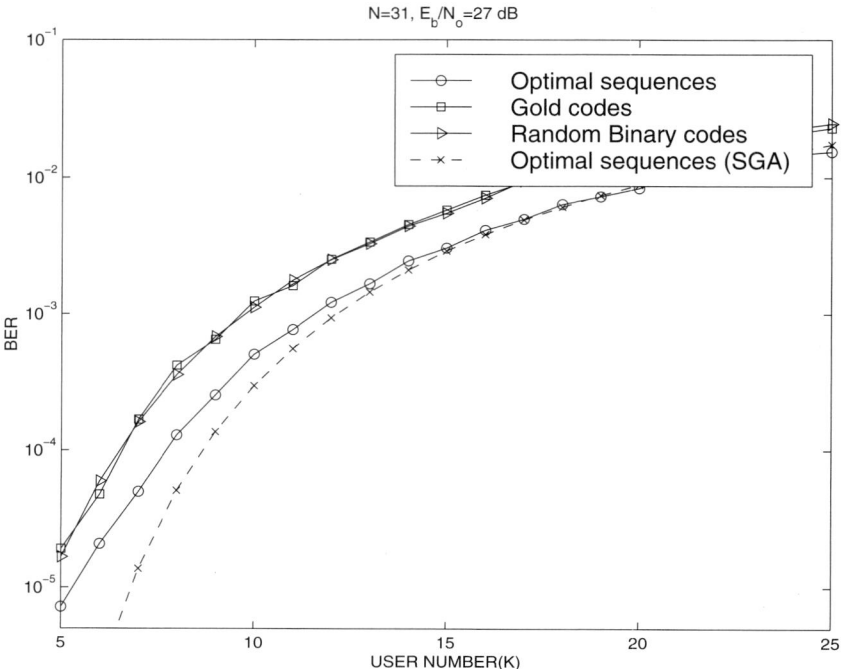

Fig. 5.12. Comparison of error probabilities of an asynchronous CDMA system using various spreading sequences (N = 31 and channel SNR E_b/N_0 = 27 dB). (Reprinted with permission from [8], ©2000 IEEE.)

local spreading sequence at the receiver is within $\pm\frac{1}{2}$ chip, the acquisition is declared. Thus, P_D is not time invariant, and (5.49) is not valid. The simulation results of the acquisition time performance are shown in Figure 5.15. From Figures 5.14 and 5.15, the acquisition time performance for the optimal sequences is better than Gold codes when employed.

Furthermore, the acquisition time performance using different detection thresholds for 8 users in the channel link are shown in Figure 5.16. In this particular case, the optimal detection thresholds for both spreading sequences are between 0.6 and 0.7, and the corresponding acquisition performance for the optimal spreading sequences is better than Gold codes. We also observed that the optimal spreading sequences are worse when larger threshold is used. The reason to this is as follows. The auto-correlation at lag one of the spreading sequences becomes more crucial for the acquisition of sequences when the detection threshold is larger. From (5.21), the peak ($\sqrt{3} - 2$) of nonzero-lag auto-correlation of the optimal spreading sequence is always at lag one whereas it is not necessary for Gold codes.

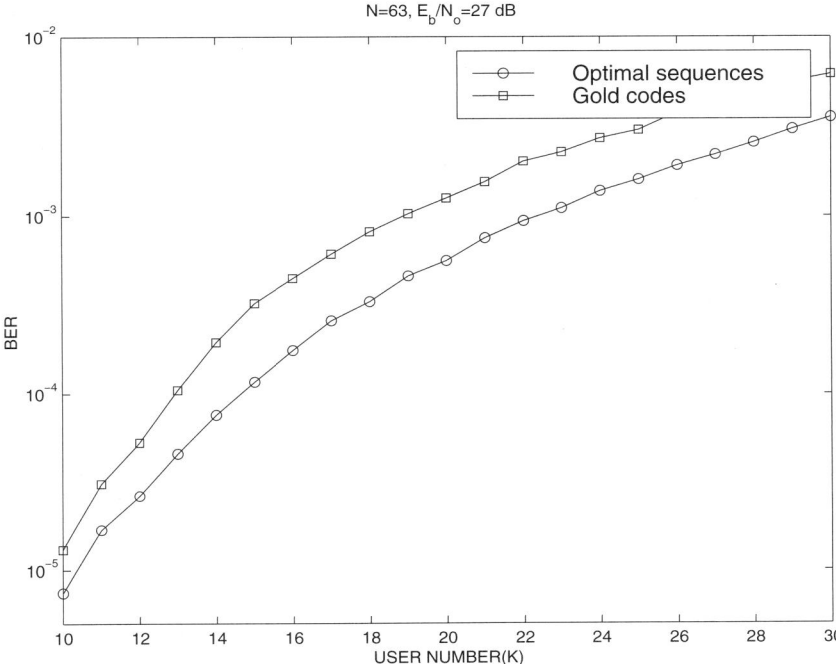

Fig. 5.13. Comparison of error probabilities of an asynchronous CDMA system using optimal spreading sequences and Gold codes ($N = 63$ and channel SNR $E_b/N_0 = 27$ dB). (Reprinted with permission from [8], ©2000 IEEE.)

Although many parameters in the acquisition scheme can be optimized, the 15% decrease of the MAI power by using optimal sequences than Gold codes is apparently an advantage of assisting the acquisition of the spreading sequences in an asynchronous CDMA system.

5.11 Conclusions

We first reviewed some aspects of chaotic communication. Then we introduced the BPSK - CDMA spread-spectrum communication system. Then we proposed a new design methodology for the design of optimal spread-spectrum sequences for asynchronous and chip-synchronous CDMA systems with respect to minimum error probability under the SGA condition. Without any assumption on spreading sequences, the optimal partial auto-correlation function of the spreading sequences is derived. Moreover, based on the ergodic theory of dynamical systems, a simple method to construct such sequences using Chebyshev polynomials, as well as analytical performance expression are provided. Using the ergodic theory of dynamical systems, a method to construct

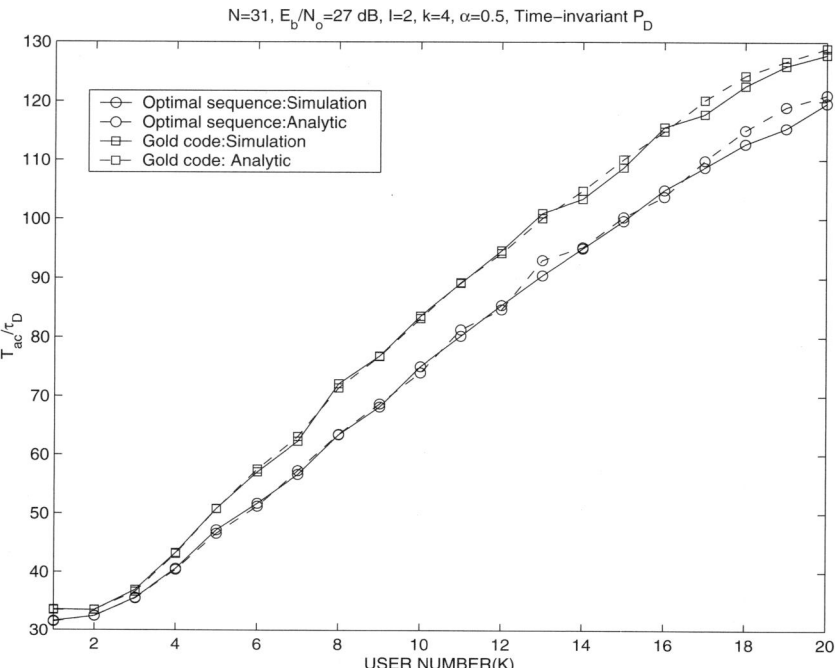

Fig. 5.14. Acquisition time performance comparison for optimal codes and Gold codes when detection probability P_D is time invariant. (Reprinted with permission from [8], ©2000 IEEE.)

and to analyze such sequences based on ergodic transformations is shown. Our method generalizes some previous approaches proposed in [14, 20, 21]. Under the SGA condition, an asynchronous CDMA system using the optimal spreading sequences allows 15% more users than when random white sequences are employed, and 73% more users of chip-synchronous systems. Simulation results also show that system performances using this family of chaotic Chebyshev spreading sequences are superior and similar to Gold codes when employed in asynchronous and chip-synchronous CDMA systems, respectively. Moreover, the Chebyshev polynomial map generator is robust against the finite-precision computational problem in terms of asynchronous CDMA system performance, as shown in simulation results. The proposed optimal spreading sequences still perform better than Gold codes in an asynchronous CDMA system over frequency-nonselective fading channels. Then some details on the generation of these optimum chaotic CDMA codes upon transformation of Gold codes are given. The acquisition time of these optimum chaotic CDMA codes are also shown to be competitive to previously known

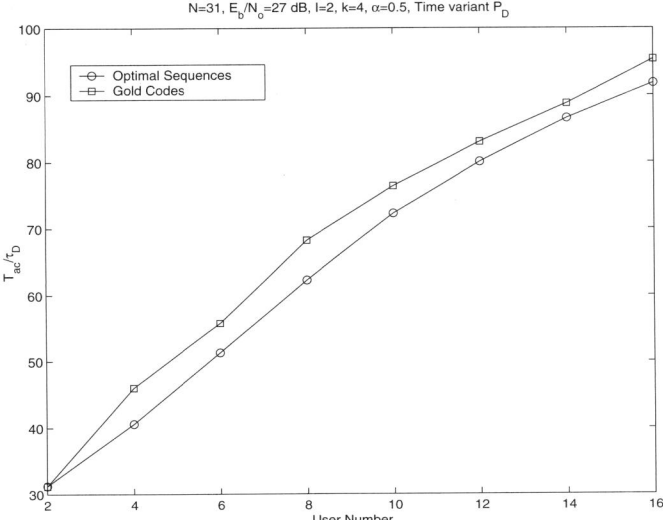

Fig. 5.15. Acquisition time performance comparison for optimal sequences and Gold codes when detection probability P_D is time variant. (Reprinted with permission from [8], ©2000 IEEE.)

CDMA codes. Extensive simulations are given to verify the performance of these chaotic CDMA codes.

Acknowledgments

This work was partially supported by the Army Research Office under MURI grant DAAG55-98-1-0269, NASA grant NCC 4-153, and CoRE grant sponsored by STM.

References

1. A. Abel, A. Bauer, K. Kelber, and W. Schwarz, Chaotic codes for CDMA applications, *Proc. ECCTD'97*, pp. 306–311, 1999.
2. R. L. Adler, and T. J. Rivlin, Ergodic and mixing properties of Chebyshev polynomials, *Proc. Amer. Math. Soc.*, vol. 15, pp. 794–796, 1964.
3. V. I. Arnold and A. Avez, *Ergodic Problems of Classical Mechanics* (W. A. Benjamin, New York, 1968).
4. D. S. Broomhead, J. P. Huke, and M. R. Muldoon, Codes for spread spectrum applications generated using chaotic dynamical systems, *Dynamics and Stability of Systems,* vol. 14, no. 1, pp. 95–105, 1999.

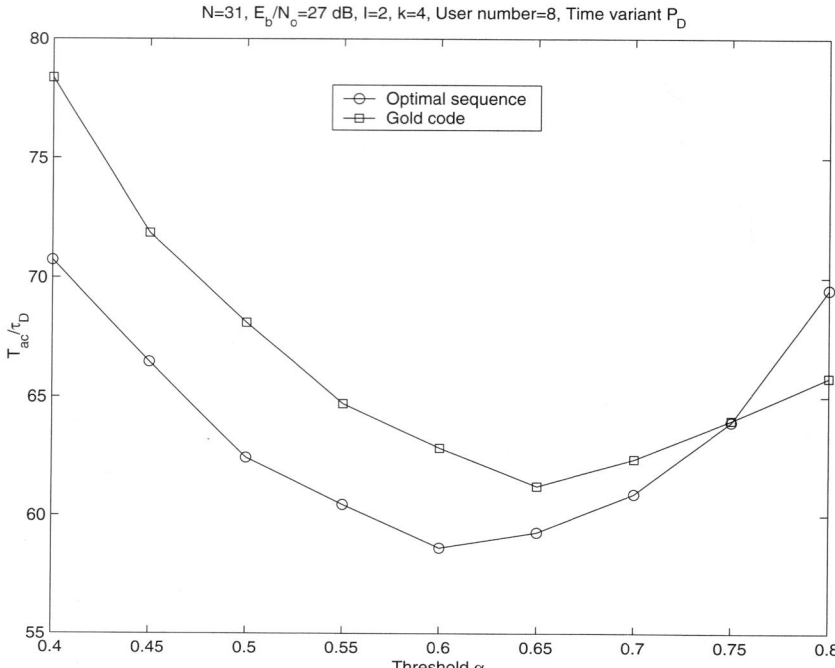

Fig. 5.16. Acquisition time performance comparison for optimal sequences and Gold codes for different detection thresholds (User number = 8). (Reprinted with permission from [8], ©2000 IEEE.)

5. C. C. Chen, E. Biglieri, and K. Yao, Design of spread spectrum sequences using Ergodic Theory, *International Symposium on Information Theory,* Sorrento, Italy, p. 379, 2000.
6. C. C. Chen, K. Yao, K. Umeno, and E. Biglieri, Design of spread spectrum sequences using chaotic dynamical systems and Ergodic Theory, *IEEE Trans. on Circuits and Systems, I,* Vol. 48, pp. 1110–1113, 2001.
7. C. C. Chen, *Chaotic signals for communication systems,* Ph. D. Dissertation, UCLA, 2000.
8. C.C. Chen, K. Yao, and E. Biglieri, Optimal spread spectrum sequences constructed from Gold codes, *Proc. IEEE Globecom,* pp. 867–871, 2000.
9. E. Geraniotis, Direct-sequence spread-spectrum multiple-access communications over nonselective and frequency-selective Rician fading channels, *IEEE Trans. Communications,* vol. 34, no. 8, pp. 756–764, 1986.
10. M. Götz and A. Abel, Design of infinite chaotic polyphase sequences with perfect correlation properties, *Proc. IEEE ISCAS'98,* pp. III 279–282, 1999.
11. J. K. Holmes and Chang C. Chen, Acquisition time performance of PN spread-spectrum systems, *IEEE Trans. Communications,* vol. 25, pp. 778–784, 1977.
12. T. Kodha and A. Tsuneda, Statistics of chaotic binary sequences, *IEEE Trans. Information Theory,* vol. 43, pp. 104–112, 1997.

13. A. Lasota, and M. C. Mackey, *Chaos, Fractals, and Noise: Stochastic Aspects of Dynamics* (Springer-Verlag, New York, 1994).

14. G. Mazzini, R. Rovatti, and G. Setti, Interference minimization by autocorrelation shaping in asynchronous DS-CDMA systems: Chaos-based spreading is nearly optimal, *Electronics Letters,* vol. 35, pp. 1054–1055, 1999.

15. L.M. Pecora and T.L. Carroll, Synchronization in Chaotic Systems, *Phys. Rev. Lett.,* vol. 64, pp. 821–824, 1990.

16. M. B. Pursley, Performance evaluation for phased-coded spread-spectrum multiple-access communication - Part I: System Aanalysis, *IEEE Trans. Communications*, vol. 25, pp. 795–799, 1977.

17. M. B. Pursley and D. V. Sarwate, Performance evaluation for phased-coded spread-spectrum multiple-access communication - Part II: Code sequence Aanalysis, *IEEE Trans. Communications*, vol. 25, pp. 800–803, 1977.

18. D. V. Sarwate and M. B. Pursley, Crosscorrelation properties of pseudorandom and related sequences, *Proc. of IEEE,* vol. 68, no. 5, pp. 593–619, 1980.

19. J. Schweizer and M. Hasler, Multiple access communications using chaotic signals, *Proc. ISCAS'96,* pp. 108–111, 1996.

20. K. Umeno, Chaotic Monte Carlo computation: a dynamical effect of randomnumber generations, *Japan Journal of Applied Physics*, vol. 39, Part 1, pp. 1442–1456, 2000.

21. K. Umeno and K. I. Kitayama, Improvement of SNR with chaotic spreading sequences for CDMA, *IEEE Information Theory Workshop*, South Africa, June, 1999.

22. K. Yao, Error probability of asynchronous spread spectrum multiple access communication systems, *IEEE Trans. Communications*, vol. 25, pp. 803–809, 1977.

6

Nonlinear Phenomena in Turbo Decoding Algorithms

Ljupco Kocarev

Summary. The turbo decoding algorithm is a high-dimensional dynamical system parameterized by a large number of parameters (for a practical realization the turbo decoding algorithm has more than 10^3 variables and is parameterized by more than 10^3 parameters). In this chapter we treat the turbo decoding algorithm as a dynamical system parameterized by a single parameter that closely approximates the signal-to-noise ratio (SNR). A whole range of phenomena known to occur in nonlinear systems, such as the existence of multiple fixed points, oscillatory behavior, bifurcations, chaos, and transient chaos are found in the turbo decoding algorithm. We develop a simple technique to control transient chaos in the turbo decoding algorithm and improve the performance of the standard turbo codes.

6.1 Introduction

Recently, it has been recognized that two classes of codes, namely *turbo codes* [1] and *low-density parity-check* (LDPC) codes [2–4], perform at rates extremely close to the Shannon limit imposed by the noisy channel coding theorem [5]. Both codes are based on a similar philosophy: constrained random code ensembles, described by some fixed parameters plus randomness, decoded using iterative decoding algorithms (or message passing decoders). Iterative decoding algorithms may be viewed as a complex nonlinear dynamical system. The aim of the present work is to contribute to the in-depth understanding of these families of error-correction codes, based on the well–developed theory of nonlinear dynamical systems [6].

Turbo codes were discovered by Berrou et al. in 1993 [1]. On the other hand, LDPC codes were originally introduced by Gallager [7] in 1962, the crucial innovation of LDPC codes being the introduction of iterative decoding algorithms. LDPC codes were rediscovered by MacKay et al. [2] in 1996. Moreover, iterative decoding of turbo codes was recognized as instances of sum-product algorithms for codes defined on general graphs [8]. The past few years have seen many new developments in the area of iterative decoding algorithms for both turbo and LDPC codes. We now briefly mention some of these achievements. The complexity, introduced by the interleaver in turbo coding, makes a rigorous analysis of the distance spectrum difficult, if not impossible, for an arbitrary interleaver instance. However, Benedetto and his co-workers [9] introduced the concept of "uniform interleaver", which proved to be a very useful tool for investigating the average distance spectrum of turbo codes. Moreover, they generalized the original parallel concatenation of convolutional codes (i.e., turbo codes) to the case of serially and hybridly concatenated codes, showing that serial concatenation results in a larger interleaver gain and, therefore, in a better average distance spectrum. Although turbo codes, on average, have poor free distance, their extraordinary performance was explained from the distance spectrum perspective by the phenomenon of spectral thinning, observed by Perez *et al.* [10].

Very recently, in a pioneering paper [11], Richardson has presented a geometrical interpretation of the turbo decoding algorithm and formalized it as a discrete-time dynamical system defined on a continuous set. This approach clearly demonstrates the relationship between turbo decoding and maximum-likelihood decoding. The turbo decoding algorithm appears as an iterative algorithm aimed at solving a system of $2n$ equations in $2n$ unknowns, where n is the block-length size. If the turbo decoding algorithm converges to a certain codeword, then the latter constitutes a solution to this set of equations. Conversely, solutions to these equations provide fixed points of the turbo decoding algorithm, seen as a nonlinear mapping. In a follow-up by Agrawal and Vardy [12] a rigorous bifurcation analysis of the iterative decoding process as a dynamical system parameterized by SNR has been carried out. These works open new research directions for analyzing and designing random coding schemes.

In this chapter, we consider the iterative decoding algorithm as a nonlinear dynamical system, where the codewords—to which the algorithm converges—correspond to fixed points in the symbol state space. We emphasize that in general the iterative decoding algorithm, being a nonlinear dynamical system, may exhibit a whole range of phenomena known to occur in nonlinear systems [13, 14]. These include the existence of multiple fixed points, oscillatory behavior, and even chaos.

The outline of this chapter is as follows. In Section 6.2 we recall Richardson's formulation of the turbo decoding algorithm as a dynamical system [11]. To this aim, we consider a classical turbo code with parallel concatenation of identical recursive convolutional codes generated by the polynomials $\{D^4 + D^3 + D^2 + D^1 + 1, D^4 + 1\}$, resulting in a rate-1/3 turbo code. The codewords are transmitted over an AWGN (additive white Gaussian noise) channel using BPSK (binary phase shift keying) modulation. With an interleaver length of $n = 1024$, the turbo decoding algorithm may be described as a dynamical system of dimension $n(= 1024)$, parameterized by $3n(= 3072)$ parameters.

In Section 6.3 we present an overview of nonlinear dynamical systems; basic definitions of chaotic set, chaotic saddle, Lyapunov exponents, and transient chaos, together with simple examples are given.

In Section 6.4 we analyze the character of the fixed points in the turbo decoding algorithm. Simulations show that at low SNRs, the turbo decoding algorithm often converges to an indecisive fixed point, which corresponds to many erroneous decisions on the information bits. On the other hand, at slightly higher SNRs, after the waterfall region, the turbo decoding algorithm converges to an unequivocal fixed point that corresponds to correct decisions on information bits.

In Section 6.5, we treat the turbo decoding algorithm as a dynamical system parameterized by a single parameter that closely approximates the SNR. By varying this parameter, we analyze the turbo decoding algorithm as a function of SNR. In each instance of the turbo decoding algorithm that we analyzed, an unequivocal fixed point was found in a wide range of SNRs: we found that this point is stable even for SNR ≈ -1.5 dB. However, at low SNR, before and at the waterfall region, the decoding algorithm often fails to converge to this fixed point, spending time, instead, on another attracting (chaotic) invariant set. The reason why the turbo decoding algorithm is unable to find the unequivocal fixed point for low SNRs, even when the fixed point is stable, is due to the fact that the basin of attraction of this fixed point can be very small. In our simulations we found that the indecisive fixed point loses its stability at low SNR, typically in the range of -7 dB to -5 dB. The region -5 dB to 0 dB is characterized by chaotic behavior: the turbo decoding algorithm as a dynamical system possesses a chaotic attractor. In the waterfall region, the turbo decoding algorithm converges either to the chaotic invariant set or to the unequivocal fixed point, after a long transient behavior. The latter indicates the existence of a chaotic non attracting invariant set in the vicinity of the unequivocal fixed point.

Section 6.6 considers an application of the theory developed here. We use a simple technique for controlling transient chaos, thereby reducing the number of iterations needed by the turbo decoding algorithm to reach the unequivocal fixed point.

6.2 Dynamics of Iterative Decoding Algorithms

6.2.1 Preliminaries

Let $\mathcal{H}$ be the set of all ordered binary strings of length n. We use $\boldsymbol{b}^0 = (0, 0, \ldots, 0)^T$, $\boldsymbol{b}^1 = (1, 0, \ldots, 0)^T$, $\boldsymbol{b}^2 = (0, 1, \ldots, 0)^T$, $\ldots$, $\boldsymbol{b}^n = (0, 0, \ldots, 1)^T$, $\boldsymbol{b}^{n+1} = (1, 1, \ldots, 0)^T$, $\ldots$, $\boldsymbol{b}^{2^n-1} = (1, 1, \ldots, 1)^T$ to denote the elements of $\mathcal{H}$ sorted in increasing order of Hamming weight, and within each weight class, sorted in reverse lexicographical order. A *density* on $\mathcal{H}$ is a positive real function defined over $\mathcal{H}$. A density f on $\mathcal{H}$ induces a probability measure Pr_f on the set of all subsets of $\mathcal{H}$, $\mathcal{P}(\mathcal{H})$, as follows,

$$Pr_f(A) := \frac{\sum_{\boldsymbol{b} \in A} f(b)}{\sum_{\boldsymbol{b} \in \mathcal{H}} f(b)},$$

for all $A \in \mathcal{P}(\mathcal{H})$.

A density p is called probability density if $\sum_{\boldsymbol{b} \in \mathcal{H}} p(\boldsymbol{b}) = 1$. A density p is normalized with respect to the all-zero binary string, $\boldsymbol{b}^0 = \boldsymbol{0}$, if $p(\boldsymbol{0}) = 1$. We say that the densities p and q are equivalent if they determine the same probability density. In each equivalence class, there is a unique probability density and a unique density normalized with respect to $\boldsymbol{0}$. For brevity, a density normalized with respect to $\boldsymbol{0}$ is simply called a normalized density.

It is useful to represent densities in the logarithmic domain. Given a density f, let $F = log \circ f$ be its logarithmic representation. We say F is a log-density on $\mathcal{H}$. A log-density F is a real valued function on $\mathcal{H}$, taking both positive and negative values. Let Φ denote the set of all log densities that correspond to the normalized densities; that is, $F \in \Phi$ if and only if $F(\boldsymbol{0}) = 0$.

Let $\mathcal{H}_i \subset \mathcal{H}$ be the set of all binary strings whose ith bit is 1. A density f is referred to as a *product density* if according to the induced probability measure Pr_f, all bits are independent of each other. For a normalized product density f, $f(\boldsymbol{b}^i)$, $i = 1, \ldots, n$, is the likelihood ratio of the ith bit according to the density f,

$$f(\boldsymbol{b}^i) = \frac{Pr_f(\mathcal{H}_i)}{Pr_f(\mathcal{H}_i^c)} = \frac{\sum_{\boldsymbol{b} \in \mathcal{H}_i} f(b)}{\sum_{\boldsymbol{b} \in \mathcal{H}_i^c} f(b)}.$$

It is clear that for a product log-density we have

$$f(\boldsymbol{b} = b_1 b_2 \ldots b_n) = \prod_{i:b_i=1} f(\boldsymbol{b}^i).$$

We refer to a log density that corresponds to a product density as a *product log-density*. Let Π be the set of all product log-densities in Φ. Using the last expression, for a product long-density $F \in \Pi$, we can write

$$F(\boldsymbol{b} = b_1 b_2 \ldots b_n) = \sum_{i:b_i=1} F(\boldsymbol{b}^i).$$

Therefore, $F(\boldsymbol{b}^i)$ is the log-likelihood ratio of the ith bit according to the density f and densities in Π are completely specified by their values on $\boldsymbol{b}^1$, $\boldsymbol{b}^2, \ldots, \boldsymbol{b}^n$. Furthermore, Π is an n-dimensional linear subspace of Φ. A basis for Π is given by a $2^n \times n$ matrix B, having $(\boldsymbol{b}^i)^T$ as its ith row.

We say that two densities p and q have the same bitwise marginal distributions if

$$Pr_p(\mathcal{H}_i) = Pr_q(\mathcal{H}_i)$$

for $i = 1, \ldots, n$. For a log-density P, we define a projection map $\pi_P : \Pi \to \Pi$ by setting $\pi_P(Q)$ to be the unique normalized product log-density that has the same bitwise marginals as $P + Q$. In another words,

$$\pi_P(Q)(\boldsymbol{b}^i) = \log \frac{\sum_{\boldsymbol{b} \in \mathcal{H}_i} p(\boldsymbol{b})q(\boldsymbol{b})}{\sum_{\boldsymbol{b} \in \mathcal{H}_i^c} p(\boldsymbol{b})q(\boldsymbol{b})} \tag{6.1}$$

for $i = 1, \ldots, n$.

6.2.2 The Turbo Decoding Algorithm as a Dynamical System

A classical turbo code is a parallel concatenation of two recursive systematic binary convolutional codes, C_1 and C_2 [1]. Let $\boldsymbol{i}$ be the information bit sequence of length n at the input to the turbo encoder, and let $\boldsymbol{c}_1(\boldsymbol{i})$ (respectively, $\boldsymbol{c}_2(\boldsymbol{i})$) be the parity bits produced by the first (respectively, second) encoder. The information bit sequence $\boldsymbol{i}$ along with the parity bit sequences $\boldsymbol{c}_1(\boldsymbol{i})$ and $\boldsymbol{c}_2(\boldsymbol{i})$ form a turbo codeword $(\boldsymbol{i}, \boldsymbol{c}_1(\boldsymbol{i}), \boldsymbol{c}_2(\boldsymbol{i}))$.

We assume that the turbo code is transmitted over a noisy binary-input memoryless channel. Let $\tilde{\boldsymbol{i}}$, $\tilde{\boldsymbol{c}}_1$, and $\tilde{\boldsymbol{c}}_2$ be the channel outputs corresponding to the input sequences $\boldsymbol{i}$, $\boldsymbol{c}_1$, and $\boldsymbol{c}_2$, respectively. Ideally, we would like to compute the posterior probability density, $p(\boldsymbol{b}|\tilde{\boldsymbol{i}}, \tilde{\boldsymbol{c}}_1, \tilde{\boldsymbol{c}}_2)$, where $\boldsymbol{b} \in \mathcal{H}$. Let us assume that the input bits are independent of each other and are equally likely to be either 0 or 1. Under this standard assumption, $p(\boldsymbol{b}|\tilde{\boldsymbol{i}}, \tilde{\boldsymbol{c}}_1, \tilde{\boldsymbol{c}}_2)$ is equivalent to the density $p_{ML}(\boldsymbol{b}) = p(\tilde{\boldsymbol{i}}|\boldsymbol{b})p(\tilde{\boldsymbol{c}}_1|\boldsymbol{b})p(\tilde{\boldsymbol{c}}_2|\boldsymbol{b})$. A direct computation of p_{ML} requires taking both sets of parity bits, $\tilde{\boldsymbol{c}}_1$ and $\tilde{\boldsymbol{c}}_2$, simultaneously into account by constructing a joint trellis of two convolutional encoders, which is computationally prohibitive.

The turbo decoder consists of two components: a decoder D_1 for the convolutional code C_1 and a decoder D_2 for the code C_2. These decoders use the BCJR [17] algorithm to compute the a posteriori probabilities of the information bits. Let q_1 be the a priori product density used by decoder D_2 and let q_2 be the a priori product density used by decoder D_1. We assume that both densities q_1 and q_2 are initialized to the uniform density. The decoding begins with the decoder D_1 computing the posterior likelihood ratios of the information bits based on the prior density q_2 and the observations $\tilde{\boldsymbol{i}}$ and $\tilde{\boldsymbol{c}}_1$. The posterior likelihood ratio of the ith information bit is given by

$$\frac{\sum_{\boldsymbol{b} \in \mathcal{H}_i} p(\boldsymbol{b}|\tilde{\boldsymbol{i}}, \tilde{\boldsymbol{c}}_1)}{\sum_{\boldsymbol{b} \in \mathcal{H}_i^c} p(\boldsymbol{b}|\tilde{\boldsymbol{i}}, \tilde{\boldsymbol{c}}_1)} = \frac{\sum_{\boldsymbol{b} \in \mathcal{H}_i} p(\tilde{\boldsymbol{i}}|\boldsymbol{b})p(\tilde{\boldsymbol{c}}_1|\boldsymbol{b})q_2(\boldsymbol{b})}{\sum_{\boldsymbol{b} \in \mathcal{H}_i^c} p(\tilde{\boldsymbol{i}}|\boldsymbol{b})p(\tilde{\boldsymbol{c}}_1|\boldsymbol{b})q_2(\boldsymbol{b})}. \tag{6.2}$$

Let p_0, p_1, and p_2 be the normalized densities equivalent to $p(\tilde{\boldsymbol{i}}|\boldsymbol{b})$, $p(\tilde{\boldsymbol{c}}_1|\boldsymbol{b})$, and $p(\tilde{\boldsymbol{c}}_2|\boldsymbol{b})$, respectively. The first decoder D_1 uses the normalized density $p_0 p_1 q_2$ to compute posterior likelihood ratios. In the logarithmic domain, $p_0 p_1 q_2$ corresponds to $P_0 + P_1 + Q_2$. Because $(P_0 + Q_2) \in \Pi$, it follows from (6.1) that the posterior log-likelihood ratio of the ith information bit is given by

$\pi_{P_1}(P_0 + Q_2)(\boldsymbol{b}^i)$. Next, the extrinsic information for the ith information bit is obtained by dividing its posterior likelihood ratios, computed with (6.2), by the product of its channel and prior likelihood ratios. This extrinsic information is then passed from the decoder D_1 to the decoder D_2 by setting the prior density q_1 in such a way that the likelihood ratios of the information bits according to q_1 equal their extrinsic information; that is, we set

$$\frac{q_1(\boldsymbol{b}^i)}{q_1(\boldsymbol{b}^0)} = \frac{\sum_{\boldsymbol{b} \in \mathcal{H}_i} p(\boldsymbol{b}|\tilde{\boldsymbol{i}}, \tilde{\boldsymbol{c}}_1) \, p(\boldsymbol{b}^0|\tilde{\boldsymbol{i}}) \, q_2(\boldsymbol{b}^0)}{\sum_{\boldsymbol{b} \in \mathcal{H}_i^c} p(\boldsymbol{b}|\tilde{\boldsymbol{i}}, \tilde{\boldsymbol{c}}_1) \, p(\boldsymbol{b}^i|\tilde{\boldsymbol{i}}) \, q_2(\boldsymbol{b}^i)}.$$

In the logarithmic domain, the last equation can be rewritten as

$$Q_1(\boldsymbol{b}^i) = \pi_{P_1}(P_0 + Q_2)(\boldsymbol{b}^i) - (P_0 + Q_2)(\boldsymbol{b}^i)$$

for $i = 1, \ldots, n$. Recall that q_1 and q_2 are initialized to induce the uniform probability distribution on $\mathcal{H}$. In the logarithmic domain, this corresponds to setting $Q_1^{(0)} = Q_2^{(0)} = \boldsymbol{0}$. Because q_1 is a product density, the likelihood ratios $q_1(\boldsymbol{b}^i)/q_1(\boldsymbol{b}^0)$ determine the density q_1 uniquely. Therefore, we have:

$$Q_1^{(l+1)} = \pi_{P_1}(P_0 + Q_2^{(l)}) - (P_0 + Q_2^{(l)}). \tag{6.3}$$

The second decoder D_2 performs a similar operation and compute the modified prior log-density Q_2:

$$Q_2^{(l)} = \pi_{P_2}(P_0 + Q_1^{(l)}) - (P_0 + Q_1^{(l)}). \tag{6.4}$$

The decoding algorithm iteratively performs the operations indicated by (6.3) and (6.4): $l = 0, 1, 2, \ldots$.

Equations (6.3) and (6.4) may be considered as a discrete-time dynamical system. The log-densities P_0, P_1, and P_2 are completely specified by the channel likelihood ratios of the codeword bits. Consequently, the turbo decoding algorithm is parameterized by $3n$ parameters. The iterated variables, Q_1 and Q_2, are product log-densities, and each of them can be specified by n log-likelihood ratios. Hence, in the above formulation, the turbo decoding algorithm is a n-dimensional dynamical system depending on $3n$ parameters. As shown in [11], this mapping depends smoothly on its variables and parameters.

What are the parameters of the turbo decoding algorithm? We assume that the turbo codewords are transmitted over an AWGN channel using BPSK modulation. Let $\mathbf{s}(\boldsymbol{b})$ be the Euclidean-space representation of the binary string $\boldsymbol{b}$ under the BPSK map, and let $\mathbf{s}_1 = \mathbf{s} \circ \boldsymbol{c}_1$ and $\mathbf{s}_2 = \mathbf{s} \circ \boldsymbol{c}_2$, where "$\circ$" denotes the composition of two functions. Without loss of generality, we consider the case when the vector $(\mathbf{s}(\boldsymbol{b}^0), \mathbf{s}_1(\boldsymbol{b}^0), \mathbf{s}_2(\boldsymbol{b}^0))$ is transmitted and the vector $(\boldsymbol{x}, \boldsymbol{y}, \boldsymbol{z})$ is received. The normalized posterior density p_1, induced on the information bits by $\boldsymbol{y}$, is given by

$$p_1(\boldsymbol{b}) = \frac{Pr(\boldsymbol{b}|\boldsymbol{y})}{Pr(\boldsymbol{b}^0|\boldsymbol{y})} = \exp \frac{||\boldsymbol{y} - \mathbf{s}_1(\boldsymbol{b}^0)||^2 - ||\boldsymbol{y} - \mathbf{s}_1(\boldsymbol{b})||^2}{2\sigma^2},$$

where σ^2 is the noise variance and $||\cdot||^2$ denotes the squared Euclidean distance in $I\!R^n$. The corresponding normalized log-density P_1 is given by

$$P_1(\boldsymbol{b}) = \frac{||\boldsymbol{y} - \mathbf{s}_1(\boldsymbol{b}^0)||^2 - ||\boldsymbol{y} - \mathbf{s}_1(\boldsymbol{b})||^2}{2\sigma^2} = -\frac{2}{\sigma^2} \sum_{j:s_{1j}(\boldsymbol{b}^0) \neq s_{1j}(\boldsymbol{b})} y_j,$$

where s_{1j} and y_j are the jth components of $\mathbf{s}_1$ and $\boldsymbol{y}$, respectively. In a similar manner, log densities P_0 and P_2 are induced by the received vectors $\boldsymbol{x}$ and $\boldsymbol{z}$, respectively. Therefore, the turbo decoding algorithm has $3n$ parameters: $x_1, \ldots, x_n$, $y_1, \ldots y_n$, and $z_1, \ldots, z_n$.

6.3 Nonlinear Dynamical Systems

We define a differentiable discrete-time dynamical system by an evolution equation of the form

$$\boldsymbol{x}_{n+1} = f(\boldsymbol{x}_n),$$

where f is a differentiable function and the variables $\boldsymbol{x}$ vary over a state-space M, which can be $I\!R^m$ or a compact manifold. Computer experiments with iterative decoding algorithms usually exhibit transient behavior followed by what appears to be an asymptotic regime. Thus iterative decoding algorithms are dissipative systems. In general, for dissipative systems, there exists a set $U \subseteq M$, which is asymptotically contracted by the time evolution to a compact set; that is, the set $\bigcap_{n \geq 0} f^n(U)$ is compact.

A subset A of M is said to be *invariant* if $f(A) = A$. We say that a compact invariant set $A \subset M$ is topologically *transitive* if there exists $\boldsymbol{x} \in A$ such that $\omega(\boldsymbol{x}) = A$, where $\omega(\boldsymbol{x})$ is the set of limit points of the orbit $f^n(\boldsymbol{x})_{n \geq 0}$. We say that A is *Lyapunov stable* if for every neighborhood U of A, there exists a neighborhood V of A such that $f^n(V) \subset U$ for all $n \geq 0$. In the following definitions, we assume that A is a compact transitive set.

Definition 1.(Basin of attraction) *The* basin of attraction $\mathcal{B}(A)$ *of a set A is the set of all $\boldsymbol{x} \in M$ with $\omega(\boldsymbol{x}) \subseteq A$.*

For nonempty A, the basin of attraction $\mathcal{B}(A)$ is always nonempty, because it includes A. For A to be an attractor, we require that $\mathcal{B}(A)$ is large in the appropriate sense.

Definition 2.(Attractor) *A set A is an (asymptotically stable) attractor if it is Lyapunov stable and $\mathcal{B}(A)$ contains a neighborhood in A.*

The tool with which one often analyzes attractors in a dynamical system is ergodic theory. For a compact invariant set A, there is a probability measure ρ invariant under the dynamics (time evolution) whose support is contained in A.

Definition 3.(Chaotic set) *A compact invariant set A is* chaotic *if A supports an ergodic measure but is not uniquely ergodic.*

In what follows we denote by $\mathcal{L}(\cdot)$ a Lebesgue measure on M. For instance $\mathcal{L}(A) = V(A)$, where $V(A)$ is the volume of A, is a Lebesgue measure on $\mathbb{R}^m$.

Definition 4.(Chaotic saddle) *A chaotic invariant transitive set A is a* chaotic saddle *if there is a neighborhood U of A such that $(\mathcal{B}(A) \cap U) \neq A$ and $\mathcal{L}(\mathcal{B}(A)) = 0$.*

Example 1.(Piece-wise linear map) Consider a piecewise linear function $f : \mathbb{R} \to \mathbb{R}$ defined by

$$f(x) = \begin{cases} a_1 x + c_1 & -\infty < x \leq b_1 \\ -a_2 x + c_2 & b_1 < x \leq b_2, \\ a_3 x + c_3 & b_2 < x \leq b_3, \\ \cdots & \cdots \\ -a_{2d} x + c_{2d} & b_{2d-1} < x < +\infty, \end{cases} \tag{6.5}$$

where the parameters $a_i > 1$, $b_i \in (0, 1)$, and c_i are chosen so that $f(0) = 0$, $f(1) = 1$, and the map $f(\cdot)$ is continuous. If $a_i b_i < 1$ for all i, the map has two attractors: a chaotic one, which is a subset of $(0, 1)$, and a fixed point located at $-\infty$. The basin of attraction of the chaotic attractor is the interval $(0, 1)$. If $a_i b_i > 1$ for all i, the map has only one attractor: the fixed point at $-\infty$. However, the map admits a chaotic saddle: this is the set of initial points which stay in the unit interval when time n goes to infinity.

6.3.1 Lyapunov Exponents

Lyapunov exponents are useful quantitative indicators of asymptotic expansion and contraction rates in a dynamical system. They, therefore, have fundamental bearing upon stability and bifurcations. In particular, the local stability of fixed points (or periodic orbits) depends on whether the corresponding Lyapunov exponents are negative. For invariant sets with more complex dynamics, such as chaotic attractors and chaotic saddles, the situation is much more subtle. Invariant measures in this case are often not unique; for instance, associated with each (unstable) periodic orbit contained in an attractor is a Dirac ergodic measure whose support is the orbit. Ergodic measures are thus not unique if there is more than one periodic orbit (as is the case with most chaotic attractors), and each ergodic measure carries its own Lyapunov exponents.

Theorem 1.(Multiplicative ergodic theorem) *Let ρ be a probability measure on the space M, and let $f : M \to M$ be a measure-preserving map such that ρ is ergodic. Let $D_{\boldsymbol{x}}f = (\partial f_i / \partial x_j)$ denote the matrix of partial derivatives of the components f_i at $\boldsymbol{x}$. Define the matrix*

$$T_{\boldsymbol{x}}^n \overset{\text{def}}{=} (D_{f^{n-1}(\boldsymbol{x})}f)(D_{f^{n-2}(\boldsymbol{x})}f) \cdots (D_{f(\boldsymbol{x})}f)D_{\boldsymbol{x}}f = D_{\boldsymbol{x}}f^n$$

and let $T_{\boldsymbol{x}}^{n*}$ be the adjoint of $T_{\boldsymbol{x}}^n$. Then for ρ-almost all $\boldsymbol{x}$, the following limit exists

$$\Lambda_{\boldsymbol{x}} \overset{\text{def}}{=} \lim_{n \to \infty} (T_{\boldsymbol{x}}^{n*} T_{\boldsymbol{x}}^n)^{1/(2n)}.$$

The logarithms of the eigenvalues of $\Lambda_{\boldsymbol{x}}$ are called *Lyapunov (or characteristic) exponents*. We denote these exponents by $\lambda_1 \geq \lambda_2 \geq \cdots$ or by $\lambda^{(1)} > \lambda^{(2)} > \cdots$ when they are no longer repeated by their multiplicity $m^{(i)}$. If ρ is ergodic, the Lyapunov exponents are almost everywhere constant.

The finite set $\{\lambda_i\}$ captures the asymptotic behavior of the derivative along almost every orbit, in such a way that a positive λ_i indicates eventual expansion (and, hence, sensitive dependence on initial conditions) whereas negative λ_i indicates contraction along certain directions.

The exponential rate of growth of distances is given in general by λ_1; if one picks a vector at random, then its growth rate is λ_1. The growth rate of a k-dimensional Euclidean volume element is given by $\lambda_1 + \lambda_2 + \cdots + \lambda_k$.

6.3.2 Ergodic Measures

Many important results of ergodic theory hold for arbitrary invariant measures ρ. However, in working with dynamical systems two measures play a central role: natural measures and Sinai–Ruelle–Bowen (SRB) measures. For the definition of SRB measures, we refer the reader to [15, 16]. We describe the natural measures next.

In general, it is exceptional that an attractor admits only one ergodic invariant measure. In typical cases, there are uncountably many distinct ergodic measures, each concentrated on an invariant set (the attractor) of Lebesgue measure zero. Nevertheless, in computer experiments in seems that one invariant probability measure ρ_{natural} can be derived more or less automatically from the time that the system spends in various parts of the attractor A. This is the so-called *natural measure* which describes how frequently various parts of A are visited by the orbit $n \to \boldsymbol{x}(n)$. More precisely, let S be a subset of the basin of attraction $\mathcal{B}(A)$ and let $\rho_\varepsilon(\boldsymbol{x}, S)$ be the fraction of time the trajectory originating at $\boldsymbol{x}$ spends in the ε-neighborhood of S. Define $\rho(\boldsymbol{x}, S) = \lim_{\varepsilon \to 0^+} \rho_\varepsilon(\boldsymbol{x}, S)$. If $\rho(\boldsymbol{x}, S)$ is the same for almost all $x \in \mathcal{B}(A)$, then we denote this value by $\rho_{\text{natural}}(S)$ and call it the natural measure of the attractor.

Example 2. For the logistic map $f : [0, 1] \to [0, 1]$ given by $f(x) = 4x(1-x)$, the whole phase-space is a chaotic set, with natural ergodic measure $\rho = dx/\pi\sqrt{x(1-x)}$ and Lyapunov exponent $\lambda = \log 2$.

Example 3.(Piecewise linear map) Consider the piecewise linear map (6.5) with $a_i b_i = 1$ for all i. In other words, every linear piece maps a part of the unit interval onto $[0, 1]$. Because $|f'(x)| > 1$ everywhere, all periodic orbits are unstable and the whole state space $[0, 1]$ is a chaotic set. For the natural measure ρ of this chaotic set, one has:

$$\lambda = h(\rho) = \sum_{k=1}^{2d} a_k^{-1} \log a_k .$$

The quantity $h(\cdot)$ above is known as the measure-theoretic entropy or the Kolmogorov–Sinai entropy.

6.3.3 Largest Lyapunov Exponent as Classification Tool

In computer experiments with iterative decoding we have encountered a number of qualitatively different attractors, each of them associated with a different type of time evolution. We found that the largest Lyapunov exponent turns out to be a good indicator for deducing what kind of state the ergodic measure ρ is describing.

Attracting Fixed Points and Periodic Points: The point $Q \in \mathbb{R}^m$ is a *fixed point* of $f : \mathbb{R}^m \to \mathbb{R}^m$ if $f(Q) = Q$. It is *attracting* if there is a neighborhood U of Q such that $\lim_{n \to \infty} f^n(x) = Q$ for all $x \in U$. Clearly, the asymptotic measure of the attractor is the measure $\rho = \delta_Q$, where δ_Q is the Dirac's delta function at Q. This measure is invariant and ergodic. The point Q is a *periodic point* of period k if $f^k(Q) = Q$. The least positive k for which $f^k(Q) = Q$ is called the prime period of Q; then Q is a fixed point of f^k. The set of all iterates of a periodic point forms a periodic trajectory (orbit). The Lyapunov exponents in both cases are simply related to the eigenvalues of the Jacobian matrix evaluated at Q and, therefore, are all negative. Thus the largest Lyapunov exponent satisfies $\lambda_1 < 0$.

Attracting Limit Cycle: Let Γ be a closed curve in $\mathbb{R}^m$, homeomorphic to a circle. If Γ is an attractor, a volume element is not contracted along the direction of Γ. Therefore, its asymptotic measure has one Lyapunov exponent equal to zero and all the others are negative. Thus $\lambda_1 = 0$.

Chaotic Attractor: An attractor A is *chaotic* if its asymptotic measure (natural measure) has a positive Lyapunov exponent. If any one of the Lyapunov exponents is positive, a volume element is expanded in some direction at exponential rate and neighboring trajectories are diverging. This property is called *sensitive dependence on initial conditions*. Thus for chaotic attractors, $\lambda_1 > 0$ and $\sum_{i=1}^m \lambda_i < 0$ (because the invariant set A is an attractor).

Example 4. Let $a \in \mathbb{R}$ be a parameter, and consider the two-dimensional system defined on the plane $\mathbb{R}^2$ as follows:

$$x_{n+1} = y_n \tag{6.6}$$

$$y_{n+1} = ay_n(1 - x_n). \tag{6.7}$$

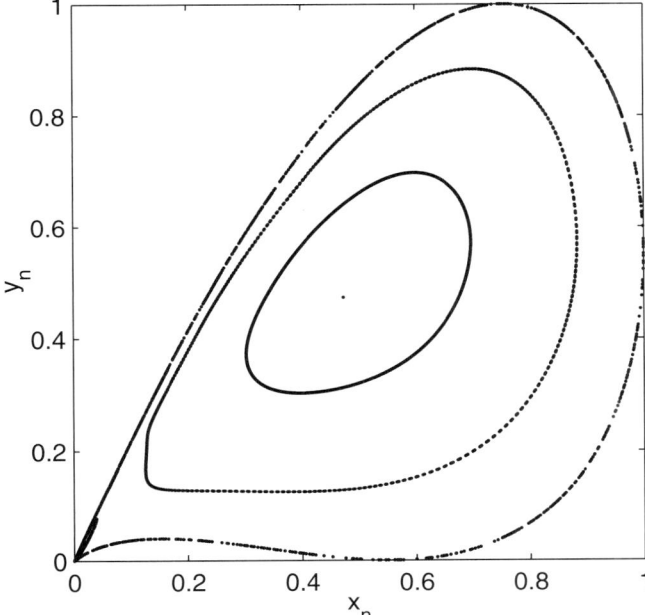

Fig. 6.1. Route to chaos for the simple 2-D map given by (6.6)–(6.7). The values of a corresponding to the invariant sets, starting from the fixed point towards chaos, are: 1.9, 2.1, 2.16, and 2.27.

The system has a fixed point at $x = y = (a - 1)/a$, which is stable for $1 < a \leq 2$. As a passes through the value 2, this fixed point loses stability and spawns an attracting invariant circle (via Neimark-Sacker bifurcation). This circle grows as the parameter a increases, becoming noticeably warped. When $a = 2.27$, the circle has completely broken down, forming a chaotic attractor. Figure 6.1 shows a typical route to chaos for this system, as well as the different attractors: fixed point, limit-cycle, and chaotic attractor.

6.3.4 Transient Chaos

Chaotic saddles are nonattracting closed invariant sets having a dense orbit. A trajectory starting from a random initial condition in a state-space region that contains a chaotic saddle typically stays near the saddle, exhibiting chaotic like dynamics, for a finite amount of time before eventually exiting the region and asymptotically approaching a final state, which is usually nonchaotic. Thus, in this case, chaos is only transient.

The natural measure for a chaotic saddle is defined as follows. Let U be the region that encloses a chaotic saddle. If we iterate N_0 initial conditions, chosen uniformly in U, then the orbits that leave U never return to U. Let N_n be the number of orbits that have not left U after n iterates. For large n, this number will decay exponentially with time:

$$\frac{N_n}{N_0} \sim \exp\left\{-\frac{n}{\tau}\right\} .$$

We call τ the *lifetime* of the chaotic transient. Let W be a subset of U. Then the natural measure of W is

$$\mu(W) \overset{\text{def}}{=} \lim_{n \to \infty} \lim_{N_0 \to \infty} \frac{N_n(W)}{N_n},$$

where $N_n(W)$ is the number of orbit points that fall in W at time n. The last two equations imply that if the initial conditions are distributed according to the natural measure and evolved in time, then the distribution will decay exponentially at the rate $\alpha = 1/\tau$. Points that leave U after a long time do so by being attracted along the stable manifold of the saddle, bouncing around on the saddle in a (perhaps) chaotic way, and then exiting along the unstable manifold. For the natural measure μ of a chaotic saddle, one has

$$\alpha = \sum_{\lambda_i > 0} \lambda_i - h(\mu) .$$

We now present an example where the characteristics of a chaotic saddle, namely the chaotic transient lifetime and the Lyapunov exponents, can be computed in a closed form.

Example 5.(Piecewise linear map) Consider the piecewise linear map (6.5) with $a_i b_i > 1$ for all i. In other words, every linear piece maps a part of the unit interval onto an interval containing the entire unit interval. This requires $|f'(x)| > 1$ everywhere, and thus all periodic orbits are unstable and the chaotic saddle is the closure of the set of all finite periodic orbits. Note that the chaotic saddle is a subset of the unit interval. For the natural measure μ of this chaotic saddle, one finds:

$$\frac{1}{\tau} = -\log \sum_{i=1}^{2d} a_i^{-1}$$

$$\lambda = \frac{\sum_{k=1}^{2d} a_k^{-1} \log a_k}{\sum_{i=1}^{2d} a_i^{-1}} .$$

$$h(\mu) = \frac{\sum_{k=1}^{2d} a_k^{-1} \log(a_k \sum_{i=1}^{2d} a_i^{-1} \log a_i)}{\sum_{i=1}^{2d} a_i^{-1}} .$$

Note that this system satisfies the relation $h(\mu) = \lambda - \alpha$ characteristic of a chaotic saddle.

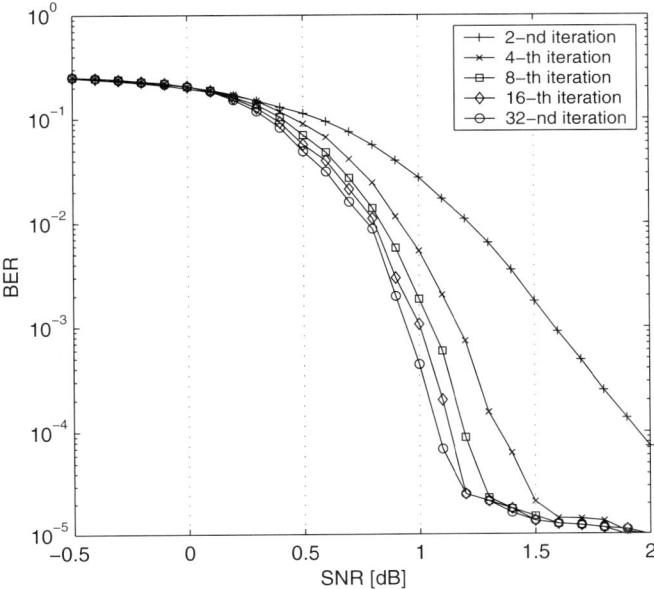

Fig. 6.2. The performance of the classical turbo code with interleaver length 1024 and rate 1/3, for increasing number of iterations. Note that the waterfall region of the turbo code corresponds approximately to the SNR region spanning 0.25 dB to 1.25 dB.

6.4 Fixed Points in the Turbo-Decoding Algorithm

6.4.1 Basic Concepts of Dynamical System Theory

The basic goal of the theory of dynamical systems is to understand the asymptotic behavior of the system itself. If the process is described by a differential equation whose independent variable is time, then the theory attempts to predict the ultimate behavior of the solutions of the equation in either the distant future $(t \to \infty)$ or the distant past $(t \to -\infty)$. If, on the other hand, the process is a discrete-time process such as iteration of the function $G : I\!R^m \to I\!R^m$, then the theory attempts to understand the eventual behavior of the set of the points $\{x, G(x), G^2(x), \ldots\}$, called a trajectory (or orbit) of x. Functions that determine discrete-time dynamical systems are also called *mappings*, or *maps*. Trajectories of points can be quite complicated sets, even for very simple nonlinear mappings. However, there are some trajectories that are especially simple and which play a central role in the study of the entire system, as described in the following.

The point $M \in I\!R^m$ is a *fixed point* for $G : I\!R^m \to I\!R^m$ if: $G(M) = M$. The point M is a *periodic point* of period k if $G^k(M) = M$. The least positive k for which $G^k(M) = M$ is called the prime period of M. The set of all

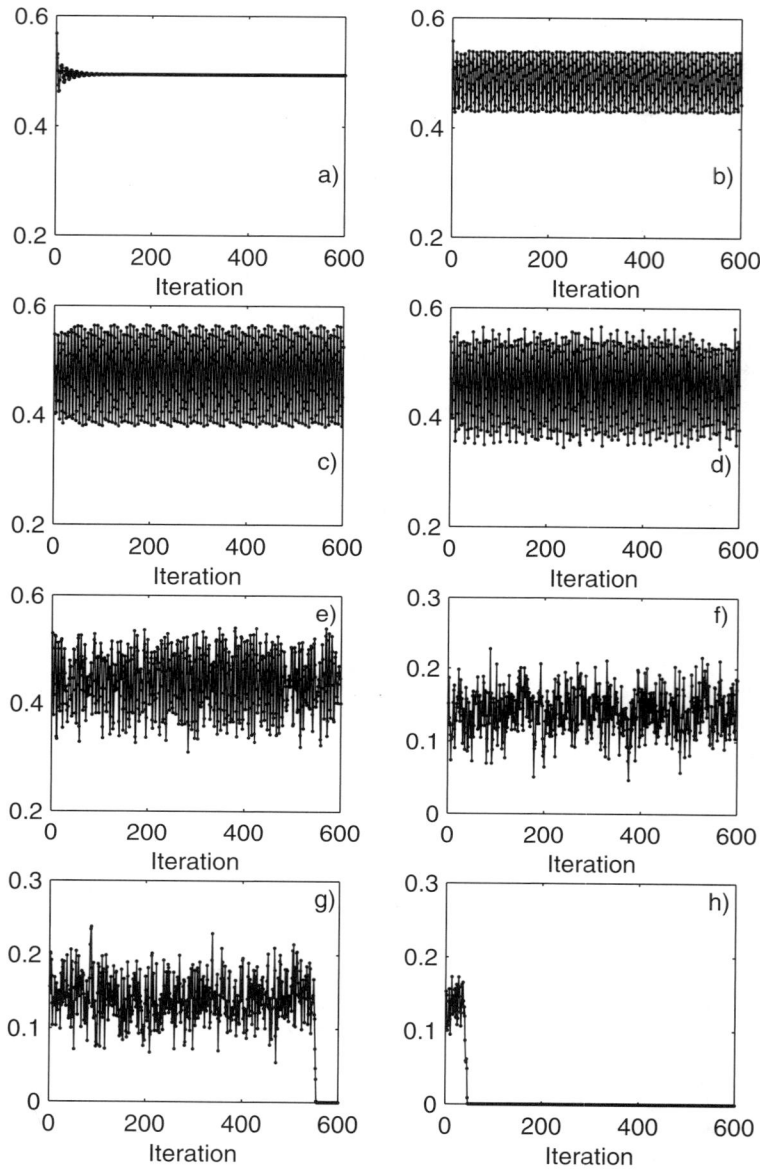

Fig. 6.3. Transition to unequivocal fixed point via Neimark–Sacker bifurcation: average entropy versus time ($E(l+1)$ vs. l). The values of SNR are: (a) -6.7 dB; (b) -6.5 dB; (c) -6.3 dB; (d) -6.1 dB; (e) -5.9 dB; (f) 0.75 dB; (g) 0.80 dB; (h) 0.85 dB. Figures (a) and (b) indicate the occurrence of Neimark–Sacker bifurcation. Note also the chaotic transients in Figures (g) and (h).

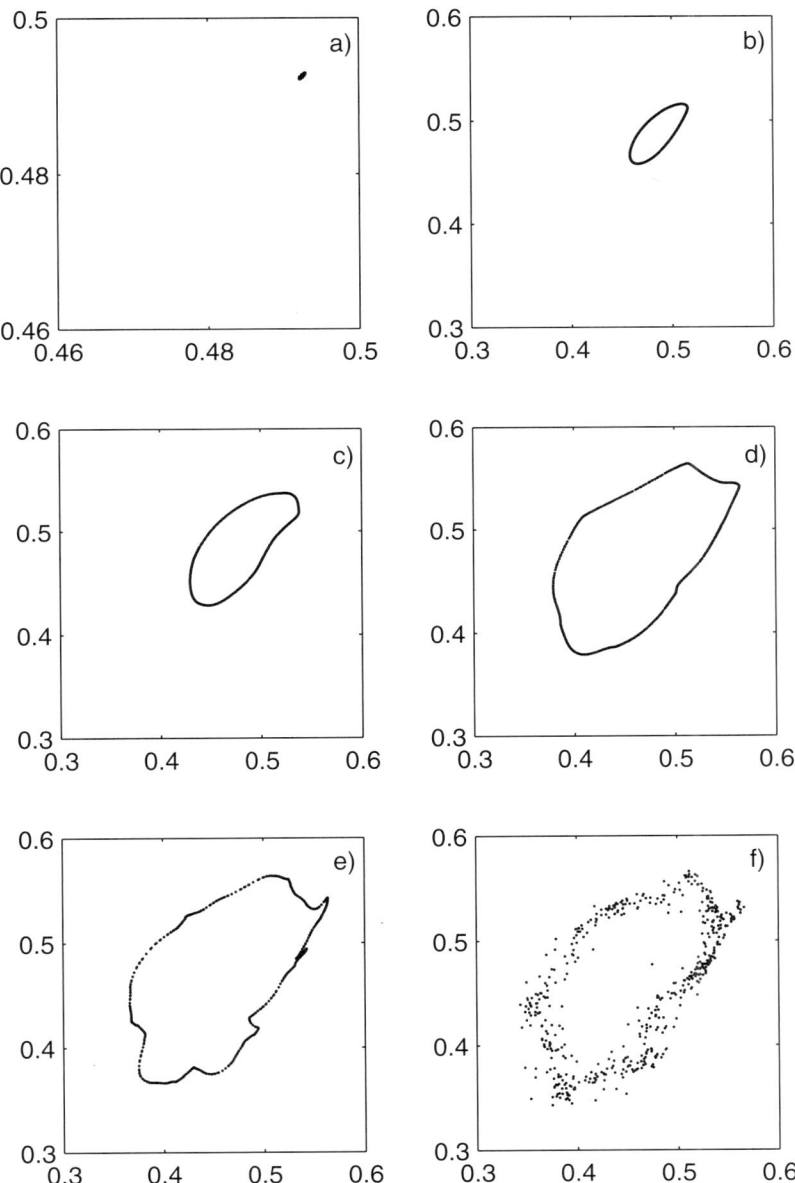

Fig. 6.4. Neimark–Sacker bifurcation and transition to chaos: $E(l+1)$ versus $E(l)$. The values of SNR are: (a) -6.7 dB; (b) -6.6 dB; (c) -6.5 dB; (d) -6.3 dB; (e) -6.2 dB; (f) -6.1 dB. Figures (a) and (b) indicate the occurrence of a Neimark–Sacker bifurcation. Note also the torus–breakdown route to chaos in Figures (d) through (f).

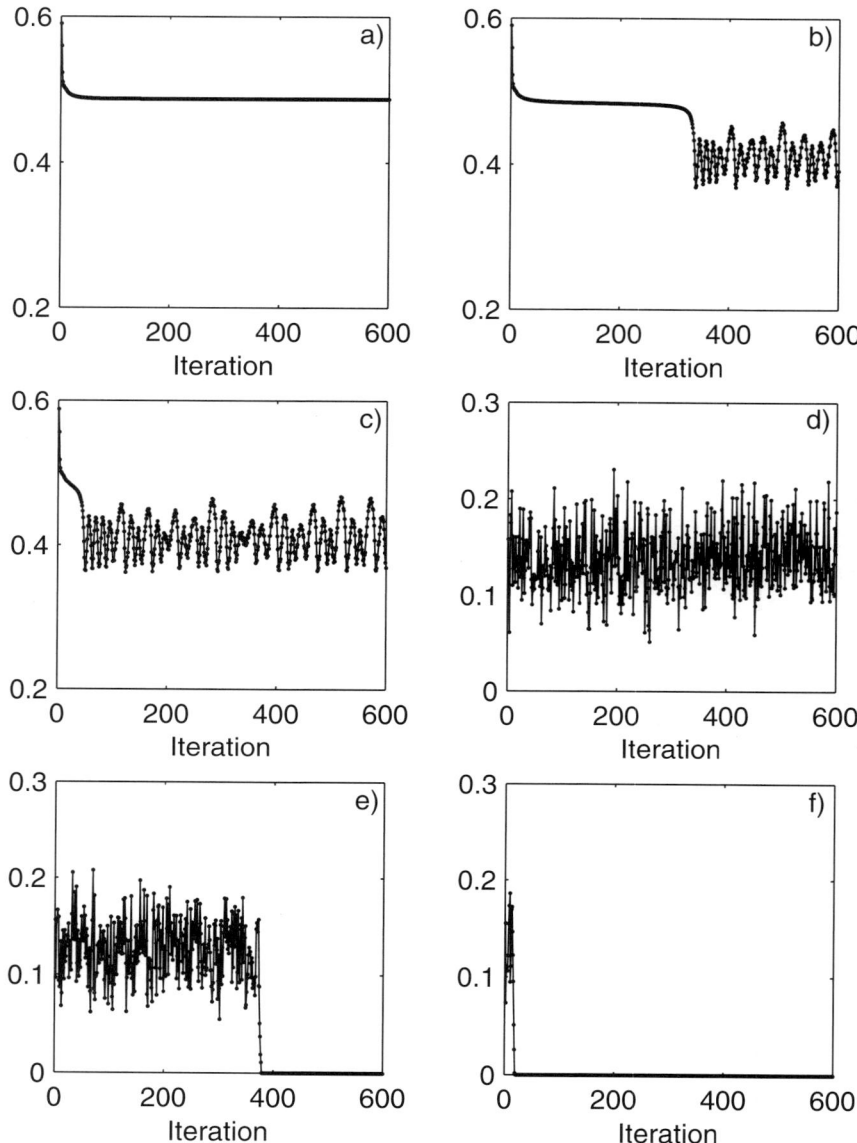

Fig. 6.5. Transition to unequivocal fixed point via tangent bifurcation: average entropy $E(l)$ versus time l. The values of SNR are: (a) -7.65 dB; (b) -7.645 dB; (c) -7.6 dB; (d) 0.30 dB; (e) 0.35 dB; (f) 0.4 dB. The indecisive fixed point (a) looses its stability via tangent bifurcation, and (b) the corresponding trajectory in state space approaches a chaotic attractor. Figure (e) indicates transient chaos. In (f) the algorithm reaches the fixed point solution in a few iterations.

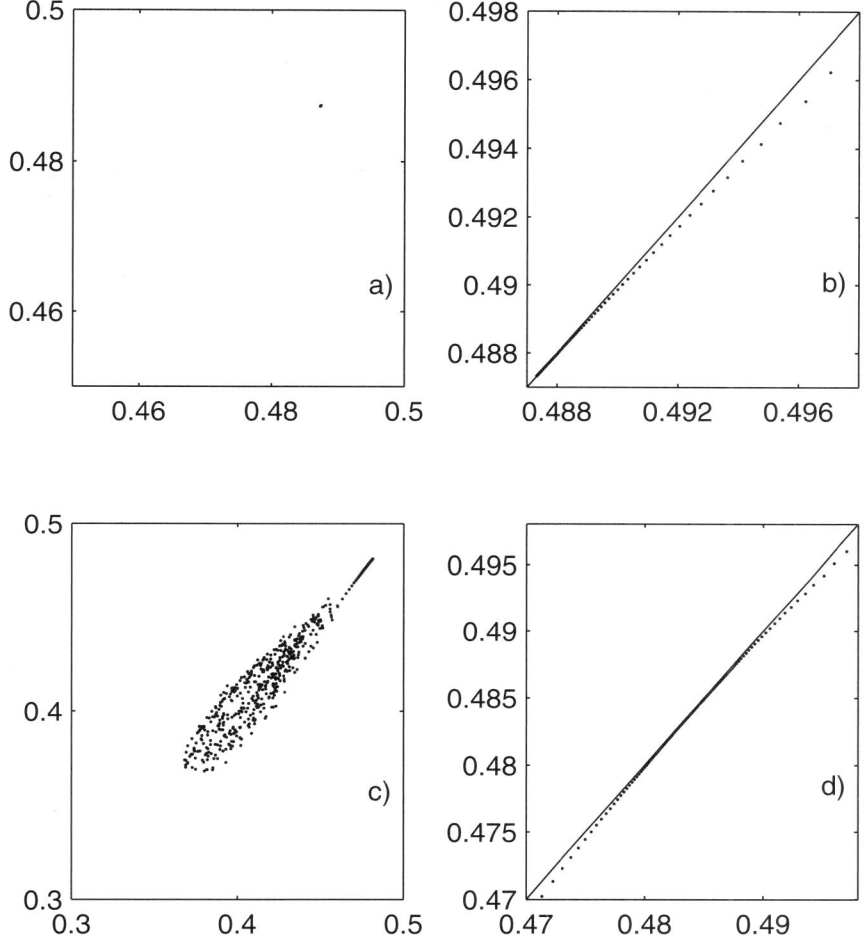

Fig. 6.6. Tangent bifurcation: $E(l+1)$ versus $E(l)$. The values of SNR are: (a)–(b) -7.65 dB; (c)–(d) -7.64 dB. Figures (b) and (d) are nothing but zooms of Figures (a) and (c), respectively.

iterates of a periodic point form a periodic trajectory (orbit). A fixed point M for $G : I\!\!R^m \to I\!\!R^m$ is called hyperbolic if $DG(M)$ has no eigenvalues on the unit circle, where $DG(M)$ is the Jacobian matrix of G computed at M. If M is a periodic point of period k, then M is hyperbolic if $DG^k(M)$ has no eigenvalues on the unit circle. There are three types of hyperbolic periodic points: sinks, sources, and saddles. M is a sink (attracting periodic point) if all of the eigenvalues of $DG^k(M)$ are less than one in absolute value. M is a source (repelling periodic point) if all of the eigenvalues of $DG^k(M)$ are greater than one in absolute value. M is a saddle point if some of the

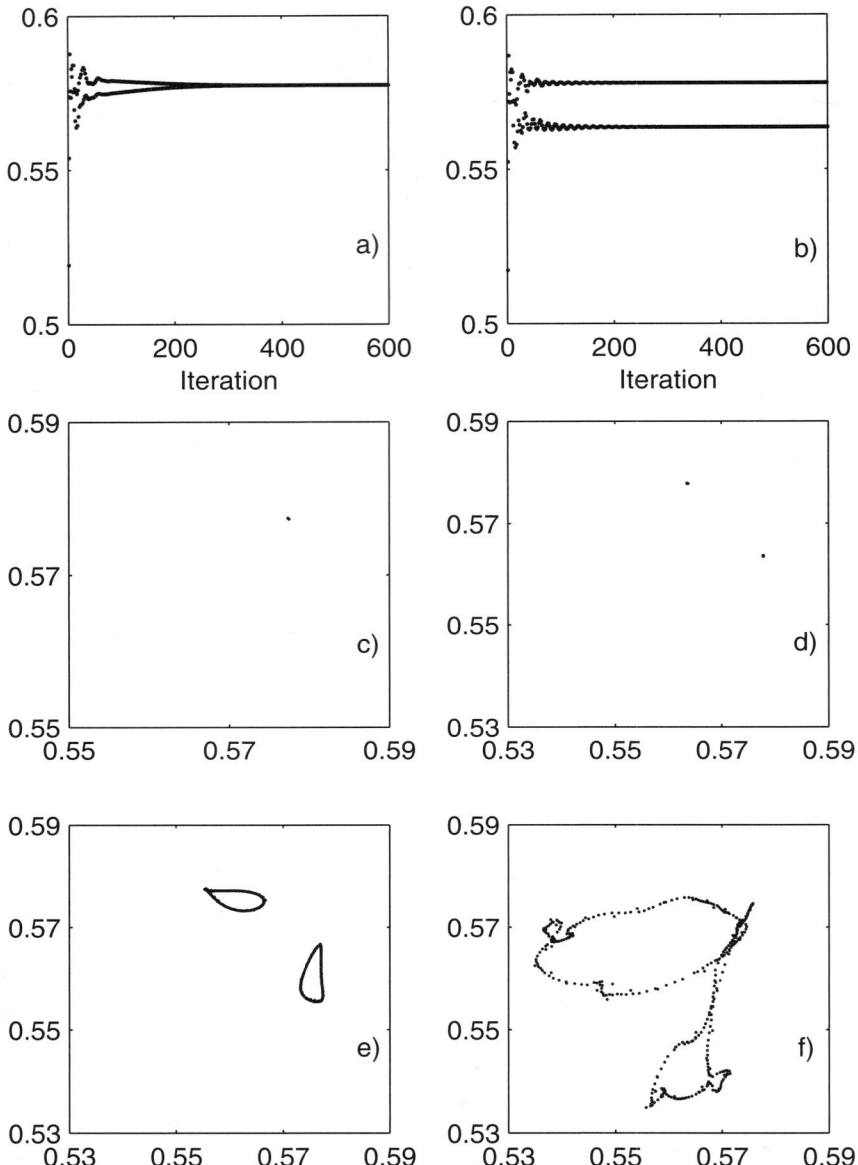

Fig. 6.7. Transition to unequivocal fixed point via flip (or period doubling) bifurcation. (a) and (b): $E(l + 1)$ versus l. (c), (d), (e) and (f): $E(l + 1)$ versus $E(l)$. The values of SNR are: (a) -6.0 dB; (b) -5.96 dB; (c) -6.0 dB; (d) -5.96 dB; (e) -5.94 dB; (f) -5.84 dB. Figures (a) and (b), or (c) and (d), show a flip bifurcation. Figures (d) and (e) indicate a Neimark–Sacker bifurcation, and (e) and (f), torus breakdown route to chaos.

eigenvalues of $DG^k(M)$ are larger and some are less than one in absolute value. Suppose that G admits an attracting fixed point at M. Then there is an open set about M in which all points tend to M under forward iterations of G. The largest such open set is called the stable set or basin (domain) of attraction of M and is denoted by $W^s(M)$.

We now consider a discrete-time dynamical system parameterized by a single parameter α, $G(\boldsymbol{x}, \alpha)$, $\alpha \in \mathbb{R}$. We assume that G is a smooth function. Because the system is smooth, the fixed point as well as the Jacobian evaluated at this point is a continuous function of the system parameter. As the system parameter α is changed, the magnitudes of the eigenvalues may also change, and it is possible that one or more eigenvalues cross the unit circle. This non-hyperbolic behavior usually indicates the occurrence of a bifurcation. In a generic system, by changing the system parameter, either a single eigenvalue will cross the unit circle through ± 1 or two complex conjugate eigenvalues will cross the unit circle together. Therefore, when a fixed point changes continuously with the parameter, it can bifurcate and lose its stability by one of the following three mechanisms: (i) *tangent* (or *fold*) bifurcation: an eigenvalue approaches $+1$; (ii) *flip* (or *period doubling*) bifurcation: an eigenvalue approaches -1; (iii) *Neimark–Sacker* bifurcation: a pair of complex conjugate eigenvalues crosses the unit circle. These mechanisms are illustrated in details on the turbo decoding algorithm in Section 6.5.

6.4.2 Indecisive and Unequivocal Fixed Points

In our analysis we considered the classical turbo code [1] with identical constituent recursive convolutional codes generated by the polynomials $\{D^4 + D^3 + D^2 + D^1 + 1, D^4 + 1\}$, producing a rate-1/3 turbo code. The codewords were transmitted over an AWGN channel using BPSK modulation. The length of the interleaver was $n = 1024$. Figure 6.2 shows the performance of the turbo code.

Assume that the log-densities P_1 and P_2 are product log-densities. We know that if P is a product log-density then $\pi_P(Q) = P + Q$. Therefore, if P_1 and P_2 are product log-densities, the turbo decoding algorithm converges to the fixed point $(Q_1^*, Q_2^*) = (P_1, P_2)$ in a single iteration, regardless of the initial conditions. The continuity of the fixed points with respect to the parameters (P_1, P_2) implies that if (P_1, P_2) is close enough to a pair of product log-densities (P_1^*, P_2^*), then the turbo decoding algorithm has a unique fixed point close to (P_1^*, P_2^*). Moreover, we expect that the unbiased initialization $(\boldsymbol{0}, \boldsymbol{0})$ will be in the domain of attraction, because for product densities, the domain of attraction is $\Pi \times \Pi$ which includes the unbiased initialization.

The turbo decoding algorithm has two types of fixed points: indecisive and unequivocal [12]. For asymptotically low SNRs, P_1 and P_2 converge to the product log-density $\boldsymbol{0}$, and therefore, the turbo decoding algorithm should have a fixed point close to $(\boldsymbol{0}, \boldsymbol{0})$. Simulations show that not only is this true but the signal-to-noise ratio required for the existence of this fixed point is not extremely low. In fact, for low SNR, the turbo decoding algorithm converges and most of the extrinsic log-likelihood ratios, $Q_1^*(\boldsymbol{b}^i)$ and $Q_2^*(\boldsymbol{b}^i)$, for $i = 1, \ldots, n$, are close to 0. In this case, the probability measures induced by Q_1^* and Q_2^* are close to 0.5. We refer to a fixed point with these characteristics as an *indecisive* fixed point [12]. At such fixed points the turbo decoding algorithm is relatively ambiguous regarding the values of the information bits. Therefore, hard decisions corresponding to these fixed points typically will not form a codeword. For high signal-to-noise ratios, the log-densities Q_1^* and Q_2^* are concentrated on the information sequence that corresponds to the codeword closest to the received vector. In other words, with high probability, $Pr_{q_1^*}(b_i = 0)$ and $Pr_{q_2^*}(b_i = 0)$ are either approximately 0 or 1, depending upon whether the i-th information bit of the codeword closest to the received vector is 1 or 0. Because the final log-likelihood ratios computed by the decoding algorithm are strong indicators of the information bits, we refer to a fixed point with these characteristics as an *unequivocal* fixed point [12]. Hard decisions corresponding to unequivocal fixed points, will usually form a codeword.

Because the turbo decoding algorithm is a high-dimensional dynamical system, we suggest the following representation of its trajectories in the state space. At each iteration l, the turbo decoding algorithm computes $2n$ log-densities Q_1 and Q_2. From these log-densities one can calculate, for each iteration, the probabilities $p_i^l(0)$ and $p_i^l(1)$ that ith bit is 0 or 1. Let us define

$$E(l) = -\frac{1}{n} \sum_{i=1}^{n} p_i^l(0) \ln p_i^l(0) + p_i^l(1) \ln p_i^l(1).$$

Thus, E represents the a posteriori average entropy, which is in a way a measure of the reliability of bit decisions of an information block with size $n = 1024$. When all bits are detected correctly or almost correctly, p_i is close to 1 for all i and, therefore, the unequivocal fixed point is represented by the point close to $E = 0$. On the other hand, for an indecisive fixed point, when all bits are equally probable, which is the case when SNR goes to $-\infty$, $E = 1$. In the following we present two types of figures: $E(l)$ versus l, and $E(l+1)$ versus $E(l)$.

6.5 Bifurcation Analysis of Turbo-Decoding Algorithm

The turbo decoding algorithm is an n-dimensional system with $3n$ parameters. This is a complex dynamical system with a large number of variables and parameters and, therefore, is not readily amenable for analysis. As a discrete-time dynamical system, the turbo decoding algorithm is parameterized by the log-densities P_0, P_1 and P_2. Given the transmitted codeword, P_0, P_1, and P_2 are completely specified by the noise values $x_1, x_2, \ldots, x_n, y_1, y_2, \ldots,$ and $y_n, z_1, z_2, \ldots, z_n$. To study the dynamics of the turbo decoding algorithm we would like to parameterize it by the SNR, that is, essentially $1/\sigma^2$ [12]. For large enough values of n, as typical for turbo codes, the noise variance σ^2 is approximately equal to: $\tilde{\sigma}^2 = \sum(x_j^2 + y_j^2 + z_j^2)/3n$. In this work, we focus on bifurcations in the turbo decoding algorithm for a single parameter, namely, the parameter $\tilde{\sigma}^2$ [12]. In particular, we fix the $3n - 1$ noise ratios $x_1/x_2, x_2/x_3, \ldots, z_{n-1}/z_n$ and treat the turbo decoding algorithm as it was depending on the single parameter $\tilde{\sigma}^2$, which is closely related to the SNR. By varying this parameter, we were able to analyze the turbo decoding algorithm as a function of SNR.

6.5.1 Bifurcation Diagram

We have performed many simulations changing the parameter SNR from $-\infty$ to $+\infty$ with different realizations of the noise (different noise ratios x_1/x_2, x_2/x_3, ..., z_{n-1}/z_n). In each instance of the turbo decoding algorithm that we analyzed, an unequivocal fixed point existed for all values of SNR from $-\infty$ to $+\infty$: this point is always represented with the point corresponding to the average entropy $E = 0$. This fixed point becomes stable at around -1.5 dB. However, the algorithm "cannot see" this point until 0–0.5 dB, when, in some cases, the initial point of the algorithm (which is always at $E = 1$) is within the basin of the attraction of the unequivocal fixed point.

The algorithm has another fixed point: the indecisive fixed point. For low values of SNR, when SNR goes to $-\infty$, this fixed point is represented by $E = 1$. In our simulations we found that the indecisive fixed point moves toward smaller values of E with increasing SNR, and loses its stability (or disappears) at low SNRs, typically in the range of -7 dB to -5 dB. The mechanisms responsible for the instability/disappearance of the indecisive fixed point of the turbo decoding algorithm are the following: tangent bifurcation, flip (or period doubling) bifurcation, and Neimark–Sacker bifurcation. Correspondingly, a transition from indecisive to unequivocal fixed points occurs in the turbo decoding algorithm. This transition corresponds to a large region of SNRs: from -7 dB to 1 dB. On the other hand, the region -5 dB to 0 dB is characterized by chaos: the turbo decoding algorithm as a dynamical system has a chaotic attractor. In the region 0 dB–1 dB chaotic transients occur. We note here, by comparison with Figure 6.2, that the waterfall region of the turbo code corresponds to the transient chaos behavior. Also, in this region the unequivocal fixed point is stable and the size of its basin of attraction gradually grows for increasing SNRs.

6.5.2 Bifurcations of Fixed Points

As explained above, the transition from indecisive to unequivocal fixed points in the turbo decoding algorithm is due to bifurcations of the indecisive fixed point. There are three ways in which a fixed point of a discrete-time dynamical system may fail to be hyperbolic: when the Jacobian matrix evaluated at the fixed point has a pair of complex eigenvalues crossing the unit circle, an eigenvalue at +1, or an eigenvalue at −1. In what follows we describe such bifurcations as they were observed in the classical turbo decoding algorithm:

- **Neimark–Sacker bifurcation.** In this case the Jacobian matrix evaluated at the fixed point undergoing the bifurcation admits a pair of complex conjugate eigenvalues on the unit circle. After the bifurcation, the fixed point goes unstable and is surrounded by an isolated, stable, close invariant curve, topologically equivalent to a circle. This bifurcation is illustrated by Figures 6.3a–b and Figures 6.4a–b. Note that the Figure 6.3 shows the average entropy E versus time l, whereas in Figure 6.4 we report $E(l+1)$ versus $E(l)$.

- **Tangent bifurcation.** Tangent (or fold) bifurcations are associated with a real eigenvalue at +1. After the bifurcation, the fixed point undergoing the bifurcation disappears, without resulting in an invariant set in its neighborhood. Figures 6.5a–b, and 6.6 illustrate this bifurcation. Note that Figure 6.5 shows the average entropy E versus time l, whereas on Figure 6.6 we report $E(l+1)$ versus $E(l)$. Figure 6.6d shows an enlargement of part of Figure c); it clearly indicates the occurrence of a tangent bifurcation. After the bifurcation the trajectory spends a long time in the vicinity of the disappeared fixed point, then approaches a chaotic attractor.

- **Flip bifurcation.** Flip (or period doubling) bifurcation is associated with an eigenvalue at −1. As a result of the flip bifurcation, a stable fixed pointa becomes unstable, and an asymptotically stable period-two orbit appears in the neighborhood of the resulting unstable fixed point. Figures 6.7a–d illustrate this bifurcation. Figures 6.7a and b show the average entropy E versus time l, whereas in Figure 6.7c and d we report $E(l+1)$ versus $E(l)$.

6.5.3 Chaotic Behavior

The turbo decoding algorithm exhibits chaotic behavior for a relatively large range of SNR values. Our analysis indicates three routes to chaos: period-doubling, intermittent, and torus breakdown. A torus breakdown route to chaos is evident, for example, in Figure 6.4, where a fixed point undergoes a Neimark–Sacker bifurcation giving rise to a periodic orbit; the later forms a torus which bifurcates leading to a chaotic attractor. Some chaotic time series are visible in Figure 6.5. The same route to chaos can be observed in Figure 6.7 where, this time, a period doubling cascade is interrupted by the occurrence of Neimark–Sacker bifurcations. The largest Lyapunov exponent of the chaotic attractor from Figure 6.4f was computed to be equal to 0.051. This chaotic attractor exists for all values of SNRs in the interval $[-6.1, 0.5]$. The values of the largest Lyapunov exponent for some parameter values are: 0.63 for SNR $= -4$dB, 1.28 for SNR $= -2$dB, 1.68 for SNR $= 0$dB, and 1.73 for SNR $= 0.5$dB.

In the waterfall region, the turbo decoding algorithm converges either to the chaotic invariant set or to the unequivocal fixed point, after a long transient behavior indicating an existence of a chaotic nonattracting invariant set in the vicinity of the unequivocal fixed point. In some cases, the algorithm spends a few thousand iterations before reaching the fixed point solution. As SNR increases, the number of iterations decreases, as can be seen, for example, from Figures6.3g–h, and Figures 6.5e–f.

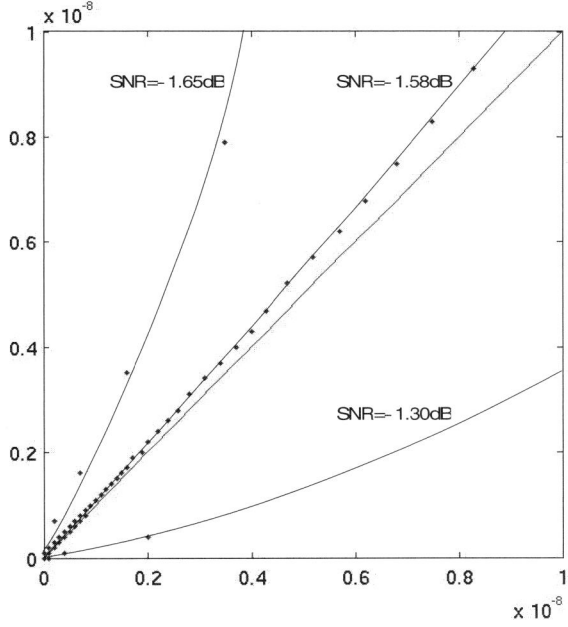

Fig. 6.8. Stability analysis of the unequivocal fixed point at the origin.

6.6 Control of Transient Chaos

In this section we consider an application of nonlinear control theory in order to speed up the convergence of the turbo decoding algorithm. First of all, we analyze the stability of the unequivocal fixed point. Figure 6.8 shows $E(l + 1)$ versus $E(l)$ for three different values of SNRs: -1.65 dB, -1.58 dB, and -1.30 dB. Note that we are very close (10^{-8}) to the fixed point at the origin. The two curves which are above the line $E(l + 1) = E(l)$ have slopes at the origin greater than 1, indicating, for these values of SNRs: -1.65 dB and -1.58 dB, that the fixed point at origin is unstable. However, the third curve,

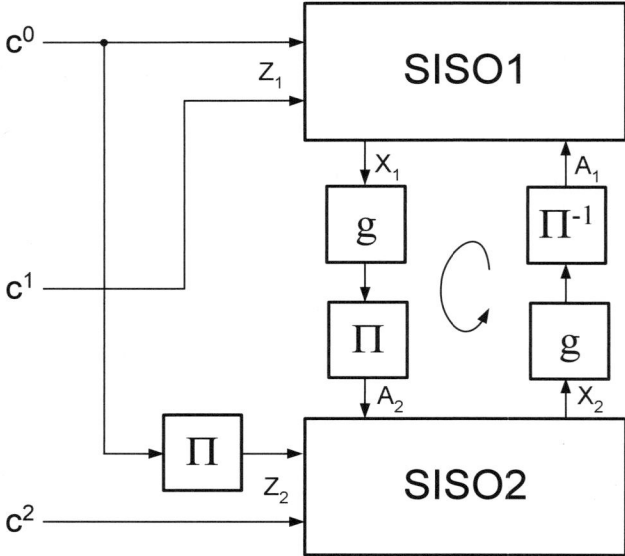

Fig. 6.9. Block diagram of the turbo decoder with control of the transient chaos. The control function is given by $g(X_i) = \alpha X_i e^{-\beta|X_i|}$ ($g(X_i) = X_i$ for classical turbo codes).

which is below the line $E(l+1) = E(l)$, has slope smaller than 1. This is a numerical confirmation that the unequivocal fixed point becomes stable for SNR ≈ -1.5 dB. Although the unequivocal fixed point is stable in the region -1.5 dB to 0 dB, the algorithm does not seem to converge to it from the initial condition. This is due to the fact that the basin of attraction of this point in this region of SNR is too small. In the region 0 dB to 1 dB, which corresponds roughly to the waterfall region, the turbo decoding algorithm converges either to the chaotic invariant set or to the unequivocal fixed point, after a long transient behavior indicating the existence of a chaotic nonattracting invariant set in the vicinity of the unequivocal fixed point. In some cases, the algorithm spends a few thousand iterations before reaching the fixed point solution.

We have developed a simple adaptive control mechanism to reduce the long transient behavior in the decoding algorithm. For example, let us consider the trajectory shown in Figure 6.3g, which converges to the unequivocal fixed point after 577 iterations. When the control is applied, the algorithm approaches the unequivocal fixed point in 13 iterations only.

A schematic block diagram of the turbo decoder with adaptive control is depicted in Figure 6.9. Our control function $g(\cdot)$ is given by:

$$g(X_i) = \alpha X_i e^{-\beta|X_i|}, \tag{6.8}$$

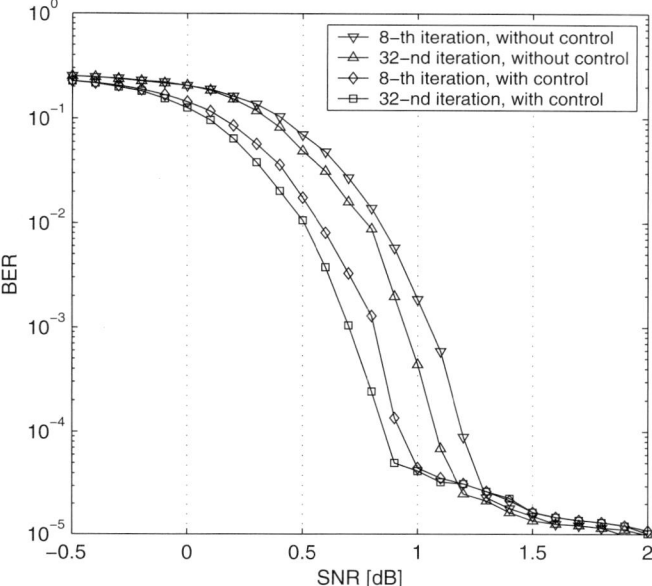

Fig. 6.10. The performance of the classical turbo code (for interleaver length 1024) with and without control of the transient chaos. (Reprinted with permission from [18]. ©IEE 2002.)

where X_1, X_2 are extrinsic information variables, and α, β are parameters. In simulations, we have used $\alpha = 0.9$ and $\beta = 0.01$, although similar results were obtained with other values of $\alpha \in [0.8, 1]$ and $\beta \in [0.01, 0.001]$. The adaptive control algorithm of Figure 6.9 is very simple, and can be easily implemented (both in software and/or in hardware) without significantly increasing the complexity of the decoding algorithm.

The performance of the control strategy is reported in Figure 6.10. On average, turbo decoding with control exhibits a gain of 0.25 dB to 0.3 dB over the conventional turbo decoding algorithm. Note that the turbo decoding algorithm with control, stopped after 8 iterations, shows better performance than the conventional turbo decoding algorithm stopped after 32 iterations. Thus adaptive control produces an algorithm that is four times faster, while providing about 0.2 dB gain over the conventional turbo decoding algorithm. On the other hand, we can see from Figure 6.10 that control is not very effective in the error-floor region. This is to be expected because the iterative decoding process does not exhibit transient chaos in this region. Finally, the error frame statistics with/without control, as a function of the SNR, are reported in Figure 6.11.

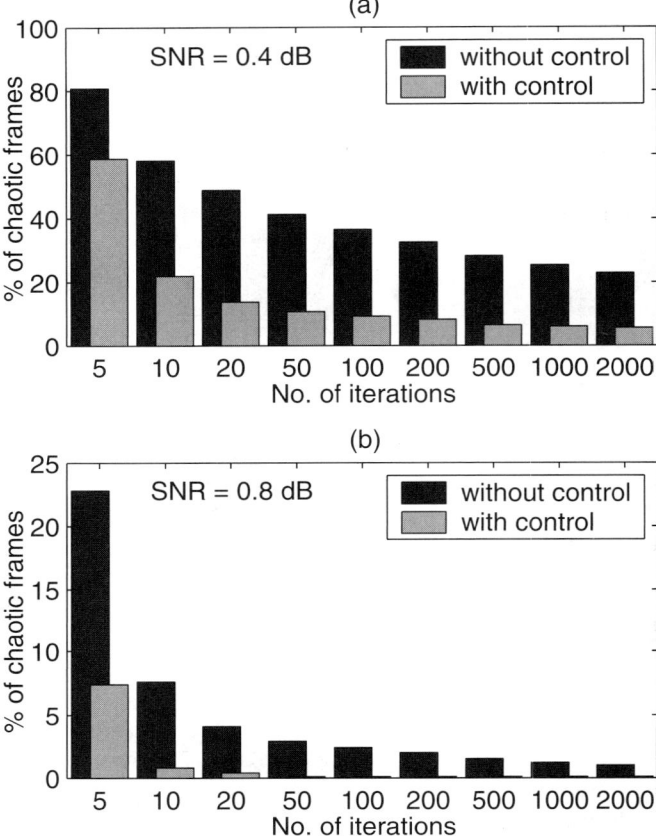

Fig. 6.11. Histogram showing the number of frames that still remain chaotic after a certain number of iterations, with/without control of transient chaos: (a) at SNR of 0.4 dB, (b) at SNR of 0.8 dB.

6.7 Conclusions

The turbo decoding algorithm can be viewed as a high-dimensional dynamical system parameterized by a large number of parameters. In this work, we have shown that the turbo decoding algorithm exhibits a whole range of phenomena known to occur in nonlinear systems. These include the existence of multiple fixed points, oscillatory behavior, bifurcations, chaos, and transient chaos. As an application of the theory developed, we have devised a simple technique to control transient chaos in the waterfall region of the turbo decoding algorithm. This results in a faster convergence and a significant gain in terms of BER performance. Part of this work has been already published in [18], and more detailed treatment of the nonlinear phenomena in iterative decoding algorithms can be found in [19].

Acknowledgments

This work was supported in part by ARO, STMicroelectronics, Inc., and by the UC MICRO program.

References

1. Berrou, C., Glavieux, A., Thitimajshima, P. Near Shannon limit error-correcting coding and decoding: Turbo-Codes. Proc. IEEE International Communications Conference, pp. 1064–1070, 1993.
2. MacKay D. J. C., Neal, R. M. Near Shannon limit performance of Low Density Parity Check Codes. Electronics Letters, vol. 32, pp. 1645–1646, 1996.
3. MacKay, D. J. C. Good error correcting codes based on very sparse matrices. IEEE Trans. Information Theory, vol. 45, pp. 399–431, 1999.
4. Richardson, T., Urbanke, R. The capacity of low-density parity check codes under message-passing decoding. IEEE Trans. Information Theory, vol. 47(2), pp. 599–618, 1999.
5. Shannon, C. E. A mathematical theory of communication. Bell Systems Technical Journal, vol. 27, pp. 379–423; pp. 623–56, 1948.
6. Wiggins S., F. John, F. Introduction to Applied Nonlinear Dynamical Systems and Chaos, 3rd ed. (Springer-Verlag, New York, 1990).
7. Gallager, R. G. Low density parity check codes. IRE Trans. Inf. Theory, vol. 8, pp. 21–28, 1962.
8. Wiberg, N. Codes and decoding on general graphs. Dissertation thesis no. 440, Dept. of Electrical Engineering, Linkoping University, Sweden, 1996
9. Benedetto, S., Montorsi, G. Unveiling turbo codes: some results on parallel concatenated coding schemes. IEEE Trans. Information Theory, vol. 42, pp. 409–428, 1996; Benedetto, S., Divsalar, D., Montorsi G., Pollara, F. Serial concatenation of interleaved codes: performance analysis, design and iterative decoding. IEEE Trans. Inform. Theory, vol. 44, pp. 909–926, 1998.
10. Perez, L., Seghers, J., Costello, D. A distance spectrum intrepretation of turbo codes. IEEE Trans. Inform. Theory, vol. 42, pp. 1690–1709, 1996.
11. Richardson, T. The geometry of turbo-decoding dynamics. IEEE Trans. Information Theory, vol. 46, pp. 9–23, 2000.
12. Agrawal, D., Vardy, A. The turbo decoding algorithm and its phase trajectories. IEEE Trans. Inform. Theory, vol. 47, pp. 699–722, 2000.
13. Kuznetsov, Y. A. Elements of Applied Bifurcation Theory (Springer-Verlag, New York, 1995).
14. Ott, E. Chaos in Dynamical Systems (Cambridge University Press, New York, 1993).
15. Sinai, Y.G. Gibbs measures in ergodic theory. Russ. Math. Surv., vol. 166, pp. 21–69, 1972.

16. Bowen, R., Ruelle, D. The ergodic theory of axiom A flows. Inventiones Math., vol. 29, pp. 181–202, 1975

17. Bahl, L. R., Cocke, J., Jelinek, F., Raviv, J. Optimal decoding of linear codes for minimizing symbol error rate. IEEE Trans. Inform. Theory, vol. 20, pp. 284–287, 1974.

18. Kocarev, L., Tasev, Z., Vardy, A. Improving turbo codes by control of transient chaos in turbo-decoding algorithms. Electronics Letters, vol. 38(20), pp. 1184–1186, 2002.

19. Kocarev, L., Lehmann, F., Maggio, G.M., Scanavino, B., Tasev, Z., Vardy, A. Nonlinear dynamics of iterative decoding systems and applications. to appear in IEEE Trans. Inform. Theory, 2006.

7

Security of Chaos-Based Communication and Encryption

Roy Tenny, Lev S. Tsimring, Henry D.I. Abarbanel, Lawrence E. Larson

Summary. During the last decade a new approach for secure communication, based on chaotic dynamics attracted the attention of the scientific community. In this chapter we give an overview and describe the research that was done at the Institute for Nonlinear Science (INLS) on this topic. We begin this chapter with a brief introduction to chaos-based encryption schemes. We then describe a new method for public key encryption that we have developed which is based on distributed chaotic dynamics. Next, we lay out a quantitative cryptanalysis approach for symmetric key encryption schemes that are based on active/passive decomposition of chaotic dynamics. We end this chapter with a summary and suggestions for future research.
[1]

7.1 Introduction

Using chaos in communication has been a very active area of research in the last decade. The main perceived benefits of using chaos in communications are related to its nonperiodic and seemingly random appearance. Therefore, there were several proposals of using chaotic waveforms for spreading the spectrum of the information signal (e.g., [5,6]). Other methods of communications based on chaotic dynamics do not provide processing gain, however they are intended to provide a certain degree of privacy or security to the transmission (e.g., [7–9]). The second group of methods falls into a new category of chaos-based encryption schemes. Chaos-based encryption schemes have several advantages over traditional encryption schemes:

- Chaos-based encryption schemes can be defined over continuous number fields and are not limited to integer number fields as traditional encryption schemes are. This provides chaos-based encryption schemes with a richer variety of functions that can be used for encryption. Also, it is possible to use chaos-based encryption schemes that do not require digitization of the

---

[1] Portions of this chapter were taken from publications [1–4].

message (traditional encryption schemes are defined over integer number fields and therefore require digitization of the data).

- Encryption can be implemented directly using high-speed analog components (optical or electrical) such as lasers and the like. Traditional encryption schemes can be implemented only by using digital hardware.
- Encoding and broadband modulation can be implemented using a single analog circuit. In traditional encryption two circuits are needed: a digital circuit for encryption, and an analog circuit for broadband modulation.
- Chaotic dynamics can generate nonperiodic pseudo-random waveforms. Pseudo-random sequences generated by traditional encryption schemes are implemented using digital hardware, and therefore are always periodic, with a period that depends on the number of bits used to represent the state of the PN sequence generator.

The main disadvantages in using chaotic encryption schemes are as follows.

- The security of chaotic encryption is difficult to quantify. The security of traditional encryption schemes based on integer number theory have been studied for a long time and is considered to be reliable. In contrast, the security of chaotic communication schemes often relies on a mixture of analytic methods and intuition. Encryption and cryptanalysis using chaotic dynamics is a relatively new field that has been studied for nearly a decade.
- Typically the power efficiency, bandwidth efficiency, and bit error rate performance of chaos-based communication schemes is inferior to that of traditional communication schemes.

In this chapter we give a brief overview of existing chaos-based secure communication schemes and describe some of the research we conducted on this topic. We begin by reviewing existing chaos based encryption schemes in Section 7.2 followed by review of cryptanalytic strategies in Section 7.3. In Section 7.4 we describe a quantitative approach to cryptanalysis of active/passive chaotic encryption schemes. In Section 7.5 we introduce a new method for public (asymmetric) key encryption based on chaotic nonlinear dynamics We end this chapter with a summary and discussion in Section 7.6.

7.2 Chaos-Based Encryption Schemes

There are two main types of chaos-based encryption schemes: The first type is similar to traditional encryption schemes that are defined over integer number fields. In those schemes chaotic dynamics is used to generate a continuous chaotic waveform which is digitized in order to generate a pseudo-random binary sequence. The resulting pseudo-random binary sequence is then used to mask a transmitted binary message. So far, this type of chaos-based encryption schemes did not prove to be more secure than traditional nonchaos-based encryption schemes [10].

The second type of chaos-based encryption schemes generates at the transmitter a continuous chaotic waveform that is used to mask a continuous message. This type of chaos-based encryption is interesting because it is defined over continuous number fields, and therefore has several advantages over traditional encryption schemes that are limited to integer number fields.

A chaotic encryption scheme is typically based on a chaotic dynamical system in which the parameters and/or the initial state serve as a key. Most of the proposed chaotic communication schemes use the following methods: chaos synchronization [7, 8, 11, 12], controlling chaos [13, 14] and chaos shift keying [5]. Chaotic encryption can be discrete in time and performed by iterating a chaotic map

$$
\begin{aligned}
\mathbf{x}(n+1) &= \mathbf{F}(\mathbf{x}(n), m(n)) \\
\mathbf{x}(0) \quad &= \mathbf{x}_0
\end{aligned}
\tag{7.1}
$$

or continuous in time and achieved by integrating a differential equation

$$
\begin{aligned}
\dot{\mathbf{x}}(t) &= \mathbf{F}(\mathbf{x}(t), m(t)) \\
\mathbf{x}(0) &= \mathbf{x}_0,
\end{aligned}
\tag{7.2}
$$

where $\mathbf{x}$ is the vector of the state variables, and m is a message. The vector field $\mathbf{F}$ and/or the initial state $\mathbf{x}_0$ can be used as an encryption key. The message m is referred as the plaintext and the transmitted signal is referred as the ciphertext. A chaos-based encryption scheme is depicted in Figure 7.1.

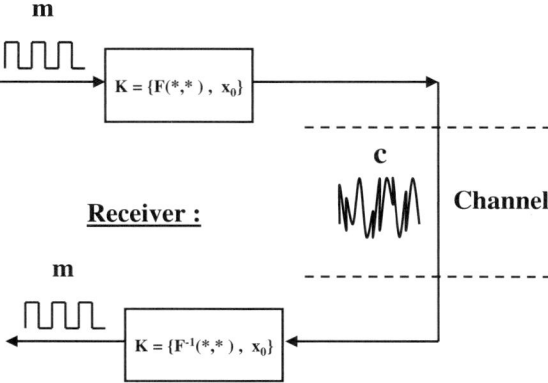

Fig. 7.1. Illustration of an encryption scheme based on chaotic dynamics.

In the following we describe some of the main types of secure communication schemes based on chaos.

7.2.1 Chaos Synchronization Using Additive Mixing

Communication schemes that are based on chaos synchronization and additive mixing of a chaotic signal with a message are described in [7] and shown in Figure 7.2. In additive mixing communication schemes a message signal is added to a chaotic signal generated by a chaotic dynamical system at the transmitter, and the sum of the two is transmitted through the channel. At the receiver (which is synchronized to the transmitter) the chaotic component is subtracted from the received signal to recover the original transmitted message. In Figure 7.2 the transmitter state evolution is governed by the dy-

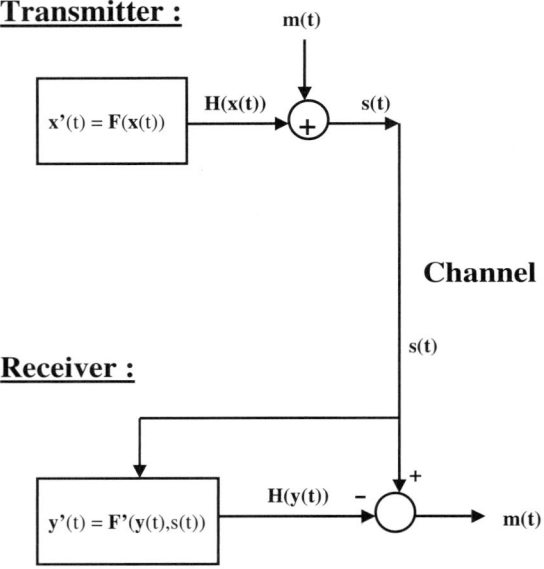

Fig. 7.2. Chaotic communication scheme based on chaos synchronization and additive mixing of a message with a chaotic component.

namical system

$$\dot{\mathbf{x}}(t) = \mathbf{F}(\mathbf{x}(t)). \tag{7.3}$$

A scalar $H(\mathbf{x}(t))$ which is a function of the transmitter state $\mathbf{x}(t)$ is added to the message $m(t)$, and the sum of the two is transmitted over the channel. The transmitted signal $s(t)$ is given by

$$s(t) = H(\mathbf{x}(t)) + m(t). \tag{7.4}$$

It is a common practice to choose the chaotic scalar added to the message as one of the components of the transmitter state $\mathbf{x}(t)$, however, in the general case $H(\mathbf{x}(t))$ can be any function of the transmitter state $\mathbf{x}(t)$, as long as the receiver synchronizes to the transmitter using the scalar $H(\mathbf{x}(t))$. The evolution of the receiver state $\mathbf{y}(t)$ is governed by the dynamical system

$$\dot{\mathbf{y}}(n) = \tilde{\mathbf{F}}(\mathbf{y}(t), s(t)). \tag{7.5}$$

The receiver state $\mathbf{y}(t)$ synchronizes to the transmitter state $\mathbf{x}(t)$ at the rate of the largest Lyapunov exponent λ, so that

$$|\mathbf{y}(t) - \mathbf{x}(t)| \propto e^{-\lambda t}. \tag{7.6}$$

At the receiver an estimation $\hat{m}(t)$ for the message $m(t)$ is calculated by subtracting the estimation $H(\mathbf{y}(t))$ from the received signal,

$$\hat{m}(t) = s(t) - H(\mathbf{y}(t)). \tag{7.7}$$

The addition of a message signal $m(t)$ to the chaotic scalar $H(\mathbf{x}(t))$ at the transmitter can degrade the quality of the synchronization between the transmitter and the receiver, and even result in loss of synchronization if the message component is too large. Also, in the presence of a large message the chaotic signal will no longer hide the message. Therefore, in order to maintain synchronization between the transmitter and the receiver, and to ensure that the chaotic signal masks the message, the dynamic range of the message has to be significantly smaller than the dynamic range of the chaotic scalar $H(\mathbf{x}(t))$ added to the message:

$$|m(n)| \ll |H(\mathbf{x}(t))|. \tag{7.8}$$

The advantage in using additive mixing is the simplicity of the circuitry used to implement the transmitter and the receiver. However, additive mixing schemes are often not efficient because only a small portion of the transmitted power is used to convey the message, and most of the transmitted power is consumed by the chaotic component $H(\mathbf{x}(t))$. Also, as we show in the next section, the security of such communication schemes is often poor, and various cryptanalysis schemes that are capable of decoding the message and reconstructing the secret dynamics have been developed.

7.2.2 Chaos Synchronization and Communication Using Active/Passive Decomposition

. In chaotic communication schemes that are based on additive mixing the added message component should be kept small compared to the chaotic component in order to maintain synchronization. This is not the case in chaos synchronization communication schemes that are based on active/passive decomposition, where the message component can be of the same order or even larger than the chaotic component. This property can be very useful for enhancing the security of an encryption scheme, as we demonstrate in Section 7.4.

In chaotic communication schemes that are based on active/passive decomposition of chaotic dynamics a dynamical system is divided into an active and a passive component [9, 15]. At the transmitter both active and passive parts are coupled to produce chaos, however, the receiver only has the passive subsystem which is driven by the signal from the active subsystem of the transmitter. By design, all the conditional Lyapunov exponents of the passive subsystem are negative, and the receiver state can synchronize to the transmitter state at the rate of the largest conditional Lyapunov exponents. An active/passive decomposition-based communication scheme scheme is sketched in Figure 7.3. In Figure 7.3 the receiver uses the transmitted signal $s(n)$ to

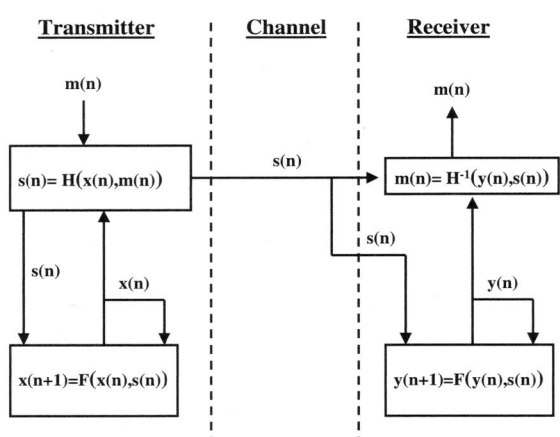

Fig. 7.3. Illustration of an active/passive decomposition based secure communication scheme. Reprinted from [3]. ©World Scientific Publishing Co.

synchronize the receiver state $\mathbf{y}(n)$ to the transmitter state $\mathbf{x}(n)$ at the rate of the conditional Lyapunov exponent λ which is conditioned on $s(n)$.

The transmitter dynamical system is given by

$$\mathbf{x}(n + 1) = \mathbf{F}(\mathbf{x}(n), s(n)). \tag{7.9}$$

The scalar signal $s(n)$ transmitted from transmitter to receiver is given by the function

$$s(n) = H(\mathbf{x}(n), m(n)), \tag{7.10}$$

or alternatively by

$$s(n + 1) = H(\mathbf{x}(n), m(n), s(n)). \tag{7.11}$$

The receiver is governed by the driven dynamical system

$$\mathbf{y}(n + 1) = \mathbf{F}(\mathbf{y}(n), s(n)). \tag{7.12}$$

The synchronization error between transmitter and receiver states decays at the rate of the largest conditional Lyapunov exponent, so that

$$\Big|(\mathbf{y}(n)|s(0),\dots,s(n)) - (\mathbf{x}(n)|s(0),\dots,s(n))\Big| \approx \Big|\mathbf{y}(0) - \mathbf{x}(0)\Big| \cdot e^{-|\lambda|n} \quad (7.13)$$

and once synchronization is achieved, the message can be decoded at the receiver by

$$\hat{m}(n) = s(n) - y(n). \tag{7.14}$$

Despite the obvious advantages of this method over the additive mixing, its security can be easily compromised. Short [16] proposed a method of breaking such schemes using local reconstruction of "average dynamics" (see Section 7.3.3). In Section 7.4 we describe a quantitative approach toward encryption schemes that employs active/passive decomposition based on the reconstruction of the average local dynamics.

7.2.3 Chaotic Shift Keying [2]

In chaos shift keying (Figure 7.4) the transmitter dynamics is dissipative and chaotic. The transmitter state trajectory converges to a strange attractor. A message is transmitted by altering one or more parameters of the transmitter dynamics which results in a change of the attractor position. At the receiver the message is decoded by estimating to which attractor the received signal belongs. Multiple attractor chaotic communication schemes that are based on three-dimensional Rössler systems are described by Carroll and Pecora [17].

A disadvantage of many chaos shift keying modulation schemes is that because the transmitter dynamics is altered, the symbol duration T_s should be long enough to allow the state trajectory to converge to the new attractor during the transmission of each symbol. The symbol duration T_s is determined by the largest negative Lyapunov exponent λ that determines the rate of convergence to the attractor. The long time spent converging to the attractor results in a lower symbol rate. Furthermore, in many cases the security of such a method is quite low, as the change in parameter values usually results in noticeable changes in the low-order statistical properties of the transmitted signal.

7.2.4 Controlling Chaos

From a control theory point of view a chaotic dynamics can be considered as a nonlinear plant, and various techniques can be used to control it. Several schemes have been proposed for controlling chaotic dynamics in order to synchronize between a transmitter and a receiver or to modulate data through control of the evolution in time of a chaotic trajectory. A synchronization scheme based on chaos control was proposed by Kapitaniak [18] and depicted in Figure 7.5.

---

[2] Not to confuse the methods described in this section with a class of differential chaos shift keying [5] and its generalizations (see Chapter 4) which are based on transmitting pieces of modulated chaotic time series together with unmodulated (reference) pieces for differential detection

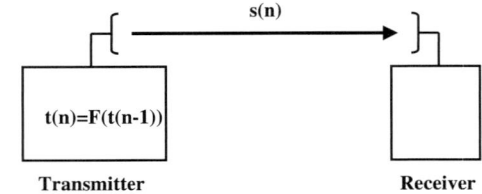

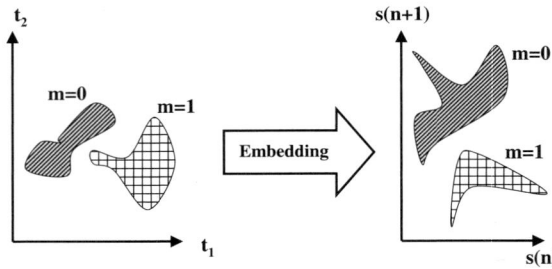

Fig. 7.4. Communication scheme using chaos shift keying [17]: By altering a parameter the transmitter dynamics converges to one of two (or more) attractors that correspond to "0" and "1". The attractors are represented by shaded surfaces in the transmitter state space t_1, t_2. The receiver cannot observe the transmitter state **t**. It needs to reconstruct the transmitter embedding phase space using time delays of the transmitted signal $s(t)$. In order to decode the message, the receiver estimates to which of the allowed attractors in the reconstructed embedding phase space the observed trajectory converged.

In Figure 7.5 a transmitter with chaotic dynamics transmits a signal to a receiver. Both transmitter and receiver have identical dynamics, however, the receiver's dynamics also depends on a control signal that is calculated based on the synchronization error between the transmitter and the receiver. Feasibility of this synchronization scheme was demonstrated in [18] using the nonautonomous Duffing equation as an underlying chaotic dynamical system. The transmitter dynamics was given by

$$\ddot{x} + a\dot{x} + x^3 = B\,cos(t). \tag{7.15}$$

The receiver dynamics are controlled using a linear function of the synchronization error,

$$\ddot{y} + a\dot{y} + y^3 = B\,cos(t) + K(x - y). \tag{7.16}$$

The control signal does not need to be continuous. In [19] Yang and Chua describe a scheme for impulsive control of chaotic synchronization communication schemes based on Chua's circuit [20].

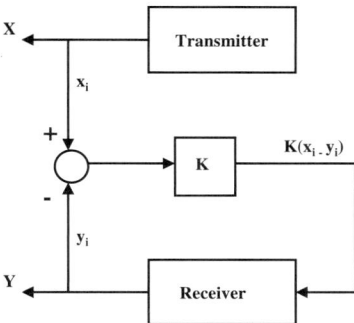

Fig. 7.5. Linear feedback is used to control the receiver dynamics in order to synchronize the receiver to the transmitter [18]. The transmitter and the receiver have internal states **x** and **y**, respectively. The scalar x_i is transmitted from the transmitter to the receiver. The synchronization error $x_i - y_i$ is used as an input to a linear controller represented by the matrix K. The output of the controller is used as a feedback to the receiver dynamics in order to synchronize the state of the receiver's dynamics to the dynamics of the transmitter.

The formulation given in [19] for impulsive control of a chaotic dynamics is

$$\begin{cases} \dot{\mathbf{x}} = \mathbf{f}(t, \mathbf{x}), & t \neq \tau_i \\ \triangle \mathbf{x} = \mathbf{U}(i, \mathbf{x}), & t = \tau_i \\ \mathbf{x}(t_0^+) = \mathbf{x}_0, & t_0 \geq 0, \; i = 1, 2, \ldots, \end{cases} \qquad (7.17)$$

where **x** is the state and $\mathbf{x}_0$ is the initial state of the chaotic system governed by the vector field $\mathbf{f}(t, \mathbf{x})$. $\triangle \mathbf{x}$ is a shift in the state **x** generated by impulsive control applied to the dynamics at times τ_i. The amount and direction of shift $\triangle \mathbf{x}$ is determined by the state **x**, and the time index i through the function U.

Chaos control can be applied to chaotic communication schemes not only in order to synchronize between transmitter and receiver, but also in order to modulate data. In [13, 14] the state trajectory of a chaotic dynamical system is controlled so that it follows a predetermined sequence of symbols, as shown in Figure 7.6. In the design of a communication scheme based on the symbolic dynamics the phase space of the transmitter is divided into regions where each region represents a discrete symbol. Often partitioning of the phase space is a heuristic process based on intuition. Conceptually it is similar to vector quantization, however, the two are not identical: in symbolic dynamics a symbol often corresponds to a portion of a chaotic attractor that has a fractal dimension whereas in vector quantization symbols typically correspond to volumes of integer dimension. Figure 7.6 demonstrates an example of a communication system based on division of a chaotic attractor into two parts that correspond to two symbols. The security of such schemes depends on the complexity of the partition of the phase space corresponding to the transmitted symbols.

When the partition is simple (as in the example of Figure 7.6) the security of the communication is very low.

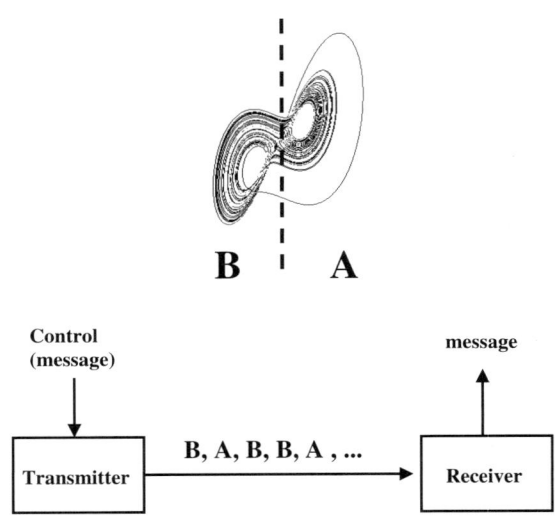

Fig. 7.6. Chaotic modulation by controlling a sequence generated by a symbolic dynamics [13]. The phase space of the system is divided into regions. In this example it was partitioned into two regions, A and B, that correspond to the right and left sides of the attractor. A control signal at the transmitter determines the sequence of the symbols generated by the chaotic system in order to modulate a message into the chaotic waveform. The message is decoded at the receiver by determining the sequence of symbols and converting that sequence to the corresponding message.

7.3 Cryptanalysis Attacks on Chaos-Based Encryption Schemes

7.3.1 Attacks That Do Not Require Searching for the Encryption Key

In some cases it is possible to break a chaos-based encryption scheme without searching for the secret key K that was used to encrypt the message. This kind of attack is usually applicable if there is dependency between the low-order statistics (mean, variance, etc.) of the transmitted signal and the message. In particular, transmission of binary data is risky, because it turns out that even high-dimensional and complicated chaotic dynamics that may be considered secure can reveal information about the transmitted bit without knowing the exact dynamics, and simple clustering of the transmitted data into two groups enables decoding of the message.

In [21] Yang et al. showed that an encryption scheme based on a Chua's circuit can broken without reconstructing the chaotic dynamics by using the spectrogram of the transmitted ciphertext. In [22] these authors showed that the same encryption scheme can also be broken by generating return maps of the transmitted ciphertext, again without a need to reconstruct the chaotic dynamics.

7.3.2 Attacks That Rely on Partial Knowledge of the Chaotic Dynamics

If an unauthorized receiver knows the general class of dynamics from which the secret chaotic dynamics was chosen, she only needs to find the specific parameters of that dynamics. For instance, if the unauthorized receiver can assume that the secret dynamics is generated by a Chua's circuit, or by a certain type of chaotic laser, then she can reduce the dimension of the key space that needs to be searched.

Low-dimensional modeling of the dynamics. An interesting example for breaking an encryption scheme knowing the general class of the chaotic dynamics was demonstrated by Geddes et al. [23]. They assumed that the unauthorized receiver knows that the chaotic system at the transmitter is an erbium-doped fiber-ring laser, but the parameters of the system are unknown. They claim that because the dimension of the transmitter dynamics is large $(D > 10)$ then attacks based on nonlinear dynamics forecasting are difficult. However, they show that by modeling the transmitter dynamics using a simple linear model with two delay loops and four parameters, they could estimate the parameters of the simplified lower-dimensional model, and decode the message.

Generalized synchronization. If the general type of dynamics is known to the unauthorized receiver, then even if the exact parameters of the system are unknown, a small parameter mismatch may still result in an ability of the eavesdropper to decode the message. In [24] Yang et al. show that an encryption scheme based on Chua's circuit can be broken using the phenomenon of generalized synchronization. The unauthorized receiver does not know the exact values of the parameters of a Chua's circuit (those parameters are the secret key K), and yet she assumes that the general form of the chaotic system is that of the Chua's circuit and assigns some parameters that are different from the the true values at the transmitter. Nonetheless, a generalized synchronization between the transmitter and the unauthorized receiver occurs, and the binary message can be decoded using variations in the synchronization error.

7.3.3 Attacks that Reconstruct the Encryption Key without a priori Knowledge About the Dynamics

The unauthorized receiver may also attempt to reconstruct the secret chaotic dynamics of the transmitter without having any a priori knowledge about the type of dynamics used. Such an approach was first proposed by by Short [16, 25, 26]. Here we illustrate this approach on the example of active/passive communication scheme introduced in the previous section. The mathematical formulation we use is somewhat different from the one used in [16, 25, 26], however, it is more suitable for the quantitative cryptanalysis we describe in Section 7.4.

We use the following notations.

- $\mathbf{s}(n) = \{s(n), \ldots, s(n - D_e + 1)\}$ is a reconstruction of the transmitter phase space $\mathbf{x}(n)$ using D_e consecutive samples of the transmitted signal $s(n)$.
- $\mathbf{s}_\epsilon$ are samples of $\mathbf{s}$ that are contained within a small D_e-dimensional hypercube neighborhood ϵ of size L_ϵ.
- $\mathbf{x}_\epsilon$ is a sample $\mathbf{x}$ preceded by a sequence $\mathbf{s}_\epsilon$. Similar definition applies to $\mathbf{x}_\epsilon$, $\mathbf{y}_\epsilon$, s_ϵ, and m_ϵ.
- $\langle s_\epsilon \rangle$ is the average of the N_ϵ available samples of s that are preceded by $\mathbf{s}_\epsilon$. $\langle s_\epsilon \rangle$ is calculated by $\langle s_\epsilon \rangle = (1/N_\epsilon) \sum_{i=1}^{N_\epsilon} s_{i,\epsilon}$ where N_ϵ is the total number of sequences $\mathbf{s}_\epsilon$ and $s_{i,\epsilon}$ is the i 'th sample.

We assume that the message $\mathbf{m}$ is added to the transmitted signal, so that the transmitted signal $s(n)$ given by Equation (7.10) is now given by

$$s(n) = H(\mathbf{x}(n)) + m(n). \tag{7.18}$$

By limiting Equation (7.18) to instances that are preceded by the driving sequence $\mathbf{s}(n - 1) = \{s(n - 1), \ldots, s(n - D_e)\}$ we obtain

$$s_\epsilon(n) = H(\mathbf{x}_\epsilon(n)) + m_\epsilon(n). \tag{7.19}$$

Equation (7.19) can be interpreted as a local manifestation of Equation (7.18) in a small D_e -dimensional hypercube of size L_ϵ. By choosing small L_ϵ we can approximate the local dynamics within the neighborhood by its local average. Based on Equation (7.19) we can estimate the average P_ϵ of the local dynamics $H(\mathbf{x}_\epsilon(n))$ by averaging of the transmitted signal $s_\epsilon(n)$:

$$\begin{aligned} P_\epsilon &\equiv E\{H(\mathbf{x}_\epsilon)\} \\ &= E\{s_\epsilon\} - E\{m_\epsilon\}. \end{aligned} \tag{7.20}$$

Assuming that the message has zero mean, we arrive at

$$P_\epsilon = E\{s_\epsilon\}. \tag{7.21}$$

Because the unauthorized receiver can obtain only a finite number of transmitted samples $s(n)$ for estimating the local dynamics, then it needs to estimate P_ϵ by the average of all samples s preceded by the driving sequence $\mathbf{s}_\epsilon$,

$$P_\epsilon \approx \hat{P}_\epsilon = \langle s_\epsilon \rangle. \tag{7.22}$$

The unauthorized can use Equation (7.19) and the estimation P_ϵ of the local dynamics given by Equation (7.22) to obtain the following estimation $\hat{m}$ of the message $m(n)$:

$$\hat{m}_\epsilon(n) = s_\epsilon(n) - \hat{P}_\epsilon. \tag{7.23}$$

In Section 7.4 we develop the cryptanalytis methods described in this section, and suggest measures that can be taken in order to enhance the security of an active/passive-based communication scheme.

7.4 Security of Chaotic Encryption Schemes Based on Active/Passive Decomposition

In this section we develop the ideas of the previous section towards a more quantitative cryptanalysis of a class of chaotic encryption schemes based on the active/passive decomposition (see Section 7.3.3). Here we lay out the main points of our cryptanalysis, and the full details of derivations and thorough discussion of our analysis can be found in [3, 4].

7.4.1 Quantifying Security Analysis

We assume that the unauthorized receiver does not have any a priori knowledge about the type of chaotic dynamics used and is therefore required to reconstruct the dynamics within each small neighborhood using time delay embedding technique.

We base our analysis on the average dynamics reconstruction approach [16, 25, 26] described in Section 7.3.3. The message reconstruction error of the unauthorized receiver is given by

$$e(n) = m(n) - \hat{m}(n), \tag{7.24}$$

where $\hat{m}(n)$ is obtained using the estimation given by Eq. (7.23).

In [3] we show that if the unauthorized receiver reconstructs the average dynamics within a hypercube of size L_ϵ using the average dynamics reconstruction, the reconstruction mean square error $E\{e^2\}$ can be approximated by

$$E\{e^2\} \approx c \cdot e^{-2|\lambda| D_e} + \frac{1}{N_\epsilon} \cdot \sigma_m^2 + kL_\epsilon^2, \tag{7.25}$$

where λ is the conditional Lyaponov exponent of the passive subsystem, D_e is the embedding dimension, and k and c are constants. N_ϵ is the number of samples used to obtain a local estimation of the dynamics within a neighborhood of size L_ϵ, and σ_m is the standard deviation of statistically independent message samples $m(n)$.

It can be shown [3] that if the transmitted message is larger than the chaotic component, that is, $|m| \gg |H(\mathbf{x})|$, then Eq. (7.25) becomes

$$E\{e^2\} = c \cdot e^{-2|\lambda|D_e} + \frac{\sigma_m^2}{N_s} \cdot \left(\frac{L_m}{L_\epsilon}\right)^{D_e} + kL_\epsilon{}^2. \qquad (7.26)$$

From Eq. (7.26) it is evident that the unauthorized receiver needs to use at least N_s samples in order to reconstruct the chaotic dynamics with the mean square error below $e_{\max}^2$. N_s can be estimated as

$$N_s = \frac{\sigma_m^2}{e_{max}^2 - kL_\epsilon{}^2 - c \cdot e^{-2|\lambda|D_e}} \cdot \left(\frac{L_m}{L_\epsilon}\right)^{D_e}. \qquad (7.27)$$

If the number of samples transmitted by the authorized receiver is less than N_s, she can conclude that the message is safe because the unauthorized receiver cannot obtain enough data to reconstruct the dynamics with accuracy required for decoding the data.

The authorized receiver can increase the number of samples N_s that can be safely transmitted by using a modulation scheme with a large dynamical range L_m. By using fine-grained modulation that is sensitive to reconstruction noise the authorized receiver can force the unauthorized receiver to reconstruct the dynamics with low reconstruction error $e_{\max}^2$ and thus increase the number of samples N_s that can be safely transmitted. Such a modulation is shown in Figure 7.7. The modulation in Figure 7.7,a is fine grained with small dynamical range. Therefore the unauthorized receiver will be required to reconstruct the chaotic dynamics with high accuracy, however, the message component may not be large enough to mask the chaotic dynamics. The modulation in Figure 7.7,b has a large dynamical range but is not fine grained. Therefore the message can be used to mask the chaotic component, however, the unauthorized receiver will not be required to reconstruct the dynamics with high accuracy. The most secure modulation scheme is shown in Figure 7.7,c. The whole dynamic range is divided into small bins corresponding to "0" and "1", and the transmitter selects one of the many bins corresponding to a given symbol randomly. The receiver knows the modulation scheme although it does not have to know the random code used to select particular bins. On the other hand, this modulation scheme presents significant difficulties for the unauthorized receiver because the large dynamical range makes the reconstruction of the chaotic dynamics difficult, and the fine-grained modulation (small bins) requires high accuracy of reconstruction for data decoding. The multilevel modulation scheme shown in Figure 7.7,d is also fine grained and has a large dynamical range. It is more efficient than the modulation shown in Figure 7.7,c because the transmitted message can take more than two values, however, it compromises security because an unauthorized receiver can determine if not the symbol itself, the subgroup of symbols to which the symbols belongs. In the modulation shown in Figure 7.7,e each time a bit needs to be transmitted, an interval corresponding to "0" or "1" is randomly chosen, and a random value within that interval is transmitted. Therefore by using this modulation technique the message component appears like a continuous uniformly distributed random signal and may be better for masking the the chaotic dynamics.

The modulation technique that combines fine-grained "randomized" symbols with a large dynamical range (Figure 7.7c) is very different from the common strategy to use a modulation which is as small as possible to enable good masking of the message by a larger chaotic component. A fine-grained "randomized" modulation with large dynamical range enables not only masking of the message by the chaotic signal but also the masking of the chaotic dynamics by the randomized message signal, making the reconstruction of the chaotic dynamics by the unauthorized receiver exceedingly difficult.

7.4.2 Simulation

We simulated an active/passive decomposition scheme using a chaotic tent map. This dynamics used is simple yet sufficient for demonstrating the main points of our quantitative security analysis.

The transmitter dynamics is described by the equations

$$
\begin{aligned}
x(n+1) &= Tent(s(n)) - b \cdot x(n) \\
s(n) &= x(n) + m(n)
\end{aligned}
\tag{7.28}
$$

with the tent map

$$
\begin{aligned}
Tent(x) &= \begin{cases} ax' - \frac{h}{2}, & if \quad x' \le \frac{w}{2} \\ \frac{3h}{2} - ax', & if \quad x' > \frac{w}{2} \end{cases} \\
w &= \frac{2h}{a} \\
x' &= x \bmod (w).
\end{aligned}
\tag{7.29}
$$

The parameters used for the tent map were $h = 0.3$, $a = 1.5$. The conditional Lyapunov exponent given by $\lambda = \ln(b)$, was controlled by the parameter b which was set to $b = 0.5$ unless stated otherwise.

The receiver dynamics was governed by the map

$$
y(n+1) = Tent(s(n)) - b \cdot y(n).
\tag{7.30}
$$

We used statistically independent message samples $m(n)$ uniformly distributed with dynamical range L_m:

$$
m \sim U[-\frac{L_m}{2}, \frac{L_m}{2}].
\tag{7.31}
$$

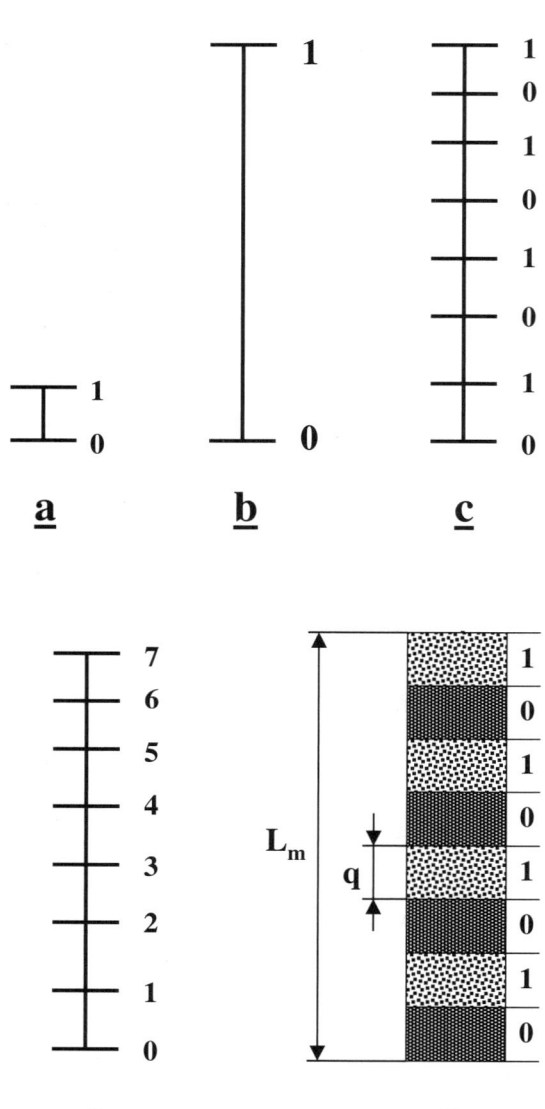

Fig. 7.7. Modulation schemes: (a) small dynamical range, fine grained; (b) large dynamical range, not fine grained; (c) large dynamical range and fine grained, binary; (d) large dynamical range and fine grained, multilevel; (e) fine grained, large dynamical range, continuous uniform distribution. (Reprinted from [3], ©2004 World Scientific Publishing Co.)

The standard deviation of $\sigma_m{}^2$ of the message is related to its dynamical range L_m by

$$\sigma_m{}^2 = \frac{L_m{}^2}{12}. \tag{7.32}$$

If no message is added to the chaotic dynamics ($m(n) = 0$), a simple time delay embedding reconstruction of the sequence $s(n)$ reveals the underlying tent map, as shown in Figure 7.8. Once the message $m(n)$ is added to the

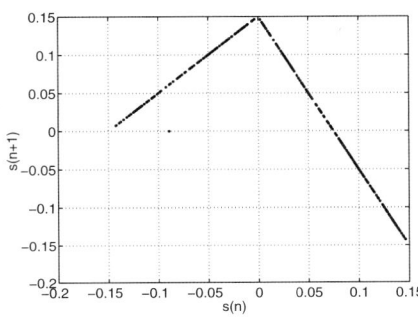

Fig. 7.8. Recovering the underlying dynamics using time delay embedding of the sequence $s(n)$. (Reprinted from [3], ©2004 World Scientific Publishing Co.)

chaotic dynamics, the map reconstructed using time delay embedding of $s(n)$ blurs, as shown in Figure 7.9. Therefore, in this communication scheme not only the chaotic component can be used to mask the message, but also the message signal can be used to mask the chaotic map and make its reconstruction more difficult. In [3] we show that in addition to masking the chaotic map, the statistically independent message samples also break the dependence between consecutive samples of the sequence $s(n)$. Therefore, even with one-dimensional tent map defined by Eq. (7.29), the addition of statistically independent message samples $m(n)$ results in a sequence $s(n)$ of infinite embedding dimension.

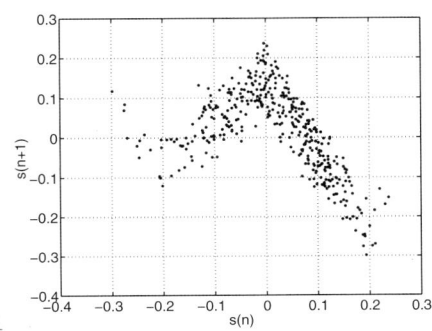

Fig. 7.9. Addition of message samples $m(n)$ perturb the reconstructed dynamics shown in Figure 7.8. (Reprinted from [3], ©2004 World Scientific Publishing Co.)

1

In the analysis of of simulation results we use the following definitions:

$$
\begin{aligned}
M &= 10\,Log_{10}[\sigma_m{}^2] \\
X &= 10\,Log_{10}[E\{(H(\mathbf{x}))^2\}] \\
N &= 10\,Log_{10}[E\{e^2\}] \\
SNR &= \frac{M}{N} \\
C &= \frac{X}{N},
\end{aligned}
\tag{7.33}
$$

where M, X, and N are the mean square in a logarithmic scale of the message $m(n)$, the chaotic component $x(n)$, and the dynamics reconstruction error $e(n)$. SNR is the reconstruction signal to noise ratio. C is the "cleaning factor" which is used to measure the ratio between the chaotic component X that was added to mask the message and the actual message reconstruction error N by the unauthorized receiver. C measures the extent to which the unauthorized receiver succeeded in cleaning the message M from the masking chaotic component X.

Shown in Figure 7.10 is the reconstruction SNR as a function of the number of samples N_s that were available to the unauthorized receiver for reconstruction. The larger N_s, the larger SNR is. Because the embedding dimension D_e and the neighborhood size L_ϵ used for reconstruction were finite ($D_e = 3, L_\epsilon = 0.02$), the reconstruction SNR reaches an asymptotic value as implied by Eq. (7.25).

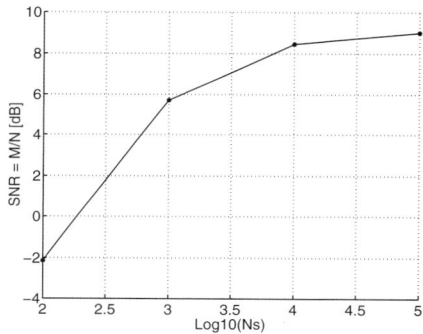

Fig. 7.10. Reconstruction signal to noise ratio versus number of samples N_s used for reconstruction. $D_e = 3$, $b = 0.5$. (Reprinted from [3], ©2004 World Scientific Publishing Co.)

In Figure 7.11 we present the cleaning factor C as a function of the reconstruction embedding dimension D_e and the parameter b that determines the conditional Lyapunov exponent $\lambda = \ln(b)$. When calculating the theoretical reconstruction error we added a constant offset Δ to the estimation, and whenever the expected number of neighbors within a neighborhood of size L_ϵ was less than one ($N_\epsilon < 1$), we set $N_\epsilon = 1$. Also, in the simulation we discarded samples for which no neighbors in a small neighborhood ϵ where found.

It is evident from Figure 7.11 that for a fixed value of b there is an optimal value D_{opt} for the embedding dimension D_e that maximizes the cleaning factor C. The unauthorized receiver may attempt to estimate this optimal value of

D_e in order to maximize the cleaning factor C and minimize the reconstruction error N.

Also, it is evident from Figure 7.11 that by using larger b (smaller $|\lambda|$) the cleaning factor decreases, and the communication scheme becomes more secure.

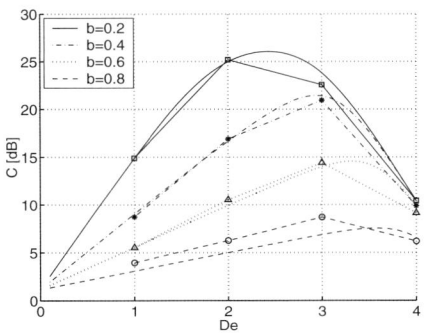

Fig. 7.11. Cleaning factor C dependence on D_e and b. Shown are theoretical values (lines) and simulation results (symbols). Simulation parameters: $N_s = 2.5e7$, $M = -13\text{db}$, $L_\epsilon = 0.02$. Theoretical estimation parameters: $c = 0.22$, $k = 0.3$, $\Delta = 9\text{dB}$. (Reprinted from [3], ©2004 World Scientific Publishing Co.)

The dependencies of the cleaning factor C on the neighborhood size L_ϵ and on the reconstruction embedding dimension D_e are shown in Figure 7.12. It is evident from Figure 7.12 that for a fixed value of the embedding dimension there is an optimal value for the estimation neighborhood size L_ϵ where a maximal value for the cleaning factor C is obtained. From Figure 7.12 it is evident that for the tent map model the unauthorized receiver can maximize the cleaning factor C by using reconstruction parameters $D_e = 2$ and $L_\epsilon = 0.016$.

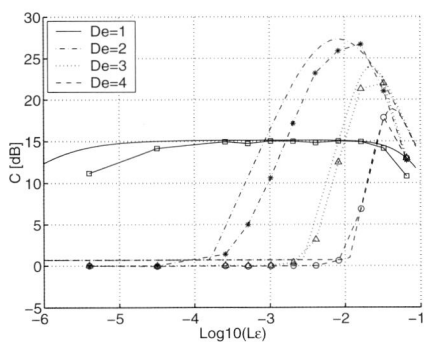

Fig. 7.12. Cleaning factor C dependence on D_e and L_ϵ. Shown are simulation (symbols) and theoretical values (lines). Simulation parameters: $N_s = 25e6$, $M = -13\text{dB}$, $b = 0.2$. Theoretical estimation parameters: $c = 0.22$, $k = 0.3$, $\Delta = 9\text{dB}$. (Reprinted from [3], ©2004 World Scientific Publishing Co.)

Let P be the fraction of the total number of samples N_s contained within a neighborhood of size L_ϵ. P is given by

$$P = \frac{N_\epsilon}{N_s}. \tag{7.34}$$

An increase in the message dynamical range L_m will result in an increase of the attractor volume in the reconstructed embedding space, and in a decrease in P, as shown in Figure 7.13. A decrease in P implies better security because for

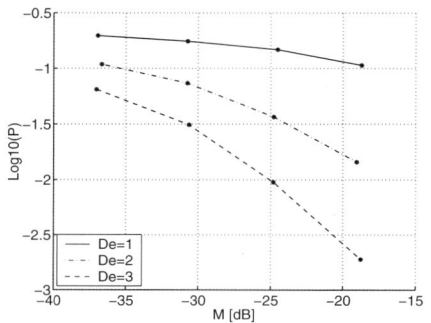

Fig. 7.13. The neighbors fraction P dependence on M and D_e. $N_s = 10^5$, $b = 0.5$, $L_\epsilon = 0.03$. (Reprinted from [3], ©2004 World Scientific Publishing Co.)

a given number of samples N_s available for reconstruction of the dynamics, the number of samples N_ϵ that are available for local reconstruction is smaller, and the unauthorized receiver will experience larger reconstruction error. Indeed, as implied by Figure 7.14, the cleaning factor decreases as we increase the message dynamical range L_m. It is important to note that an increase in the message dynamical range L_m will result in an increase in security only if the modulation is kept fine grained (by keeping small L_q) in order to force the unauthorized receiver to reconstruct the local dynamics with high accuracy.

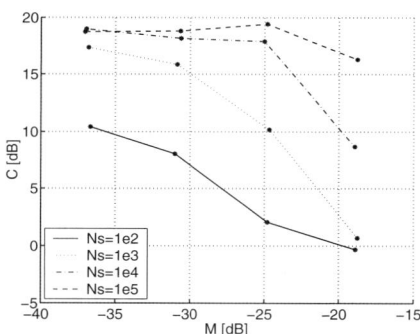

Fig. 7.14. Cleaning factor C dependence on M and N_s for $D_e = 3$, $b = 0.5$, $L_\epsilon = 0.03$. (Reprinted from [3], ©2004 World Scientific Publishing Co.)

Parameter Settings of Encryption Scheme

In the previous sections we showed that the security of an encryption scheme can be maintained by using fine-grained modulation (small L_q), with large dynamical range L_m and small value of the parameter b which results in a small $|\lambda|$. Also, the number of samples N_s that can be safely transmitted for a fixed set of system parameters should be limited, so that the unauthorized receiver will not be able to reconstruct the chaotic dynamics with accuracy necessary for decoding the message. Parameters of the encryption scheme can be determined using the following steps:

- Decide what is the maximal number of samples N_s that need to be transmitted.
- Choose the smallest bin size L_q that will allow the authorized receiver to decode the message with a low decoding error rate in the presence of channel and circuit noise.
- Determine the minimum value of the unauthorized receiver dynamics reconstruction mean square error $e_{\min}$ that can be allowed without compromising the message.
- Choose a small enough value for the parameter b that will ensure that the unauthorized receiver will not be able to reconstruct the dynamics with a reconstruction error lower than $e_{\min}$ by using N_s samples for dynamics reconstruction.

We now demonstrate the proposed procedure for determining the parameters of an encryption scheme using a more specific example.

- An encryption scheme is required to safely transmit up to $N_s = 25e6$ samples.
- We assume that the transmission power P_T is limited to $P_T - 13\mathrm{dB}$. By using an active/passive scheme in which the message component M is larger than the chaotic component X the power P_t of the transmitted signal is approximately the power M of the message component, so we choose $M = -13\mathrm{db}$.
- In our example we assume that the transmission channel adds Gaussian noise with standard deviation $\sigma_{channel} = 1e-3$ ($N_{channel} = 20\mathrm{Log}10(\sigma_{channel}) = -60\mathrm{dB}$).
- Our (arbitrary) design requirement is that the decoding error probability P_a encountered by the authorized receiver will be $P_a = 0.04$. Because of the presence of channel noise with standard deviation $\sigma_{channel} = 1e-3$ we need to set the distance L_q between adjacent bins to $L_q = 4e-3$ in order to ensure decoding error rate $P_a = 0.04$.
- We assume that in order to maintain message security, the probability P_u that the unauthorized receiver will decide that the wrong message bin has been transmitted should be at least $P_u = 0.6$. Therefore the standard deviation of the unauthorized receiver reconstruction error should be $4e-3$ ($e_{\min} = -48\mathrm{dB}$).
- From Figure 7.12 we find that if we use $b = 0.2$ then the largest cleaning factor the unauthorized receiver can obtain by reconstructing the dynamics using $N_s = 25e6$ samples is $C = 27\mathrm{dB}$. This largest value of the cleaning factor is obtained by using reconstruction embedding dimension $D_e = 2$

and averaging in a neighborhood of size size $L_\epsilon = 0.016$. In our specific implementation the chaotic component power is $X = -21\text{dB}$ therefore the power N of the reconstruction error is $N = X - C = -21 - 27 = -48\text{dB}$. This reconstruction error equals e_{min} and satisfies our requirements for security.

7.5 Public Key Encryption Using Distributed Dynamics

Public key encryption schemes are designed to solve the problem of sharing a secret encryption/decryption key between transmitter and receiver through a secure channel. In public key encryption there is no need to use a secure channel to share a key, and it is assumed that all data exchange between transmitter and receiver is public. In public key encryption schemes a public key is used to encrypt a message at the transmitter and a private key which is different from the public key is used to decode the message at the receiver. In classical cryptography which is based on the integer number theory several public key encryption schemes have been developed: RSA [27], Elliptic curves [28], El-Gamal [29], Knapsack [30], McAlliece coding theory-based system [31], and others.

Until recently, only private key chaos-based encryption schemes have been developed. In [1] and [2] we proposed the first public key encryption scheme that is based on continuous state chaotic distributed dynamics rather than on integer number theory. In this section we give an overview of this new method.

7.5.1 Overview of Distributed Dynamics Encryption

The Distributed Dynamics Encryption (DDE) scheme is illustrated by Figure 7.15. A chaotic nonlinear system is distributed between transmitter and receiver that are coupled bidirectionally through a communication channel. The entire dynamics, which are comprised of the receiver and transmitter, is dissipative and converges to an attractor. The subsystem at the transmitter is public and so are the coupling signals. The receiver subsystem is kept private and is assumed to be known only to the authorized receiver. A message is modulated by altering the dynamics of the transmitter which in turn results in a shift in the attractor position (Figure 7.16). The authorized receiver can simulate the entire dynamics offline, before the real transmission begins, and establish the attractor positions that correspond to the transmission of specific data symbols, (in binary transmission, "0" or "1"). When the real transmission takes place, the dynamical system is initialized with a random state, and then is iterated long enough to allow the state trajectory to converge to one of the allowed attractors. The authorized receiver that knows the positions of all allowed attractors can decode the message by estimating to which of them the trajectory has converged (Figure 7.17). It is important to note that because the authorized receiver has no access to the transmitter state, it must

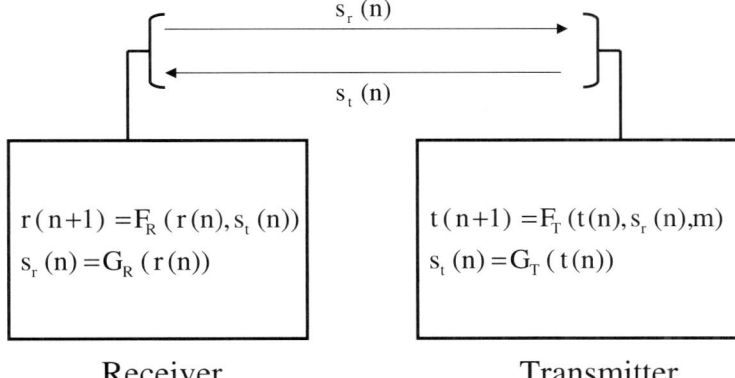

Fig. 7.15. Public and private keys of DDE. Private key: $\mathbf{F}_R(\bullet), G_R(\bullet), \mathbf{r}(n)$. Public key: $\mathbf{F}_T(\bullet), G_T(\bullet), s_r(n), s_t(n)$. Known only to transmitter: $\mathbf{t}(n), m$. From [2].

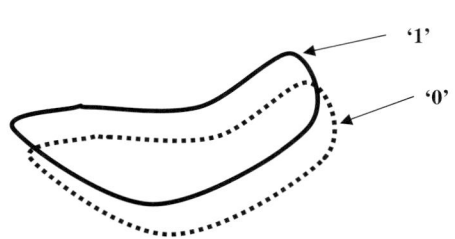

Fig. 7.16. A message is modulated by altering the parameters of the transmitter which results in a change in the attractor position. For instance, in the case of binary transmission the transmitter parameters are altered between two sets of values. The entire dynamical system converges to one of two attractors that correspond to transmission of "0" or "1."

perform decoding in a reconstructed embedding space generated from time-delayed values of its internal variables and/or transmitted signals [32]. The

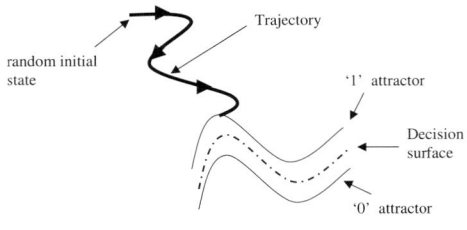

Fig. 7.17. Message decoding in a reconstructed embedding space: A trajectory starts at a random initial state and converges to one of two attractors that correspond to transmitted "0" or "1." Message is decoded by choosing the attractor closest to the trajectory after transients. From [2].

unauthorized receiver does not know the private receiver dynamics, hence she does not know the positions of the allowed attractors. Therefore, she cannot tell to which of the allowed attractors the trajectory converged, and cannot decode the transmitted bit.

In Figure 7.15, the D_R dimensional receiver state $\mathbf{r}(n) = [r_1(n), \ldots, r_{D_R}(n)]$ is given by

$$\mathbf{r}(n+1) = \mathbf{F}_R(\mathbf{r}(n), s_t(n)). \tag{7.35}$$

The receiver dynamics $\mathbf{F}_R$ and state $\mathbf{r}(n)$ are parts of the private key and are known only to the authorized receiver.

The transmitted scalar $s_r(n)$ is given by the function

$$s_r(n) = G_R(\mathbf{r}(n)). \tag{7.36}$$

Signal $s_r(n)$ is public and can be observed by an unauthorized receiver. The function G_R is a part of the private key and is known only to the authorized receiver.

The D_T dimensional transmitter state $\mathbf{t}(n) = [t_1(n), \ldots, t_{D_T}(n)]$ is given by

$$\mathbf{t}(n+1) = \mathbf{F}_T(\mathbf{t}(n), s_r(n), m(n)). \tag{7.37}$$

The dynamics $\mathbf{F}_T$ is a part of the public key, therefore the unauthorized receiver can use it to decode the message. $\mathbf{t}(n)$ and $m(n)$ are explicitly known only to the transmitter, and both authorized and unauthorized receivers need to estimate these variables. The authorized receiver modulates the parameter m ($m = 0, 1$ in the case of binary data transmission) which results in a change in the position of the attractor.

The transmitted scalar signal $s_t(n)$ is given by

$$s_t(n) = G_T(\mathbf{t}(n)), \tag{7.38}$$

where $s_t(n)$ and G_T are public and are known to both authorized and unauthorized receivers.

The authorized receiver does not know the transmitter state $\mathbf{t}(n)$ and needs to reconstruct the transmitter state in a time delay embedding phase space $\mathbf{s}(n) = (s_t(n), \ldots, s_t(n - (d-1)))$ in order to recover the attractor position in the reconstructed phase space $\mathbf{e}(n) = \left(\mathbf{r}(n), \mathbf{s}(n)\right)$.

The initial state of the transmitter is initialized with a random value at the beginning of each transmitted bit. The system is iterated long enough to allow the trajectory to converge to one of the allowed attractors. The authorized receiver decodes the message by determining to which of the allowed attractors the trajectory converged (Figure 7.17).

The receiver dynamics is altered frequently (preferably at the beginning of each transmitted bit), so the positions of the attractors corresponding to "0" and "1" changes. By altering the receiver private dynamics various cryptanalysis attacks that are discussed in Section 7.5.3 can be thwarted.

The main distinction of DDE from the traditional public key encryption scheme is that it allows an analog implementation. It is assumed that at least the transmitter is implemented using analog components. In such a case the public key is determined at the manufacturing phase. The public key is the structure of the transmitter that is available to all, and is not necessarily a stream of bits that needs to be transmitted. Of course, the hardware transmitter can have several public parameters that may determine the transmitter dynamics, but this is only an option.

The coding efficiency of DDE is measured in the number of information bits per transmitted bit: $E = \log2(M)/(n_1 * T_{bit})$, where M is the alphabet size (we can use more than two attractors), n_1 is the number of bits representing a single sample of s_r or s_t, and T_{bit} is the number of samples until the system converges to the attractor. Note that the system does not need to accurately converge to the attractor. It only has to get closer to the correct attractor than to the wrong one(s). Large separation between the attractors will require smaller T_{bit}. T_{bit} is also determined by the rate of convergence of the dynamical system, which is determined by the largest negative Lyapunov exponent.

7.5.2 Advantage of the Authorized Receiver over Unauthorized Receivers

In DDE as in all public key encryption schemes, the advantage of the authorized receiver over the unauthorized receiver is computational and not information theoretic. An unauthorized receiver has the same amount of information about the message as the authorized receiver, however, in order to use the available information to decode the message the unauthorized receiver is required to use computationally unfeasible algorithms whereas the authorized receiver can use the private key to extract the message information using a computationally feasible algorithm. In Section 7.5.3 we show that in DDE the unauthorized receiver can in principle decode the message without knowing the private key which is the receiver part of the distributed system, however, she is forced to used computationally unfeasible algorithms to do so.

In DDE the advantage of the authorized receiver over the unauthorized receiver is that the authorized receiver knows the entire system, so it can simulate the system offline before transmission begins for both $m = 0$ and $m = 1$ and determine the corresponding attractors. Knowing the position of the attractors enables the authorized receiver to decode the message by determining which of them is closest to the trajectory after convergence, as shown in Figure 7.17.

Because the positions of the attractors depend on both transmitter and receiver parts of the system, an unauthorized receiver that does not know the dynamics of the receiver cannot simulate the system offline and estimate the attractor positions. During the actual message transmission, the system is initialized with a random state and is allowed to converge to one of the allowed attractors, however, once convergence is obtained, the bidirectional transmission is stopped. For transmission of the next bit the dynamics is again initialized with a random state, and the receiver dynamics is altered so that previous attractor positions are no longer relevant. Therefore, no matter how long the transmission is, the unauthorized receiver cannot reconstruct the entire attractor structure. Furthermore, unlike the authorized receiver, the eavesdropper can observe only a single transient which corresponds to a particular value of m, and not observe transients for other values of m at a given set of system parameters.

7.5.3 Cryptanalysis of DDE

In this section we describe various methods that can be used to attack DDE, and suggest strategies to protect against such attacks. In most of the attacks described, it is assumed that the unauthorized receiver will attempt to decode the message by using methods that do not rely on the knowledge of the receiver private dynamics.

Known Ciphertext Attack: Detecting Shifts in the Attractor

The unauthorized receiver may attempt to reconstruct the whole system attractor from the (public) transmitted signal $s_t(n)$ based on the time-delay embedding method. By detecting the shift of the attractor due to the message encryption ($m = 0$ and $m = 1$), it is possible to extract the message without the knowledge of the receiver dynamics.

Protection:
In traditional chaotic switch keying, the receiver and transmitter have identical structure that has to remain the same from bit to bit. That allows an unauthorized receiver to reconstruct the attractor in the embedding space of $s_t(n)$ using familiar phase space embedding technique. The DDE scheme allows to change the secret dynamics of the receiver at any time, in particular at the beginning of every transmitted bit. Because for a single bit only a short transient from a random initial condition to the attractor is transmitted, if the receiver parameters change for every transmitted bit, an eavesdropper simply does not have sufficient data to reconstruct the dynamics. Furthermore, even if the attractor position could be identified, an eavesdropper is not able to tell whether the end point lies on the "0" attractor or the "1" attractor.

Known Ciphertext Attack: Duplicating the Transmitter Subsystem

Attack:
An unauthorized receiver may attempt to build two systems duplicating the transmitter subsystem and feed them with the observed signal $s_r(n)$ while setting the message parameter of one system to "0" and another to "1", hoping that the output of one of these two auxiliary systems will match the observed signal $s_t(n)$. That would uniquely determine the bit being transmitted.

Protection:
The transmitter dynamics is chaotic, and neither the transmitter state nor

the signal $s_t(n)$ is completely determined by $s_r(n)$. In order to reproduce the transmitted signal $s_t(n)$, an eavesdropper in addition to $s_r(n)$ needs to know the initial condition $\mathbf{t}(0)$ with impractically large precision. The state of the transmitter is randomized at the beginning of every bit, and this state is unknown to everybody (including the receiver). For different random initial states of the transmitter, the transmitter will produce different outputs $s_t(n)$ thus making this attack impossible.

Known Ciphertext Attack: Reconstructing Receiver Dynamics

Attack:
The unauthorized receiver can use $s_t(n)$ and $s_r(n)$ to reconstruct the receiver dynamics that constitutes the private key. After doing so it can simulate the dynamics offline and find the location of the attractors, and use the same methods that the authorized receiver use to decode the message.

   Protection:
There are two cases to consider. In one case, the receiver dynamics is assumed to be completely unknown. Reconstruction of the receiver dynamics can be made unfeasible by changing frequently the parameters at the receiver. This will not allow the unauthorized receiver to collect enough data to reconstruct of the secret dynamics of the receiver. By using high-dimensional receiver dynamics D_r the number of samples required to reconstruct the private dynamics can be increased. In the second case, the attacker knows the general structure of the secret receiver dynamics, and the private key would be the specific parameters used to implement the known "type" of dynamics. The difficulty in reconstructing the secret dynamics of the receiver will depend on the sensitivity of the dynamics to changes in parameters, and the size of the possible parameter space. Cryptanalysis in this case will be similar to the cryptanalysis of the traditional chaotic secret key encryption scheme. Yet, the DDE scheme is more secure, because the receiver can alter the system parameters at will, whereas in the traditional secret key chaotic encryption scheme the dynamics remains constant.

Known Ciphertext Attack: Solving Transmitter Equations

Attack:
The receiver dynamics $\mathbf{F}_T, G_T$ and the transmitted signals $(s_t(n), s_r(n))$ are public and known to the unauthorized receiver. So she may attempt to solve the following set of equations in order to recover the random initial state $\mathbf{t}(0)$ of the transmitter and the message bit m:

$$\begin{cases} s_t(0) & = G_T\left(\mathbf{t}(0), m\right) \\ \mathbf{t}(1) & = \mathbf{F}_T\left(\mathbf{t}(0), s_r(0), m\right) \\ \quad\vdots \\ s_t(D_T) & = G_T\left(\mathbf{t}(D_T), m\right) \\ \mathbf{t}(D_T+1) = \mathbf{F}_T\left(\mathbf{t}(D_T), s_r(D_T), m\right). \end{cases} \tag{7.39}$$

Protection:

By using high-dimensional transmitter dynamics $\mathbf{F}_T$ the unauthorized receiver will need to solve a set of D_T equations that are high-order polynomials of the unknown initial state $\mathbf{t}(0)$. The set of equations will contain expressions of the form $[t_i(0)]^{p^{D_T}}$ where p is the order of the transmitter state component t_i in the dynamics $\mathbf{F}_T$. For instance, by using $p = 4$ and $D_T = 10$ the unauthorized receiver will need to solve a set of equations with polynomials of order $(t_i)^{1,048,576}$. By further increasing p or D_T this calculation can be computationally unfeasible.

Ciphertext Attack: State Quantization and Maximum Likelihood Estimation

Attack:

The chaotic dynamics may contain channel and circuit noise that make the dynamics stochastic. An unauthorized receiver may attempt to generate a Hidden Markov Model (HMM) of the transmitter public dynamics for each possible value of the message m, and obtain a Maximum Likelihood (ML) estimation $\hat{m}$ of the message m. The decoded message will then be given by

$$\hat{m} = \max_{m \in (0,1)} p\left(\mathbf{s}_t^{T_{bit}} \mid m\right). \tag{7.40}$$

In order to generate the hidden Markov model the unauthorized receiver will need to quantize the transmitter state in a time delay reconstructed embedding space (Figure 7.18), and to estimate the state transition probabilities as well as the observation probabilities of the model (Figure 7.19).

Protection:

In the process of generating a hidden Markov model for the receiver dynamics, the unauthorized receiver will need to quantize the transmitter state, which in turn will result in quantization noise. The quantization noise will blur the separation between the transmitted waveforms that correspond to the transmission of "0" or "1" as illustrated by Figure 7.20. In order to decode the message the quantization noise should be kept small by using small quantization bins. However, using small quantization bins will result in an impractically large number of states in the hidden Markov model. In [2] and [4]

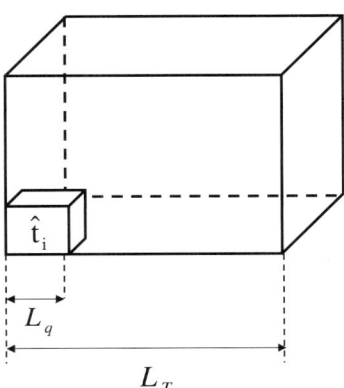

Fig. 7.18. Quantization of a D_e dimensional reconstructed embedding phase space of size L_T using quantization bins of size L_q. From [2].

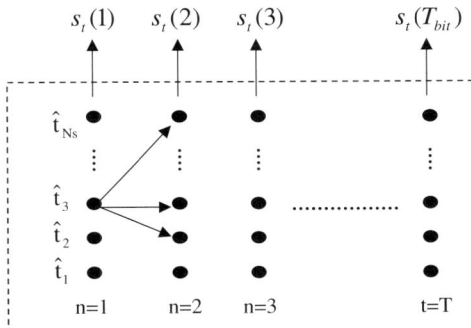

Fig. 7.19. An unauthorized receiver may estimate state transition and observation probabilities of the transmitter public dynamics and generate a hidden Markov model of the transmitter dynamics to obtain a maximum likelihood estimation of the message using a Viterbi algorithm. From [2].

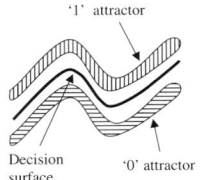 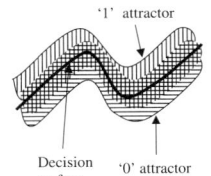

Fig. 7.20. A clear separation between the attractors in a continuous transmitter phase space **t** can be blurred by quantization noise in a quantized phase space $\hat{t}$ used for generating a hidden Markov model. From [2].

we obtained a lower bound for the number of states N_s in the hidden Markov model that is required in order to maintain quantization noise below a level that will enable message decoding by the unauthorized receiver. The number of states N_s is given by

$$N_s \geq \left[\frac{L_T \cdot w}{A} \cdot \sqrt{\frac{D_T}{3 \cdot T_{bit}}} \cdot Q^{-1}\left(1 - P_u\right) \right]^{D_T}, \qquad (7.41)$$

where L_T is the transmitter state dynamical range, D_T is the transmitter state dimension, P_u is the decoding error rate encountered by the unauthorized receiver, T_{bit} is the number of iterations the dynamics is allowed to converge from a random initial state to the attractor, A is modulation amplitude, and $Q(x) = (2\pi)^{-1/2} \int_{-\infty}^{x} e^{-z^2/2} dz$.

In Section 7.5.5 it is shown that the number of states N_s can be made large enough to make decoding using quantization of transmitter state computationally unfeasible. Large N_s can be achieved by choosing small modulation parameter A and large transmitter dimension D_T.

Known Plaintext Attack: Attractor Reconstruction Using Trajectory Ends

Attack:

Each converged trajectory end lies on one of the allowed attractors. If the unauthorized receiver possesses a collection of pairs of trajectories and the corresponding transmitted bit (plaintext), it can use the trajectory ends to reconstruct the positions of the attractors that correspond to the transmission of "0" or "1" as illustrated by Figure 7.21. In the following transmissions the unauthorized receiver can decode the message by determining to which of the reconstructed attractors an observed trajectory converged.

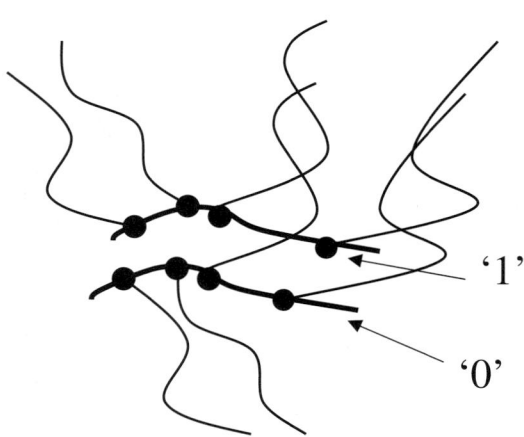

Fig. 7.21. Plaintext attack: Reconstruction of attractors positions using end points of the trajectories that lie on the attractors. From [2].

Protection:

Such an attack can also be avoided by altering the receiver dynamics at the beginning of each transmitted bit. By doing so, the trajectory ends that have been collected by the unauthorized receiver during transmission of previous bits cannot be used to decode future bits.

7.5.4 Effect of Time Delays and Unsynchronized Samples

A lack of proper synchronization may affect the decoding capability by altering the position of the attractor. For some dynamics, a severe lack of synchronization may even cause the system state trajectory to diverge. There are two main reasons for desynchronization: loss of symbols due to noise and round-trip time delay. The former can be mitigated by adding a synchronization protocol that makes sure that no symbols are lost in the transmission, or if a symbol is lost, forces the system to retransmit it. The latter can be improved either by including the time delay in the simulated dynamics of the system to determine the positions of attractor, or by limiting the bandwidth so the time delay becomes negligible. In practical implementations, the transmitter can transmit a block of data, then wait for reception of the block of data from the receiver before transmitting another block of data.

7.5.5 Simulation Results

The DDE scheme was simulated using a coupled map lattice (CML). The transmitter dynamics are described by

$$
\begin{aligned}
t_j(n+1) \quad &= d_{j,j-1} \cdot t_{j-1}^2(n) + d_{j,j} \cdot t_j^2(n) \\
&\quad + d_{j,j+1} \cdot t_{j+1}^2(n) + e_{j,j} \cdot |t_j(n)| \\
&\quad + f_j \cdot s_r^2(n) + g_j \quad, \\
j &= 1, \ldots, D_T,
\end{aligned} \tag{7.42}
$$

and the receiver dynamics are given by

$$
\begin{aligned}
r_i(n+1) \quad &= a_{i,i-1} \cdot r_{i-1}^2(n) + a_{i,i} \cdot r_i^2(n) \\
&\quad + a_{i,i+1} \cdot r_{i+1}^2(n) + b_i \cdot s_t^2(n) + c_i \quad, \\
i &= 1, \ldots, D_R.
\end{aligned} \tag{7.43}
$$

The receiver and transmitter are coupled through the signals

$$
s_r(n) = \sum_{i=1}^{D_R} h_i \cdot r_i^2(n) \quad, \tag{7.44}
$$

and

$$
s_t(n) = w \cdot \sum_{j=1}^{D_{tr}} |t_j(n)| + A \cdot m \quad. \tag{7.45}
$$

We used the transmitter and receiver dimension $D_T = 12$, $D_R = 2$, and the parameters $a, b, c, d, e, f, g, h, q$, and A, were chosen such that the dynamics was chaotic. All details of our implementation and values of parameters can be found in [4] and [33].

We allow convergence (typical) from a random initial state to the attractor (Figure 7.22) for $T_{bit} = 50$ iterations during the transmission of each message

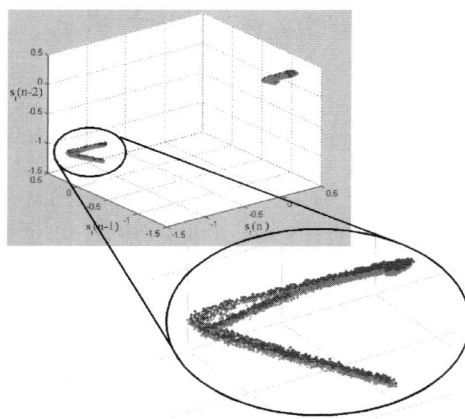

Fig. 7.22. Projection of the attractors that correspond to transmission of "0" and "1" (gray and black) on a $3D$ reconstructed embedding phase space generated using time delays of the transmitted public sequence $(s_t(n), s_t(n - 1), s_t(n - 2))$.

bit. The decision to which of the allowed attractors the system has converged was based on the last ten endpoints of the converging trajectory (Figure 7.17).

As mentioned in the previous section, the protection against reconstruction of the attractor using the endpoints of many transient trajectories is obtained by altering the receiver dynamics frequently. Plotted in Figure 7.24 are three different attractors obtained by altering the receiver dynamics while keeping the transmitter dynamics constant. Even if the transmitter parameters are kept constant while transmitting the same bit the attractor will be different for each bit due to the change in the receiver dynamics. Also, initializing the state of the transmitter with a random value at the beginning of each transmitted bit enhances security, as illustrated by Figure 7.23.

Figure 7.25 shows the dependence of the decoding error rate P_a encountered by the authorized receiver on the modulation parameter A in Eq. (7.45). P_a was estimated by simulation. The parameter A should be large enough in order to maintain low decoding error rate P_a encountered by the authorized receiver, yet not too large so that the number of hidden Markov model states N_s (estimated using Eq. (7.41)) will be exceedingly large.

Using analog components to implement the dynamics results in component noise, Figure 7.26 illustrates the the number of states N_s that the unauthorized receiver needs to use in a hidden Markov model in order to maintain low-level quantization noise as a function of the transmitter circuit component noise. Transmitter circuit component noise was simulated by adding a Gaussian noise with variance V to each of the transmitter state components t_i.

Larger component noise will require the use of a larger parameter A in order to ensure sufficient separation between the attractors corresponding to the transmission of "0" and "1". The modulation parameter A was chosen to ensure that the decoding error rate P_a encountered by the authorized receiver is less than 0.01. N_s was calculating using Eq. (7.41) assuming that the error rate P_u encountered by the unauthorized receiver should be at least $P_u =$

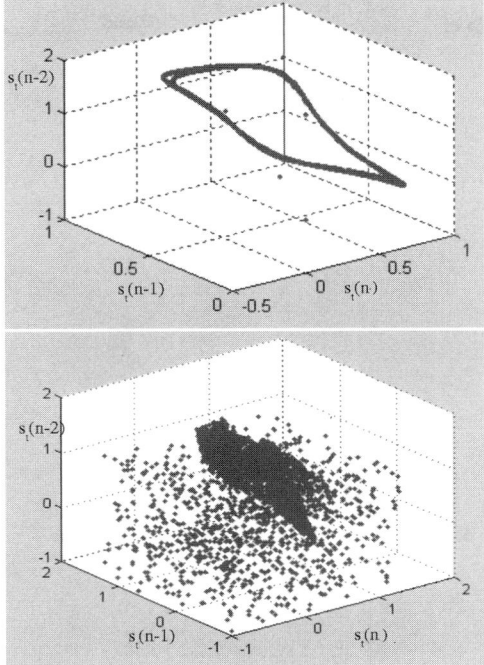

Fig. 7.23. At the beginning of each bit the transmitter state is initialized with a random value. Top: a limit cycle attractor. Bottom: the superposition of multiple trajectories starting at a random initial state and converging to the same attractor shown at the top plot.

0.2. The noise added to each transmitter state component t_i had Gaussian distribution with variance V.

It is evident from Figure 7.26 that the noise in the analog components negatively affects the security of the DDE scheme.

7.5.6 DDE Versus Traditional Public Key Encryption Schemes

- **Underlying concept:**
  Traditional public key encryption schemes rely on one-way functions that are relatively simple to calculate in one direction yet their inverse can be made exceedingly difficult to compute. DDE relies on a completely different concept: We can design a nonlinear dynamics that is simple to simulate and difficult to reconstruct analytically/numerically from a set of observations. By keeping part of the dynamics private an unauthorized receiver cannot simulate the system and is required to use computationally expensive analytic and numeric methods.

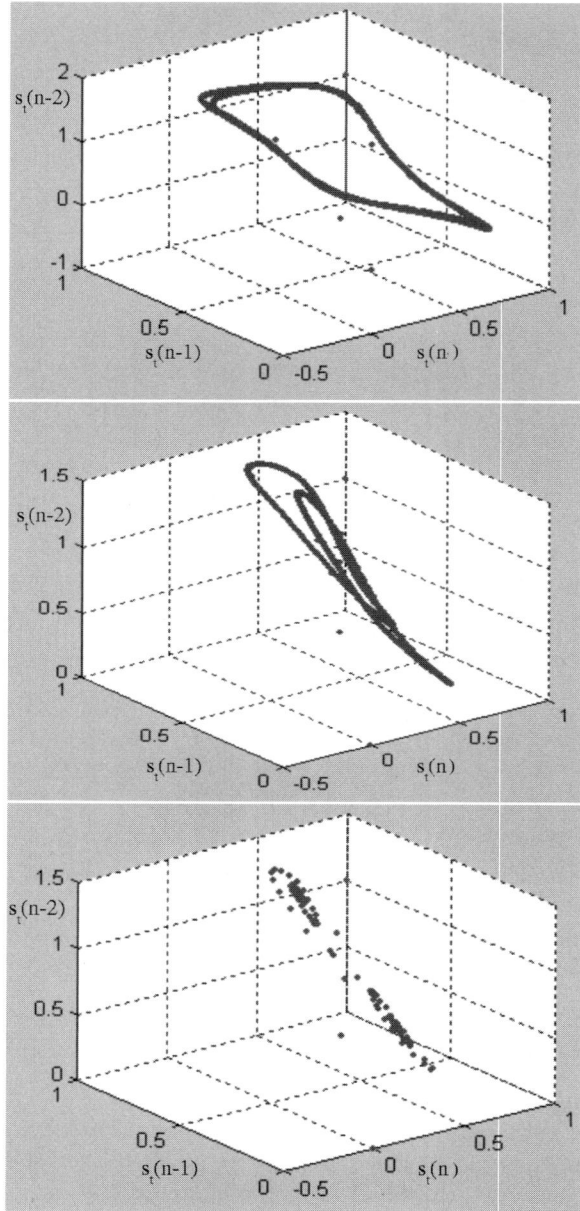

Fig. 7.24. A protection against attractor reconstruction by the unauthorized receiver is obtained by altering the receiver dynamics frequently (each bit) which results in a change in the attractor position. Three different attractors obtained by altering the receiver parameters while keeping the transmitter parameters constant.

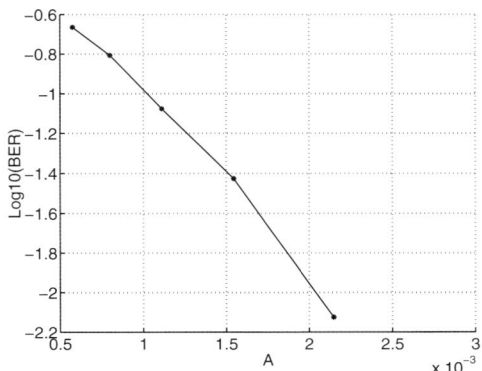

Fig. 7.25. Dependence of Bit Error Rate (BER) encountered by the authorized receiver on the modulation parameter A which affects the amount of separation between the "0" and "1" attractors. (Reprinted from [1], ©2003 American Physical Society.)

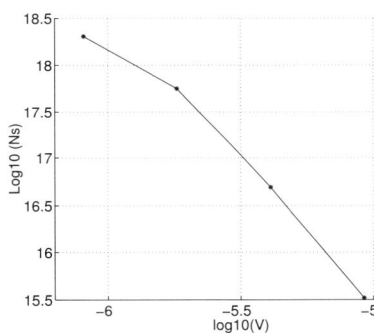

Fig. 7.26. The number of states N_s required by a Viterbi algorithm based attack as a function of the variance V of the component noise at the transmitter circuit. (From [1].)

- **Continuous number field:**
  Traditional public key encryption schemes are defined over integer number fields whereas DDE is defined over continuous number fields (simulated digitally using float or double numbers). This results in a very different underlying mathematics for cryptanalysis.
- **Hardware implementation:**
  Traditional public key encryption schemes can only be implemented using digital hardware. DDE is defined over continuous number fields and therefore can be implemented directly using analog components. Direct implementation using electrical/optical analog components may result in fast/inexpensive communication devices.
- **Algorithmic variety:**
  Traditional public key encryption schemes are based on a limited pool of algorithmic procedures. Most schemes are based on some kind of exponentiation over Galois fields. DDE represents a novel concept that can be implemented using a rich variety of different functions .
- **Block merging of encryption and modulation:**
  In traditional public key encryption schemes the digital encryption block is implemented separately from the analog modulation block. In DDE both

encryption and modulation can be implemented using analog hardware as a single block. The input to this block would be a message stream and the output would be an encrypted broadband modulated signal.

- **Operation under noisy conditions:**
  Traditional public key encryptions schemes do not allow any noise to interfere with the ciphertext. An error in a single ciphertext bit (after error correction stage) will result in a change in most (typically all) corresponding message bits. In DDE noise is assumed to be part of the encryption and decryption processes, and a certain amount of noise is allowed.

- **Key management:**
  In traditional public key encryption schemes the public key is distributed before transmission starts. In DDE the signal S_r from receiver to transmitter is a part of the public key and therefore in fact the public key is continuously updated during transmission.

- **Coding efficiency:**
  In the present implementation DDE is less efficient than traditional public key encryption schemes. Because we need to generate the whole transient toward an attractor to transmit one information bit, the amount of information needed to be transmitted far exceeds the number of message bits. However, because DDE is conceptually different than traditional public key encryption schemes, it may prove to be more secure and easier to implement for specific applications.

7.6 Conclusions

The history of chaotic secure communication is short, and its future is uncertain. In spite of their problematic security, chaotic encryption schemes definitely provide a certain degree of privacy, and a large range of applications require privacy and not necessarily a very high level of security. A clear advantage in using a chaotic encryption scheme is that it does not require digitization of data and can be implemented using analog (electrical/optical) components. The rapid growth in wireless communications may create a new type of applications that will require cheap encryption of undigitized continuous waveforms using simple analog hardware.

In this chapter we summarized the current status of algorithms of chaos-based communications with an emphasis on their security. Cryptanalysis of chaos-based encryption schemes so far has mostly been qualitative and not sufficient to guarantee security. In this chapter, we described a more quantitative approach to the cryptanalysis of a particular class of chaos-based algorithms using so-called active/passive decomposition. The cryptanalysis is based on the reconstruction of the local average dynamics as described in [16, 25, 26]. We quantify the security of an encryption scheme by estimating the number of samples that can be safely transmitted without allowing an unauthorized receiver to reconstruct the dynamics with an accuracy that is sufficient to compromise the message. Our quantitative analysis shows that even the chaos-based communication schemes that have been broken can still be considered secure as long as the number of transmitted samples N_s does not exceeds a certain threshold. We showed that the number of samples that can be safely transmitted can be increased by using a fine-grained modulation with large dynamical range. This is in contrast to the typical approach which attempts to modulate a message that is as small as possible, so it can be better masked by the chaotic component. In our approach, not only the chaotic component masks the message, but also the randomized message component is used to mask the chaotic component. The larger the dynamical range of the message, the more samples the unauthorized receiver needs in order to accurately reconstruct the chaotic dynamics. We ensure that the large dynamical range of the message will not allow the unauthorized receiver to reconstruct the masking chaotic dynamics with lower accuracy by keeping the message modulation fine grained. Furthermore, we show that using a conditional Lyapunov exponent λ with smaller absolute value increases the number of samples that can be safely transmitted, and requires the unauthorized receiver to reconstruct the chaotic dynamics in a reconstructed embedding phase space of larger embedding dimension D_e.

The quantitative approach presented here is only the first step toward building confidence in the security of chaos-based encryption schemes by putting the cryptanalysis of these schemes on a rigorous mathematical foundation. Much work remains to be done to reach the stage of cryptanalysis typical for modern digital encryption schemes based on integer number theory.

In this chapter we also introduced the first public key encryption scheme that is based on chaotic dynamics. It is defined over continuous number fields. As such, it does not require digitization of the data, and can be implemented using analog (electrical/optical) components.

We showed that various cryptanalytic attacks can be avoided by initializing the dynamics with a random state and by altering the receiver private dynamics at the beginning of each transmitted bit. Also, the larger the attractor embedding dimension, the more secure the scheme is. Possible candidates for a simple high-security implementation can be dynamical systems with long time-delayed feedback or large coupled map lattices (see, e.g., [34]). These systems may exhibit chaos of potentially very high dimension and may prove to be useful for future DDE schemes with extremely high level of security.

Proving the security of a new encryption scheme may take years of breaking and reinforcing cycles. DDE has been recently proposed and therefore it may take time until its security is thoroughly tested. It may happen that the specific implementation we used in our simulation to demonstrate the concept of DDE will be broken by some cryptanalysis attack, however, we believe that due to the rich variety of dynamics that can be used to implement DDE it is likely that schemes that are robust to a specific attack can be found. We encourage the reader who is interested in detailed derivations and reasoning of our analysis to read our publications [1, 2, 4, 33].

Acknowledgments

This work was partially supported by the Army Research Office under MURI grant DAAG55-98-1-0269, by the U.S. Department of Energy, Office of Basic Energy Sciences, Division of Engineering and Geosciences, under Grants No. DE-FG03-90ER14138, No. DE-FG03-95ER14516, and No. DE-FG03-96ER14592, by a grant from the National Science Foundation, NSF PHY0097134, and by a grant from the Office of Naval Research, N00014-00-1-0181.

References

1. R. Tenny, L. S. Tsimring, L. E. Larson, H. D. I. Abarbanel, Using Distributed Nonlinear Dyunamics for Public Key Encryption, *Phys. Rev. Lett.*, vol. 90, 047903, 2003.
2. R. Tenny, L. S. Tsimring, H. D. I. Abarbanel, L. E. Larson, Asymmetric Key Encryption using Distributed Chaotic Nonlinear Dynamics, Proc. of the IASTED International Conference on Communications Internet and Information Technology, St. Thomas, US Virgin Islands, pp. 338–345, 2002.
3. R. Tenny, L. S. Tsimring, H. D. I. Abarbanel, L. E. Larson, Steps towards quantifying cryptanalysis of chaotic active/passive decomposition encryption schemes using average dynamics estimation, *Int. J. Bifurcation and Chaos*, vol. 14, 3949–3968, 2004.
4. R. Tenny, Symmetric and Asymmetric Secure Communication Schemes using Nonlinear Dynamics, Ph.D. Dissertation, University of California, San Diego 2003.
5. H. Dedieu, M. P. Kennedy, M. Hasler, Chaos shift keying: Modulation and demodulation of a chaotic carrier using self-synchronizing Chua's circuits, *IEEE Trans. on Circuits and Systems–II*, vol. 40, pp. 634–642, 1993.
6. M. Itoh, Spread spectrum communication via chaos, *Int. J. of Bifurcation and Chaos*, vol. 9, pp.155–213, 1999.
7. K. M. Cuomo and A. V. Oppenheim, Circuit implementation of synchronized chaos with application to communications, *Phys. Rev. Lett.*, vol. 71, pp. 65–68, 1993.
8. U. Parlitz, L. Kocarev, T. Stojanovski, and H. Preckel, Encoding messages using chaotic synchronization *Phys. Rev. E.*, vol. 53, pp. 4351–4361, 1996.
9. L. Kocarev and U. Parlitz, General Approach for chaotic synchronization with applications to communication, *Phys. Rev. Lett.*, vol. 74, pp. 5028–5031, 1995.
10. L. Kocarev, Chaos-based cryptography: A brief overview. *IEEE Circuits and Systems Mag.*, vol. 1, No.3, pp. 6–21, 2001.
11. T. L. Carroll and L. M. Pecora, Cascading synchronized chaotic systems, *Physica D*, vol. 67, pp. 126–140, 1993.
12. A. R. Volkovskii and N. F. Rulkov, Synchronouns chaotic response of a nonlinear ocsillating system as a principle for the detection of the information component of chaos. *Tech. Phys. Lett.* vol. 19, pp. 97–99, 1993.
13. S. Hayes, C. Grebogi, E. Ott, and M. Spano, Experimental control of chaos for communication, *Phys. Rev. Lett.*, vol. 73, pp. 1781–1784, 1994.

14. Y.-C. Lai, E. Bollt, and C. Grebogi, Communicating with chaos using two–dimensional symbolic dynamics. *Phys. Lett. A*, vol. 255, pp. 75–81, 1999.
15. Y. Zhang, M. Dai, Y. Hua, W. Ni, and G. Du, Digital communication by active–passive decomposition synchronization in hyperchaotic systems, *Phys. Rev. E*, vol. 58, pp. 3022–3027, 1998.
16. K. M. Short and A. T. Parker., Unmasking a hyperchaotic communication scheme, *Phys. Rev. E* , vol. 58, No.1, pp. 1159–1162, 1998.
17. T. L. Carroll and L. M. Pecora, Using multiple attractor chaotic systems for communication, *Chaos*, vol. 9, pp. 445–451, 1999.
18. T. Kapitaniak, Synchronization of chaos using continuous control, *Phys. Rev. E*, vol. 50, pp. 1642–1644, 1994.
19. T. Yang and L. O. Chua, Impulsive stabilization for control and syncrhonization of chaotic systems: Theory and application to secure communication, *Trans. IEEE, on Circuits and Systems–I*, vol. 44, pp. 976–988, 1997.
20. J. M. Cruz and L. O. Chua, An IC chip of Chua's circuit, *IEEE Trans. on Circuits and Systems–II*, vol. 40, pp. 614–625, 1993.
21. T. Yang, L. B. Yang, and C. M. Yang, Breaking chaotic secure communication using a spectrogram, *Phys. Lett. A*, vol. 247, pp. 105–111, 1998.
22. T. Yang, L. B. Yang, and C. M. Yang, Cryptanalyzing chaotic secure communications using return maps, *Phys. Lett. A*, vol. 245, pp. 495–510, 1998.
23. J. B. Geddes, K. M. Short, and K. Black, Extraction of signals from chaotic laser data, *Phys. Rev. Lett.*, vol. 83, pp. 5389–5392, 1999.
24. T. Yang, L. B. Yang, and C. M. Yang, Breaking chaotic switching using generalized synchronization: Examples, *IEEE Trans. on Circuits and Systems–I*, vol. 45, pp. 1062–1067, 1998.
25. K. M. Short, Steps toward unmasking secure communications, *Int. J. of Bif. and Chaos*, vol. 4, pp. 959–977, 1994.
26. K. M. Short, Unmasking a modulated chaotic communications scheme, *Int. J. of Bif. and Chaos*, vol. 6, pp. 367–375, 1996.
27. R. Rivest, A. Shamir, and L. Adleman, A method for obtaining digital signatures and public key cryptosystems, *Comm. ACM*, vol. 21, pp.120–126, 1978.
28. V. S. Miller, Use of elliptic curves in Cryptology, *Proc. of Crypto 85*, pp. 417–426, 1986.
29. T. El-Gamal, A public key cyryptosystem and signature scheme based on discrete logarithms, *Advances in Cryptography: Proc. of Crypto 84*, Springer-Verlag, Berlin, pp. 10–18, 1985.
30. B. Chor, and R. L. Rivest, A knapsack type public key cryptosystem based on arithmetic in finite fields, *Advances in Cryptography: Proc. of Crypto 84*, 196, pp. 54–64, 1985.
31. R. J. McEliece, A public-key cryptosystem based on algebric coding theory. *JPL DSN Progress report*, vol. 42–44, pp. 114–116, 1978.
32. H. D. I. Abarbanel, R. Brown, J. J. Sidorowich, and L. S. Tsimring, The analysis of observed chaotic data in physical systems. *Rev. Mod. Phys.*, vol. 65 pp. 1331–1393, 1993.
33. DDE overview, images and updates may be found at: http://inls.ucsd.edu/~roy/DDE/MainPage/
34. K. Kaneko, Theory and Applications of Coupled Map Lattices, Nonlinear Science Theory and Applications (Wiley, New York, 1993).

8

Random Finite Approximations of Chaotic Maps

Jesús Urías, Eric Campos, and Nikolai F. Rulkov

8.1 Introduction

Utilizations of chaotic signals in communication systems [1] and radars [2, 3] are the most promising engineering applications of the dynamical systems with a chaotic behavior. Thanks to chaotic dynamics such systems generate aperiodic signals with adequate noiselike properties and, at the same time, two identical chaotic systems can synchronize in the sense that one system follows the chaotic trajectory of the other system [6]. Such synchronization is a key element of many chaos-based communication schemes. The inexpensive chaos generators designed for these applications usually rely on the analog circuitry that implements one of the known ODE systems or nonlinear maps demonstrating a chaotic behavior. Because the parameters of analog electronic elements that control the parameters of the dynamical system have some dispersion and are susceptible to the temperature and other fluctuations, the design of two analog chaotic circuit with identical parameters is quite a challenge.

In order to overcome the problem of the parameter fluctuation many engineering applications use finite (digital) systems which are more robust and reliable than continuous (analog) ones. Now many analog systems have been replaced by a digital one to accomplish an equivalent task. Such an approach would be also very attractive for implementation of chaos generators. However, a dynamical system approximated with a finite-state (digital) system loses the ability to produce a chaotic behavior. Therefore, it is impossible to use finite systems for applications similar to those implemented with chaotic systems unless one can induce a "chaoticlike" behavior in the dynamics of the finite system, for instance, by randomly perturbing it.

From an application point of view it is desirable to develop an approach for the design of inexpensive digital system that can replicate the chaotic dynamic of an analog system in the high-frequency range. Such a design can be transported to Field Programmable Gate Arrays (FPGA) or a similar technology assuming that the number of the finite states is not very large, but sufficient for the replication of chaos in a given application. This chapter discusses the theoretical issues related to the finite-state implementation of chaotic one-dimensional maps paying particular attention to the case of a large but limited number of states.

Our method to design chaoticlike digital systems specifies a Markov chain whose stationary distribution approximates a given invariant measure μ for a map f of the interval $I = [0, 1)$. We show how to construct finite partitions $\{J_i\}$ of I such that the transition probabilities $p(i|j) = \mu(f(J_j) \cap J_i)/\mu(f(J_j))$ define a Markov chain with $\{i\}$ the set of states and a stationary distribution with probabilities that coincide with the the the weights given by the f-invariant measure μ to the subintervals J_i. The method further provides a coding function that translate states of the finite system to points in the interval. In this way, the random sequences generated by the finite system produce pseudo-orbits with the same statistical properties than a μ-typical orbit of f.

In Section 8.2 partitions for the Markov chain are constructed for maps that have a generating partition. The random transition function is expressed as a simple arithmetic formula that can be implemented as a shift register. Coding functions are explicitly constructed in Section 8.4 for the one-parameter family of tent maps.

8.2 Random Finite Approximations

First, we propose a sampling scheme to discretize (I, f, μ). Let $P := \{x_0 = 0 < x_1 < \cdots < x_s = 1\}$ be a finite set of partition points in the interval $I = [0, 1)$. The subintervals $J_i = [x_i, x_{i+1})$, $i = 0, 1, \ldots, s-1$, provide a partition of I. For a point $x \in [0, 1)$, denote by $[x]_P$ the index of the subinterval in I where x lays in, for example, $[x]_P = i$ if $x \in J_i$: $[x]_P$ is the value of x as sampled by P. The set of sample indices is $\Delta := \{0, 1, \ldots, s-1\}$ and addition (modulo s) on Δ is assumed (denoted $+_s$). The sampling function $[\cdot]_P : I \to \Delta$ is not invertible and given $[x]_P = i$ our guess about x will be x_i, denoted by $\chi(i) = x_i$. The sampled version of function f, $f_P : \Delta \to \Delta$, is defined to be $f_P(i) = [f(x_i)]_P$. Functions that are continuous and bounded on I can be approximated by its sampled versions by letting $|P| := \max\{x_{i+1} - x_i : i = 0, \ldots, s-1\} \to 0$.

Lemma 1. *Let $f : I \to I$ be continuous on I. Then, for every $\varepsilon > 0$ there exists P such that for each $i \in \Delta$, $|f(x) - \chi \circ f_P(i)| < \varepsilon$ whenever x is such that $[x]_P = i$.*
Proof. Let $[x]_P = i$. By triangle inequality $|f(x) - \chi \circ f_P(i)| \leq |f(x) - f(\chi[x]_P)| + |f \circ \chi[x]_P - \chi(f \circ \chi[x]_P)|$. Then, by uniform continuity of f there is a $|P|$ such that $|f(x) - f(\chi[x]_P)| \leq \varepsilon/2$, for every $x \in I$. Thus, $|f(x) - \chi \circ f_P[x]_P| \leq \varepsilon/2 + |P|$. Lemma 1 is proved because there exists a partition P of diameter $|P|$ such that $\varepsilon/2 + |P| < \varepsilon$ is also satisfied. $\square$

Due to the finite character of the state set Δ, all orbits

$$i_0, f_P(i_0), \ldots, f_P^n(i_0), \ldots \tag{8.1}$$

of the discrete system (Δ, f_P) are eventually periodic. However, the sampling of a chaotic orbit

$$x, f(x), \ldots, f^n(x), \ldots \tag{8.2}$$

of f, yields the seemingly random sequence of integers

$$r_0, r_1, \ldots, r_n, \ldots, \tag{8.3}$$

where $r_n = [f^n(x)]_P$. Thus, the dynamics of the discretized version (Δ, f_P) is far from approximating the dynamics of the original system (I, f, μ), for example, after a while the orbits (8.1) and (8.3) will have completely different statistical behaviors (even though f_P approximates f closely in the sense of Lemma 1).

Regular motions are avoided if the deterministic finite system is perturbed randomly at every time step. Instead of the deterministic transition $i \mapsto f_P(i)$ we consider a random transition $i \mapsto j$ with transition probability $p(j|i)$. We express the random transition from state i as the random transition function $F(i)$ that takes values on Δ and such that $\mathcal{P}[F(i) = j] = p(j|i)$. Thus, the random finite system (Δ, F, p) is a Markov chain. The stochastic matrix of the chain is denoted by $M = (m_{ij})$ with entries $m_{ij} = p(j|i)$. When the matrix M is aperiodic, the stationary distribution of the chain is the only positive vector $\rho = (\rho_0, \ldots, \rho_{s-1})$ such that $\rho M = \rho$ [4].

We want any sample sequence $i_0, i_1, \ldots, i_n, \ldots$ generated by the random finite system (Δ, F, p) to "look like" a typical orbit of (I, f, μ), in the following sense. First, the sequence $\{i_n\}$ is to be interpreted as the pseudo-orbit $\{x_n = \chi(i_n)\}$ in I and we want each pair (x_n, x_{n+1}) close to the graph of f. Thus, we ask (Δ, F, p) to satisfy the following condition.

> (w1) For any $\varepsilon > 0$ there is a partition P such that the random transition function satisfies $|f(x) - \chi \circ F(i)| < \varepsilon$ with probability 1 for each $i \in \Delta$ and $[x]_P = i$.

Second, the frequency at which points in the pseudo-orbit $\{x_n\}$ touch the subinterval $J_k \in P$ should coincide with the weight given to J_k by the f-invariant measure μ. For this we ask (Δ, F, p) to satisfy the following condition.

> (w2) For every k and $i \in \Delta$

$$\mu(J_k) = \lim_{t \to \infty} \frac{1}{t} \sum_{j=0}^{t-1} \delta_k(F^j(i)), \quad \text{with probability 1.} \tag{8.4}$$

A discrete random system (Δ, F, p) is said to constitute a random finite approximation for the system (I, f, μ) if it satisfies conditions (w1) and (w2).

Condition (w1) ensures that the random transition function $i \mapsto F(i)$ is ε–close to the actual map $f(x)$, with probability 1. A sufficient condition on transition probabilities $p(i|j)$ that makes condition (w1) valid is given in Lemma 2 below.

In condition (w2) we require the stationary distribution of the Markov chain to approximate μ, up to a resolution given by the mesh $|P|$. To satisfy condition (w2) the problem is to find a partition P of I and transition probabilities $p(F(k)|k)$ such that equality in (8.4) certainly holds. Observe that condition (w2) is not a condition valid in the limit $|P| \to 0$ only but it should be satisfied by partitions with positive diameter $|P|$.

Let us write the random transition function in the form $F(i) = f_P(i) +_s r_i$, with r_i a random variable such that $\mathcal{P}[f_P(i) +_s r_i = j] = p(j|i)$. Condition (w1) to hold requires that the strength of the random perturbation r_i be small enough, that is, as the sampling mesh $|P|$ is made finer the strength of the random perturbation r_i has to become weaker. This condition on the random variable r_i, sufficient to make (w1) true, is expressed in the following.

Lemma 2. *Let $f : I \to I$ be continuous. For each P let there exist a smallest integer $R = R(P) \in \Delta$ such that $\lim_{|P| \to 0} |P| R = 0$ and $\mathcal{P}[r_i \geq R] = 0$, for each i. Then, (w1) holds.*
Proof. By definition $F(i) = [f(\chi(i))]_P + r_i$ for each $i \in \Delta$. Thus, by Lemma 1, for any $\varepsilon > 0$ there exists a partition P such that

$$
\begin{aligned}
|f(x) - \chi(F(i))| \quad &= \quad |f(x) - \chi([f(\chi(i))]_P + r_i)| \\
&\leq |f(x) - \chi([f(\chi(i))]_P)| + |\chi([f(\chi(i))]_P) - \chi([f(\chi(i))]_P + r_i)| \\
&< r_i|P| + |f(x) - \chi(f_P(i))| \quad < \quad |P| R + \varepsilon/2
\end{aligned}
$$

whenever $[x]_P = i$. The assumption on the random variable r_i allows us to take $|P|$ small enough as to have $|P| R < \varepsilon/2$, sufficient to make (w1) true. $\square$

Thus, condition (w1) is satisfied if the support of transition probabilities becomes sufficiently small (relative to $|P|^{-1}$) as $|P| \to 0$. Condition (w2) is the main matter of the following sections.

8.2.1 Pertinence of the Random Finite Approximation

Assume that (Δ, F, p) approximates (I, f, μ) in the sense of conditions (w1) and (w2). Let $i_0, i_1, \ldots, i_j, \ldots$ be a random orbit of F. Then, the pseudo-orbit $x_0, x_1, \ldots, x_j, \ldots$, with $x_j = \chi(i_j)$, mimics a typical orbit of (I, f, μ).

First, by condition (w1), for any $\varepsilon > 0$ there is a partition P such that $|f(x_j) - x_{j+1}| < \varepsilon$, for each $j = 0, 1, \ldots$. In other words, the pseudo-orbit (x_j) follows closely an actual typical orbit of f.

Second, the time average along a pseudo-orbit provides a good estimate of the averages with measure μ. We have the following

Lemma 3. *Let x_0, x_1,... be a pseudo-orbit of (I, f, μ). Let the function φ be continuous and bounded on I. Then, for every $\varepsilon > 0$ there exists P such that*

$$\left| \lim_{t \to \infty} \frac{1}{t} \sum_{j=0}^{t-1} \varphi(x_j) - \int_I d\mu\, \varphi \right| < \varepsilon\ .$$

Proof. Observe that $\varphi(x_j) = \varphi_P(i_j)$, where $\varphi_P := \varphi \circ \chi$ is defined on Δ. Considering that (a) any function $\varphi_P : \Delta \to \mathbb{R}$ can be written as $\varphi_P = \sum_{k \in \Delta} \alpha_k \delta_k$ where $\alpha_k = \varphi_P(k) \in \mathbb{R}$ and that (b) $\delta_k[\cdot]_P = i_{J_k}$ is the indicator of the interval J_k, then by assumption (w2) we have that

$$\sum_{k \in \Delta} \alpha_k \lim_{t \to \infty} \frac{1}{t} \sum_{j=0}^{t-1} \delta_k(i_j) = \sum_{k \in \Delta} \alpha_k \int_I d\mu\, i_{J_k}.$$

Thus,

$$\lim_{t \to \infty} \frac{1}{t} \sum_{j=0}^{t-1} \varphi(x_j) = \lim_{t \to \infty} \frac{1}{t} \sum_{j=0}^{t-1} \varphi_P(i_j) = \int_I d\mu\, \varphi_P \circ [\cdot]_P.$$

Then, by Lemma 1 there exists a partition P with sufficiently small diameter $|P|$ such that

$$\left| \lim_{t \to \infty} \frac{1}{t} \sum_{j=0}^{t-1} \varphi(x_j) - \int_I d\mu\, \varphi \right| \le \int_I d\mu\, |\varphi - \varphi_P \circ [\cdot]_P| < \varepsilon.$$

$\square$

Lemma 3 is illustrated by Figure 8.1. The invariant density for the logistic map, $f(x) = 4x(1 - x)$, is $\rho(x) = \left(\pi \sqrt{x(1 - x)}\right)^{-1}$ [7]. Figure 8.1 shows, in asterisks, the approximate density computed by using a pseudo-orbit that is 10^5 time-steps long. For comparison, in Figure 8.1 we also show (as empty circles) the numerical estimates for $\mu(J_k)$, obtained with an actual orbit of f (10^5 time-steps long, too). The continuous line in Figure 8.1 is the graph of the exact invariant density $\rho(x)$. Details of the random finite approximation of the logistic map are given in the next section.

8.3 Maps with a Generating Partition

The logistic map $f(x) = 4x(1 - x)$, with the invariant density $\rho(x) = \left(\pi \sqrt{x(1 - x)}\right)^{-1}$, is used to illustrate how to obtain the approximating Markov chains.

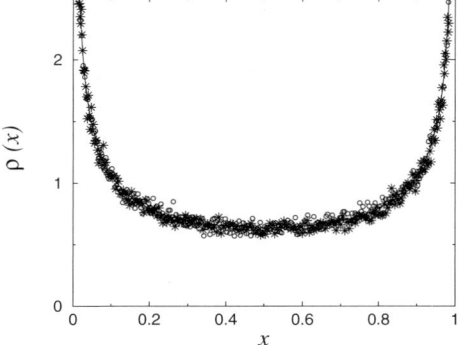

Fig. 8.1. Invariant density of the logistic map approximated by (empty circles) averages along a deterministic orbit and (asterisks) along a random time series.

As sampling mesh we use the dynamical refinement P_k, defined recursively by

$$P_{k+1} = \bigcup_{x \in P_k} f^{-1}(x) , \tag{8.5}$$

of an initial set of generating partition points $P_1 = \{0, 0.5, 1\}$. In (8.5) $f^{-1}(x) = \{(1 - \sqrt{1-x})/2, (1 + \sqrt{1-x})/2\}$. If 0 and 1 are excluded from P_1, the number of partition points is doubled at each iteration (8.5). Thus, $\#P_k = 2^k + 1$.

At the kth refinement, the set of states for the Markov chain is $\Delta = \{0, \ldots, 2^k - 1\}$. Points in P_k are indexed in increasing order: $0 = x_0 < x_1 \cdots < x_s = 1$, where $s = 2^k$, specifying the coding function $i \mapsto \chi(i) = x_i$. Correspondingly, $J_i = [x_i, x_{i+1})$, $i = 0, \ldots, s - 1$. Next, we determine transitions.

Denote by K_j, $j = 0, 1, \ldots, 2^{k-1} - 1$, the subintervals of the $(k-1)$th refinement. For each K_j there exists just one $\ell \in \Delta$ such that $K_j = J_\ell \cup J_{\ell+1}$. Because $f(J_j)$ is an interval of the $(k-1)$th refinement, then there exists only one integer $0 \le \ell < 2^k$ such that for every $j \in \{0, 1, \ldots, 2^k - 1\}$ we have that $f(J_j) = J_\ell \cup J_{\ell+1}$. This determines the allowed transitions to be $j \mapsto \ell$ and $j \mapsto \ell + 1$ if and only if $f(J_j) = J_\ell \cup J_{\ell+1}$. In terms of partition points, there are two cases to determine the value of ℓ for a given j. One is when $f'(x_j) > 0$. The value of ℓ is given by $f(x_j) = x_\ell$. The only other case is when $f'(x_j) \le 0$. The value of ℓ is given in this case by $f(x_i) = x_{\ell+2}$. Thus, the random perturbation is $F(j) = \ell + r$ with transition probabilities determined by the weight of each interval $J_{\ell+r}$, $r = 0, 1$, relative to the weight of the interval $f(J_j)$: $p(\ell + r|j) = \mu(J_{\ell+r})/\mu(f(J_j))$ and $p(i|j) = 0$ otherwise. In terms of the density function ρ the positive transition probabilities are given by

$$p(\ell + r|j) = \frac{\int_{x_{\ell+r}}^{x_{\ell+r+1}} dx \, \rho(x)}{\int_{x_\ell}^{x_{\ell+2}} dx \, \rho(x)} , \qquad r = 0, 1 . \tag{8.6}$$

The logistic map is locally eventually onto [8]. Thus, the Markov chain with transition probabilities (8.6) is aperiodic and the stationary distribution ρ has entries

$$\rho_j = \int_{x_j}^{x_{j+1}} dx \, \rho(x) = \mu(J_j) , \quad j = 0, \ldots, 2^k - 1 , \tag{8.7}$$

as a direct verification of the equation $\rho M = \rho$ shows us.

The Markov chain satisfies conditions (w1) and (w2). Indeed, the transition probabilities (8.6) are such that $\mathcal{P}[r_i \geq 2] = 0$ and then conditions in Lemma 2 that make (w1) valid are satisfied. On the other hand, the stationary distribution (8.7) is just condition (w2). Then, all the implications in Section 8.2.1 are valid for this random finite approximation of the logistic map, for example, Figure 8.1.

Observe that as partition P_k is made finer (larger values of k) the transition probabilities in (8.6) get closer to 1/2 each. Indeed, for any $\varepsilon > 0$ there exists $K > 0$ such that

$$\left| \rho(x_i) - \frac{1}{2^k(x_{i+1} - x_i)} \right| < \varepsilon$$

for each $x_i \in P_k$ with $k > K$.

Next, we systematize the previous construction for an arbitrary map f with a generating partition $\{J_i : i \in S\}$. Let $w := w_0 \cdots w_{k-1} \in S^k$ be a word of length $k > 1$. Each subinterval J_{w_0} is refined to

$$J_{w_0 \cdots w_{k-1}} := \bigcap_{i=0}^{k} f^{-i}(J_{w_i}) .$$

When $J_{w_0 \cdots w_{k-1}} \neq \emptyset$ we say that word w is admissible and we denote by Δ the set of all admissible words of length k.

For $w \in \Delta$ the point $\chi(w) = x_w = \inf J_w$ is a partition point (χ is called the coding function) . The successor of word w is the word $w^+ \in \Delta$ such that x_{w^+} is the smallest of partition points for which $x_w < x_{w^+}$. Then $J_w = [x_w, x_{w^+})$ and the set of partition points is $P = \{x_u : u \in \Delta\}$.

To specify the transition function $f_P : \Delta \to \Delta$ observe that any partition point x_w is mapped by f to another partition point $x_v = f(x_w)$. The corresponding transition in Δ is $w \mapsto v$. Let symbol $r \in S$ be such that $\inf f(J_{w_0 w_1 \cdots w_{k-1}}) = \inf J_{w_1 \cdots w_{k-1} r}$. Thus, the transition function is $f_P(w) := [f(x_w)]_P = w_1 \cdots w_{k-1} r = v$.

In the chain transitions $w \mapsto v$ are allowed if and only if $r \in \Delta$ is such that $f(J_{w_0 w_1 \cdots w_{k-1}}) \cap J_{w_1 \cdots w_{k-1} r} \neq \emptyset$. The random transition function $F(w)$ is specified by letting r (the last symbol of word v) to be a random variable with conditional probability

$$p(v|w) = \begin{cases} \mu(I_{v_0 \cdots v_{k-1}})/\mu(I_{w_1 \cdots w_{k-1}}) & \text{if } v = w_1 \cdots w_{k-1} v_{k-1}, \\ 0 & \text{otherwise,} \end{cases} \qquad (8.8)$$

for every pair, w and v, of admissible words. The stationary distribution ρ of the Markov chain with transition probabilities (8.8) has components $\rho_w = \mu(J_w)$, $w \in \Delta$. Conditions (w1) and (w2) are satisfied and the chain approximates (I, f, μ).

In conclusion, if the map is locally eventually onto and it has a generating partition, then the steps above lead us to an aperiodic Markov chain that approximates the map (in the sense of (w1) and (w2)). Then, the main problem in approximating any map by a Markov chain is to find a generating partition. In the case of Markov maps this is immediate.

8.3.1 An Arithmetic Version of the Random Transition Function

A representation of admissible words $u = u_0 \cdots u_{k-1}$ that is adequate to express arithmetically the random transition function F is to denote word u by the integer $\underline{u} := \sum_{j=0}^{k-1} q^j u_{k-1-j} = s\, x_u$, where q is the cardinality of the generating partition (e.g., $q = 2$ for the logistic map). The transformation of words $u_0 u_1 \cdots u_{k-1} \mapsto v := u_1 \cdots u_{k-1} r$ corresponds to the transformation $\underline{u} \mapsto \underline{v} = q\underline{u} +_s r$. Then

$$F(\underline{u}) := q\,\underline{u} +_s r, \qquad (8.9)$$

where $r := v_{k-1}$ has probability $\pi(r|u) := p(v|u)$. Formula (8.9) can be implemented as a shift register. A sample sequence $\underline{u}_0, \ldots, \underline{u}_n, \ldots$ of (8.9) produces the pseudo-orbit $\chi(\underline{u}_0), \ldots, \chi(\underline{u}_n), \ldots$ of (I, f, μ).

8.4 Approximations for the Tent Maps

The general construction of the previous section is applied to the tent maps

$$T_\lambda(x) = \begin{cases} x/\lambda_0, & \text{if } 0 \le x \le \lambda_0 \\ 1/\lambda_1 - x/\lambda_1, & \text{if } \lambda_0 \le x \le 1 \end{cases}, \qquad (8.10)$$

where parameter $\lambda_0 \in (0, 1)$ and $\lambda_1 = 1 - \lambda_0$. The Lebesgue measure is invariant.

For the tent maps the set of points $P_1 = \{0, \lambda_0, 1\}$ specifies a generating partition with basic subintervals $J_0 := [0, \lambda_0)$ and $J_1 := [\lambda_0, 1)$. Partition points are $x_0 = 0$ and $x_1 = \lambda_0$.

8.4.1 The Coding Function

Given the binary word $u = u_0 \cdots u_{n-1} u_n$, let us determine the location of point $\chi(u) =: x_u$ in I, relative to the point $x_{u_0 \cdots u_{n-1}}$. To do it we observe that $x_u \in J_{u_0 \cdots u_{n-1} u_n} \subset J_{u_0 \cdots u_{n-1}}$ and then the position of x_u relative to $x_{u_0 \cdots u_{n-1}}$

is determined as follows. Let $\theta_{n-1} := \theta(u_0 \cdots u_{n-1}) = 0$ if the number of 1's in word $u_0 \cdots u_{n-1}$ is even and $\theta_{n-1} = 1$ otherwise. If $\theta_{n-1} = 0$ then the total number of inversions in the dynamical refinement of the basic interval J_{u_0} up to $J_{u_0 \cdots u_n}$ is even and so there is no net inversion. When this is the case, the two subintervals that refine $J_{u_0 \cdots u_{n-1}}$ have the same order as the two basic subintervals in I. Then

$$x_{u_0 \cdots u_{n-1} u_n} = x_{u_0 \cdots u_{n-1}} + u_n \lambda_0 |J_{u_0 \cdots u_{n-1}}|, \quad \text{if} \quad \theta_{n-1} = 0.$$

When $\theta_{n-1} = 1$ the total number of inversions is odd and the order of the two subintervals is inverted. Then

$$x_{u_0 \cdots u_{n-1} u_n} = x_{u_0 \cdots u_{n-1}} + (1 - u_n) \lambda_1 |J_{u_0 \cdots u_{n-1}}|, \quad \text{if} \quad \theta_{n-1} = 1.$$

The value of x_u in either case is written as

$$x_{u_0 \cdots u_{n-1} u_n} = x_{u_0 \cdots u_{n-1}} + \begin{cases} 0 & \text{if} \quad u_n = \theta_{n-1} \\ \lambda_{\theta n-1} |J_{u_0 \cdots u_{n-1}}| & \text{if} \quad u_n \neq \theta_{n-1} . \end{cases} \quad (8.11)$$

Let $w < u$ if $x_w < x_u$. A direct consequence of (8.11) is that $u_0 \cdots u_{n-1} \theta_{n-1} < u_0 \cdots u_{n-1}[1 - \theta_{n-1}]$. This relation puts on order in Δ. For instance, $000 < 001 < 011 < 010 < 110 < 111 < 101 < 1000$. Thus, lexicographical order is not followed. The effect of the inverting action of the negative-slope branch of T_λ on encoding is illustrated in Figure 8.2.

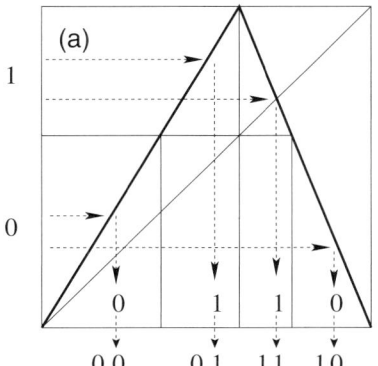
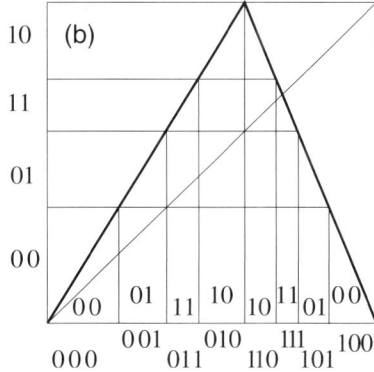

Fig. 8.2. The interval J_{11} precedes J_{10} by the effect on coding of the inverting action of the negative slope branch of the tent map (lexicographical order is not obeyed).

The length of intervals is determined by the contraction rate of each of the branches of T_λ^{-1} and is not affected by the inverting action of the branch with a negative slope: $|J_{u_0 \cdots u_{n-1}}| = \prod_{k=0}^{n-1} \lambda_{u_k}$. Thus, the coding map is

$$x_{u_0\cdots u_{n-1}u_n} = x_{u_0\cdots u_{n-1}} + a_n\lambda_{\theta_{n-1}} \prod_{k=0}^{n-1} \lambda_{u_k}$$

$$= \lambda_0 u_0 + \sum_{j=0}^{n-1} a_{j+1}\lambda_{\theta_j} \prod_{k=0}^{j} \lambda_{u_k}, \tag{8.12}$$

where $a_n = u_n$ if $\theta_{n-1} = 0$ and $a_n = 1 - u_n$ otherwise.

For each $k > 0$, the set $P_k(\lambda) := \{x_u : u \in \{0,1\}^k\} \cup \{1\}$ has $|P_k(\lambda)| = \max\{\lambda^k, (1-\lambda)^k\}$ and $\#P_k(\lambda) = 2^k + 1$. For instance, a direct application of (8.12) yields $P_3(\lambda) = \{0 < \lambda^3 < \lambda^2 < \lambda^3 - \lambda^2 + \lambda < \lambda < \lambda^3 - 2\lambda^2 + 2\lambda < \lambda^2 - \lambda + 1 < \lambda^3 - \lambda^2 + 1 < 1\}$. For $\lambda = 1/2$, $P_3(1/2) = \{j/2^3 : j = 0, 1, \ldots, 8\}$.

8.4.2 The Symmetric Tent Map

Consider map (8.10) with $\lambda = 1/2$. The coding map (8.12) reduces to $\chi(w) = \sum_{i=0}^{k-1} 2^{-i-1}w_k$. As invariant measure μ consider Lebesgue measure. For a word u of length k the weight of the interval J_u is given by its diameter. Because the dynamical partition happens to be a uniform one, we have that $\mu(J_u) = (1/2)^k$. Thus, according to (w2), the stochastic matrix M of the Markov chain on Δ gives the same probability to all transitions. The random transition function that approximates the symmetric tent map, with Lebesgue measure, is

$$F(\underline{w}) = 2\,\underline{w} + r, \tag{8.13}$$

where word w is represented by the integer value $\underline{w} := s\,x_w$, $s = 2^k$ and r is selected at random from $\{0,1\}$, with probability $1/2$ each, independent of state w. Pseudo-orbits are provided by coding function χ.

8.4.3 Singular Invariant Measures

We may proceed in the opposite way. First we specify a stochastic matrix M for the chain and then make use of (w2) to get the estimation of a measure on I for T_λ. In the limit $|P| \to 0$ the measure is invariant but not necessarily continuous.

A great variety of invariant measures for T_λ can be explored numerically by just changing matrix M. To illustrate this scheme, consider the nonsymmetric tent map (8.10) with $\lambda = 0.75$ and consider a matrix M that gives the same weight 0.5 to 0s and 1s. With this choice of M all allowed transitions $w \mapsto v$ are given the same probability $p(v|w) = 0.5$. Typical numerical results are shown in Figure 8.3. For an initial state $w \in \Delta = \{0,1\}^6$, a randomly generated sequence w, $F(w)$, $F(w)$, $\ldots$, $F^{t-1}(w)$, with $t = 10^5$, was used to compute the numbers $M_v = \sum_{j=0}^{t-1} \delta_v(F^j(w))/t$ as estimates of the second member in (8.4). The asterisks in Figure 8.3 mark, for each $u \in \Delta$, the points (x_u, N_u) with $N_u = \sum_{v: x_v < x_u} M_v$.

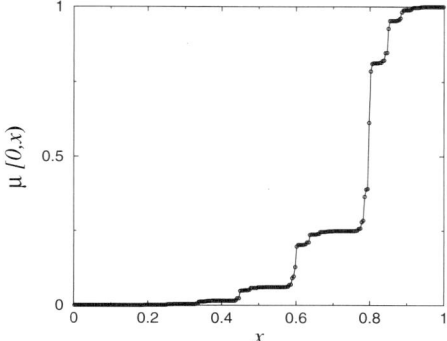

Fig. 8.3. An invariant distribution, $\mu[0, x)$, for the nonsymmetric tent map with $\lambda = 0.75$ (Example 8.4.3).

8.5 Conclusions

We have shown how to specify Markov chains that approximate a chaotic map (I, f, μ). The method enables one to design a digital chaos generator that replicates the behavior of a chaotic map. Designs use a sufficiently large, but limited number of finite states and, as the result, support an inexpensive implementation of the generator suitable for utilization in chaos-based communication systems and other applications.

The digital implementation of the chaotic systems requires an external source of noise to perturb the finite-state system. In some communication systems such noisy perturbations can be given by a properly precoded information data stream. An example of such a chaos-based communication system is considered in [5].

Acknowledgments

This work received partial financial support through an UC–SD/IICO collaboration in the MexUS program. JU received partial financial support form CONACYT (28424E). EC is a CONACyT doctoral fellow at IICO. NR was supported in part by U.S. Department of Energy (grant DE-FG03-95ER14516), the U.S. Army Research Office (MURI grant DAAG55-98-1-0269).

References

1. See, for example, recent special focus issues devoted to chaos-based communication systems: *IEEE Trans. Circuits Syst.*-I, vol. 48, 2001; *IEEE Trans. Circuits Syst.*-I, vol. 47, 2000; *Int. J. Circ. Theory Appl.*, vol. 27, 1999.

2. A. Bauer, Generation and processing of chaotic signals in range measurement systems, AIP Conference Proceedings (Applied Nonlinear Dynamics and Stochastic Systems near the Millennium, San Diego, CA, USA) No.411 pp. 249–254, 1997.

3. A. Bauer, Chaotic signals for CW-ranging systems. A baseband system model for distance and bearing estimation, Proceedings of the 1998 IEEE International Symposium on Circuits and Systems, (ISCAS '98, Monterey, CA, USA), vol. 3, pp. 275–278, 1998.

4. P. Billingsley, Probability and Measure, 3rd ed. (Wiley, NY, 1995).

5. G.M. Maggio, N.F. Rulkov, and L. Reggiani, Pseudo-chaotic time hopping for UWB impulse radio, IEEE Trans. Circuits Syst.-I, vol. 48, pp. 1424–1435, 2001.

6. L.M. Pecora, T.L. Carroll, G.A. Johnson, D.J. Mar, and J.F. Heagy, Fundamentals of Synchronization in chaotic systems, concepts, and applications, Chaos, vol. 7, pp. 520–543, 1997.

7. J. von Neumann and S.M. Ulam, On combination of stochastic and deterministic processes, Bull. AMS, vol. 33, 1120, 1947.

8. M. Pollicott and M. Yuri, Dynamical systems and ergodic theory (Cambridge University Press, 1998).

9

Numerical Methods for the Analysis of Dynamics and Synchronization of Stochastic Nonlinear Systems

How-Foo Chen and Jia-Ming Liu

Summary. The most important numerical tools needed in the analysis of chaotic systems performing chaos synchronization and chaotic communications are discussed in this chapter. Basic concepts, theoretical framework, and computer algorithms are reviewed. The subjects covered include the concepts and numerical simulations of stochastic nonlinear systems, the complexity of a chaotic attractor measured by Lyapunov exponents and correlation dimension, the robustness of synchronization measured by the transverse Lyapunov exponents in parameter-matched systems and parameter-mismatched systems, the quality of synchronization measured by the correlation coefficient and the synchronization error, and the treatment of channel noise for quantifying the performance of a chaotic communication system. For a dynamical system described by stochastic differential equations, the integral of a stochastic term is very different from that of a deterministic term. The difference and connection between two different stochastic integrals in the Ito and Stratonovich senses, respectively, are discussed. Numerical algorithms for the simulation of stochastic differential equations are developed. Two quantitative measures, namely, the Lyapunov exponents and the correlation dimension, for a chaotic attractor are discussed. Numerical methods for calculating these parameters are outlined. The robustness of synchronization is measured by the transverse Lyapunov exponents. Because perfect parameter matching between a transmitter and a receiver is generally not possible in a real system, a new concept of measuring the robustness of synchronization by comparing the unperturbed and perturbed receiver attractors is introduced for a system with parameter mismatch. For the examination of the quality of synchronization, the correlation coefficient and the synchronization error obtained by comparing the transmitter and the receiver outputs are used. The performance of a communication system is commonly measured by the bit-error rate as a function of signal-to-noise ratio. In addition to the noise in the transmitter and the receiver, the noise of the communication channel has to be considered in evaluating the bit-error rate and signal-to-noise ratio of the system. An approach to integrating the linear and nonlinear effects of the channel noise into the system consistently is addressed. Optically injected single-mode semiconductor lasers are used as examples to demonstrate the use of these numerical tools.

9.1 Introduction

In developing a nonlinear dynamical system for chaos synchronization and chaotic communication, several important issues ranging from basic dynamical characteristics to system performances have to be studied. Besides experimental measurements and characterization, various numerical tools are generally needed for a thorough study of the system. Numerical simulation is usually a necessary approach to gaining a complete picture of various dynamical characteristics of a nonlinear system and predicting the dynamics of the system under certain operating conditions. Many parameters and operating conditions, although not easily controllable precisely in experiments, can be precisely prescribed in numerical simulations. Even in the experimental approach, it is also very important to use appropriate numerical methods to analyze the experiment data for a variety of purposes, such as the verification of the chaotic states, the analysis of the complexity of the observed chaos, the qualification of the achieved chaos synchronization, and the characterization of the system performance. The required numerical methods for these studies cover several aspects, including the nonlinear dynamics of a chaotic oscillator generating a chaotic output, the characteristics of the chaotic output, the feasibility of chaos synchronization for a proposed system, the robustness of synchronization, the quality of synchronization, and the performance of the system.

Any physical system always contains some intrinsic noise. Noise in a nonlinear dynamical system is not simply additive to the system dynamics but is an integral part of the dynamics. The dynamical state of a nonlinear system can often vary with the level of the noise in the system. When considering a communication system in reality, the effect of noise on many system performance indicators, such as the bit-error rate (BER), is a key issue that needs to be addressed. For a communication system that utilizes nonlinear dynamics, the effect of noise on the system performance can neither be fully understood nor correctly quantified without considering the nonlinear interactions between noise and system dynamics. Both intrinsic noise of the nonlinear devices and noise of the communication channel have to be considered. Because noise is random and nondeterministic, the Riemann−Stieltjes integrals for solving deterministic differential equations cannot be used for the numerical analysis and simulation of a noisy nonlinear system that is described by differential equations containing stochastic noise terms. Thus, the numerical methods for the analysis and simulation of the dynamics of a stochastic nonlinear system are nontrivial and have to be carefully formulated. In Section 9.2, the general concepts of a stochastic differential equation and its correct integral are first discussed, followed by the introduction of proper numerical algorithms for implementing the analysis and simulation of a system described by stochastic differential equations. The numerical method for the integration of channel noise in a communication system is discussed in Section 9.5.

Because a chaotic waveform generated by a transmitter is used in a chaotic communication system as the message carrier, the verification of its chaotic nature and the characterization of its complexity are both needed in this application. Various numerical methods for the analysis of chaos in the time domain, the frequency domain, or the phase space have been developed. In the time domain, the Lyapunov exponents can be calculated for a chaotic state to quantify the sensitivity of the state to the initial conditions. These parameters indicate the predictability, or rather the unpredictability, of the chaotic trace evolving in time. In the frequency domain, a chaotic waveform is characterized by a broadband spectrum that can be found by taking the Fourier transform of its time series. In the phase space, a chaotic state is characterized by a strange attractor of a fractal dimension. Several different definitions of chaotic attractor dimension are used to measure the complexity of a chaotic waveform. One is the correlation dimension that increases with the complexity of the chaotic state. The numerical methods for calculating the Lyapunov exponents in the time domain and the correlation dimension of the chaotic attractor in the phase space are discussed in Section 9.3.

Although the complexity of a chaotic waveform is important, the ability to stably synchronize two nonlinear dynamical systems serving as transmitter and receiver, respectively, is a key issue for synchronized chaotic communications. Theoretically, perfect synchronization requires a perfect match in the parameters of the transmitter and the receiver being synchronized. Although allowing for perfect synchronization, perfect parameter matching does not guarantee the stability of synchronization against noise or any other form of perturbation. The robustness of chaos synchronization can be examined by finding the transverse Lyapunov exponents of the synchronized chaotic trace. Perfect parameter matching is generally not possible for real systems. Therefore, the concept of transverse Lyapunov exponents for perfect synchronization has to be modified for real systems that have parameter mismatch. These concepts and the numerical methods developed for quantifying the robustness and quality of synchronization are discussed in Section 9.4.

All of the numerical methods discussed in this chapter focus on nonlinear systems that are described by differential equations though some of them can be generalized to systems described by mapping equations, such as the tent map or the logistic map systems. Optically injected semiconductor lasers are used as examples to demonstrate the numerical methods addressed in this chapter. The dynamics and synchronization of such lasers are thoroughly discussed in Chapter 10. The examples in this chapter demonstrate the application of the numerical methods to these lasers.

9.2 Numerical Simulation of Stochastic Differential Equations

It is necessary to consider noise in the simulation of a realistic system because noise always exits in any physical system. For a nonlinear dynamical system, such as a semiconductor laser, noise is not simply a linear addition to the dynamics but is an integral part of the dynamics. The expression of noise is subject to the nonlinearity in the system, but noise also affects the dynamics of the system on the other hand. For a communication system, noise is a key factor that determines the performance of the system. Specially, because the communication systems discussed here are nonlinear systems, the effect of noise on the system performance is also nonlinear. The method of dealing with channel noise in traditional communication systems cannot be used in chaotic communications.

The difference between the simulation of a regular differential equation and that of a stochastic differential equation is that noise is always random and nondeterministic. When solving a stochastic differential equation, the actual value of noise at any moment is not known. The effect of noise can only be evaluated through an ensemble average. Therefore, the rule of the Riemann−Stieltjes integral that treats the integrand as a deterministic function of time does not apply in the integration of a stochastic differential equation that contains one or more noise terms. For this reason, the numerical method for solving stochastic nonlinear differential equations has to be treated with extra care. Instead of the Riemann−Stieltjes integral, a stochastic integral is considered in the numerical analysis.

In general, there are two different methods to calculate a stochastic integral, namely, the Ito integral and the Stratonovich integral. The Ito integral assumes that the correlation time between random processes is zero; it is designed for calculating a stochastic integral with white noise. The Stratonovich integral assumes that a correlation exists between random processes at different moments; it is designed for color noise. The difference between the Ito integral and the Stratonovich integral can be very significant [1]. Which integral has to be chosen depends on the correlation characteristic of the noise under consideration. Detailed discussions regarding these integrals are presented in this section.

Semiconductor laser systems are used as examples in this chapter to demonstrate the numerical concepts. Because the semiconductor lasers functioning as the chaotic oscillators have spontaneous emission as the noise source, all of the equations that describe the nonlinear dynamics, the chaos synchronization, and the chaotic communication are stochastic nonlinear differential equations. In general, the noise sources of semiconductor lasers include the optical noise from spontaneous emission and the carrier noise. The optical noise is white Gaussian noise, and the carrier noise is color noise. The carrier noise is ignored in the numerical analysis of a single-mode laser because this narrowband color noise is contributed mainly by the partition noise of the side modes [2,3]. Therefore, when single-mode semiconductor lasers are considered, only optical noise from spontaneous emission with a white Gaussian nature is included. The discussions in the following are thus limited to white Gaussian noise sources.

9.2.1 Langevin Equation

The rate equations of a system with white noise sources can be generalized in the following form of stochastic differential equations, known as the vector Langevin equation [4–10],

$$\frac{dx_i}{dt} = f_i(\boldsymbol{x}, t) + \sum_{j=1}^{k} g_{ij}(\boldsymbol{x}, t) n_j(t), \tag{9.1}$$

where $i = 1, \ldots, m$ is the dimension index of the differential equation, and $j = 1, \ldots, k$ is the index of the stochastic sources. In this equation, both $f_i(\boldsymbol{x}, t)$ and $g_{ij}(\boldsymbol{x}, t)$ are deterministic functions but $n_j(t)$ represents a stochastic process. This equation was provided by Langevin to explain the Brownian motion besides Einstein's explanation, which is a special case of the Fokker−Planck equation. The stochastic term $n_j(t)$ is independent white Gaussian noise that has the following properties,

$$\langle n_j(t) \rangle = 0, \tag{9.2}$$
$$\langle n_i(t) n_j(t - \tau) \rangle = \delta_{ij} \delta(\tau), \tag{9.3}$$

where $\langle \cdot \rangle$ means ensemble average. Because of δ_{ij} on the right-hand side of (9.3), $\langle n_i(t) n_j(t - \tau) \rangle$ is an autocorrelation function. For demonstration purposes, the discussions are focused on a one-dimensional Langevin equation with a single white Gaussian noise term, that is, a scalar Langevin equation, as follows,

$$\frac{dx}{dt} = f(x, t) + g(x, t) n(t). \tag{9.4}$$

The discussions on the scalar Langevin equation can be generally applied to the vector Langevin equation.

For the differential equation written in the form of (9.4) above, we would suppose that $n(t)$ is integratable and thus the integral

$$W(t) = \int^{t} n(s) ds \tag{9.5}$$

exists. This integral is actually a Wiener process, denoted by $W(t)$, which belongs to a larger subclass of the stochastic process known as the Markov process, or simply Markovian. This means that a Wiener process is memoryless. The Wiener process has the following properties,

$$W(0) = 0, \tag{9.6}$$

$$\langle W(t) \rangle = 0, \tag{9.7}$$

$$\text{Var}[W(t) - W(s)] = t - s, \quad \text{for all } 0 \le s \le t, \tag{9.8}$$

where $\text{Var}[\cdot]$ stands for variance. From this standard definition of the Wiener process, we obtain the relationship that $\text{Var}[W(t) - W(s)] = \text{E}\{[W(t) - W(s)]^2\} = t - s$ because $\langle W(t) \rangle = 0$, where $E\{\cdot\}$ stands for expectation value. Although random variables are not differentiable in the sense of ordinary calculus, one of the important characteristics of the Wiener process is that it is not differentiable even in the sense of the mean-square limit [8, 9]. This property can be seen by examining the convergence of its derivative in the mean-square limit [9]:

$$\left\langle \left[\frac{W(t+h) - W(t)}{h} \right]^2 \right\rangle = \frac{\langle [W(t+h) - W(t)]^2 \rangle}{h^2} = \frac{1}{h}, \tag{9.9}$$

and, therefore,

$$\lim_{h \to 0} \left\langle \left[\frac{W(t+h) - W(t)}{h} \right]^2 \right\rangle = \infty. \tag{9.10}$$

As we can see here, the derivative of $W(t)$ diverges. The paradox here is that $W(t)$ is the integral of $n(t)$, but itself is not differentiable even in the sense of the mean-square limit.

Because $x(t)$ is a real physical observable but the time derivative of $x(t)$ is not, the fact that $W(t)$ is not differentiable means that, mathematically speaking, a Langevin equation does not exist. Its use is only in the sense of physical intuition. The question here is the following: If a Langevin equation does not exist in the mathematical sense, can it be modified to be self-consistent? By examining some properties of the Wiener process $W(t)$, it is important to understand that the Wiener process is not a stationary process in both the strict and the wide senses: As we know, the conditional probability of the Wiener process is [10]

$$p(W_i, t_i | W_{i-1}, t_{i-1}) = \frac{1}{\sqrt{4\pi D(t_i - t_{i-1})}} \exp\left[-\frac{(W_i - W_{i-1})^2}{4D(t_i - t_{i-1})} \right], \tag{9.11}$$

where D specifies the increasing rate of its variance [10], which is $2Dt_i$. For the Wiener process discussed here, we have $D = 1$. By choosing the initial condition $p(W_0, t_0 = 0) = \delta(W_0)$, the probability density for $t \ge 0$ is

$$p_1(W, t) = \int p(W, t | W_0, t_0 = 0)\delta(W_0)dW_0 = \frac{1}{4\pi Dt} \exp\left(-\frac{W^2}{4Dt} \right), \tag{9.12}$$

which is dependent on time. Therefore, it is not a stationary process in the strict sense. From the probability density $p_1(W, t)$, we can also prove that the first moment vanishes, $\langle W(t) \rangle = 0$, and that $\langle W(t_1)W(t_2) \rangle = 2D \min(t_1, t_2)$.

From here, it can also be concluded that it is not a stationary process in the wide sense, either.

However, the increments, $\Delta W_{t_1,t_2} \equiv W_{t_2} - W_{t_1}$, of the Wiener process are stationary. The Wiener process can be represented as a continuous sum over subsequent independent increments:

$$W(t) = \sum_{i=1}^{n} (\Delta W_{t_{k+1}-t_k}). \tag{9.13}$$

The integral $W(t) = \int^t dW(s)$ as defined by the sum of the increments of the Wiener process is then well-defined. Then, the integral of (9.4) in the form of

$$x(t) - x(t_0) = \int_{t_0}^{t} f(x,s)ds + \int_{t_0}^{t} g(x,s)n(s)ds \tag{9.14}$$

can be interpreted consistently by a replacement of $n(t)dt$, which is made by directly following the interpretation of the integral of $n(t)$, as the Wiener process $W(t)$:

$$dW(t) \equiv W(t+dt) - W(t) = n(t)dt. \tag{9.15}$$

The integral equation in (9.14) can then be rewritten as

$$x(t) - x(t_0) = \int_{t_0}^{t} f(x,s)ds + \int_{t_0}^{t} g(x,s)dW(s), \tag{9.16}$$

where $f(x,t)$ is called the *drift* term of the stochastic integral equation, and $g(x,t)$ is called the *diffusion* term. Under this replacement, the integral for $n(t)$ can be defined through the integration of $dW(t)$. Equation (9.16) is named as a stochastic integral equation.

The stochastic integral expressed in (9.16) is very different from a deterministic Riemann–Stieltjes integral. To simplify the discussion, let us consider the integral in the form of

$$x(t_0 + h) - x(t_0) = \int_{t_0}^{t_0+h} u(s)ds, \tag{9.17}$$

where h is a very small time interval and the integrand $u(t)$ can be either a deterministic function or a stochastic process. When it is a deterministic function, $u(t) = f(t)$, then

$$x(t_0 + h) - x(t_0) \simeq f(t_0)h. \tag{9.18}$$

That is to say that $\Delta x \propto h$. However, when $u(t)$ is a stochastic process, the result is very different. For example, let us assume that $u(t)$ is a white Gaussian process with a unity standard deviation that $u(t) = n(t)$. The displacement $x(t_0 + h) - x(t_0)$ is then obtained as the following,

$$x(t_0 + h) - x(t_0) = \int_{t_0}^{t_0+h} n(s)ds = \int_{t_0}^{t_0+h} dW(s)$$
$$\simeq \Delta W_{t_0+h,t_0}$$
$$\simeq \sqrt{h}\, Y, \tag{9.19}$$

where Y is a Gaussian variable with a zero mean and a unity standard deviation. This result can be realized by the fact that the average of $\Delta W_{t_0+h,t_0}$ is zero, and

$$\mathrm{Var}(\Delta W_{t_0+h,t_0}) = \langle (\Delta W_{t_0+h,t_0})^2 \rangle = h. \tag{9.20}$$

As is obtained in (9.19), the displacement, $\Delta x = x(t_0 + h) - x(t_0)$, is proportional to $\sqrt{h}$ in the case of a stochastic process. The integral proportional to $\sqrt{h}$ is the result of the integral in the sense of the expectation value. An example that demonstrates the difference between using $\sqrt{h}$ and h to solve a stochastic integral equation can be found in [1]. In contrast, the Riemann–Stieltjes integral yields the same result proportional to h in both the regular convergence limit and the mean-square convergence limit.

In this subsection, we clarify the critical difference between a deterministic integral and a stochastic integral. Based on the different definitions to calculate the stochastic integral, however, we have a stochastic integral in the Ito sense and that in the Stratonovich sense. Which one has to be used depends on the characteristics of the noise source, which in turn depends on the problem being considered.

9.2.2 Stochastic Integral

The first integral on the right-hand side of (9.16) is a deterministic integral, which is a Riemann–Stieltjes integral, and it is not of concern here. Thus, we only focus on the second integral in (9.16) at this moment, which is stochastic. A discussion to cover the entire evaluation of the integral equation is addressed in the next subsection when the concept of the stochastic integral is implemented in numerical analysis.

Now suppose that the function $u(t)$ is an arbitrary function, and $W(t)$ is the Wiener process. The stochastic integral can be defined in a manner similar to the Riemann–Stieltjes integral, but not the same in the sense of convergence criteria. We divide the interval $[t_0, t]$ into n subdivisions with the order $t_0 \leq t_1 \leq t_2 \leq \ldots \leq t_{n-1} \leq t_n = t$ and define intermediate points τ_i, where the function $u(t)$ is evaluated, such that

$$t_{i-1} \leq \tau_i \leq t_i. \tag{9.21}$$

The stochastic integral can be defined as the limit of the partial sums:

$$S_n = \sum_{i=1}^{n} u(\tau_i)[W(t_i) - W(t_{i-1})]. \tag{9.22}$$

The limit applied here is in the sense of mean-square limit, which is defined as [9]

$$\lim_{n\to\infty} \left\langle \left[S_n - \int_{t_0}^{t} u(s)dW(s) \right]^2 \right\rangle = 0. \tag{9.23}$$

This definition of the stochastic integral is more general than the second integral term in (9.16). It allows $u(t)$ to be a stochastic process.

The value of the stochastic integral depends on the choice of the intermediate points τ_i. As a common example, let the function $u(t)$ be replaced by the Wiener process $W(t)$ [1, 8, 9]. When $\tau_i = t_{i-1}$,

$$\int_{t_0}^{t} W(s)dW(s) = \frac{1}{2}[W^2(t) - W^2(t_0) - (t - t_0)]. \tag{9.24}$$

However, when $\tau_i = \frac{1}{2}(t_{i-1} + t_i)$,

$$\int_{t_0}^{t} W(s)dW(s) = \frac{1}{2}[W^2(t) - W^2(t_0)]. \tag{9.25}$$

The stochastic integral (9.24) that is evaluated at $\tau_i = t_{i-1}$ is an Ito integral, whereas the stochastic integral (9.25) that is evaluated at $\tau_i = (t_{i-1} + t_i)/2$ is a Stratonovich integral. These two integrals clearly have different results.

The Ito integral, denoted by (I) in front of the integral, is defined as

$$(I) \int_{t_0}^{t} u(s)dW(s) \equiv \lim_{n\to\infty} \sum_{i=1}^{n} u(t_{i-1})(W(t_i) - W(t_{i-1})). \tag{9.26}$$

The Stratonovich integral, denoted by (S) in front of the integral, is defined as

$$(S) \int_{t_0}^{t} u(s)dW(s) \equiv \lim_{n\to\infty} \sum_{i=1}^{n} u\left(\frac{t_{i-1} + t_i}{2}\right)(W(t_i) - W(t_{i-1})). \tag{9.27}$$

The major differences between the Ito integral and the Stratonovich integral are summarized in Table 9.1.

Table 9.1. Characteristics of Ito and Stratonovich Integrals

Characteristics	Ito	Stratonovich
Relationship between $u(t_{i-1})$ and $\Delta W_{i,i-1}$	Independent	Dependent
Noise characteristics	True white noise	Color noise
Operation method	Ito calculus	Ordinary calculus

It is important to know that the Ito integral is mathematically more satisfactory, but it is not always the physically natural choice. When the noise $n(t)$

is realistic noise with a finite correlation time, not a white noise, it is more natural to choose the Stratonovich integral. Furthermore, the Stratonovich integral enables us to use ordinary calculus, which is not possible for the integral in the Ito sense. However, there is the following simple relationship between the integral equation solved in the Ito sense and that solved in the Stratonovich sense [8,9], in which $u(t) = g(x, t)$ is a deterministic function,

$$x(t) = x(t_0) + \int_{t_0}^t f(x, s)ds + (\text{I}) \int_{t_0}^t g(x, s)dW(s), \qquad (9.28)$$

$$x(t) = x(t_0) + \int_{t_0}^t \bar{f}(x, s)ds + (\text{S}) \int_{t_0}^t g(x, s)dW(s), \qquad (9.29)$$

with the modified drift term $\bar{f}(x, s)$ defined as follows

$$\bar{f}(x, s) = f(x, s) - \frac{1}{2}g(x, s)\frac{\partial g(x, s)}{\partial x}. \qquad (9.30)$$

This relationship is only valid in the integral equation, but not in the stochastic integral itself. When the diffusion term, $g(x, t)$, is a constant, the Ito integral and the Stratonovich integral are the same. This conclusion can be observed in (9.30) or, alternatively, from a logical intuition that a constant $g(x, t)$ is always independent of $\Delta W(t)$. This is a very important characteristic that serves as an important guideline for programming the simulation of the differential equations of a noisy system.

9.2.3 Numerical Algorithm

It is usually easier to start the computer algorithm for the stochastic integral in the Stratonovich sense than in the Ito sense because the integral in the Stratonovich sense follows the ordinary calculus. However, it is still usually complicated even in the Stratonovich sense. In order to apply the theory of the stochastic integral to the differential equations of a noisy nonlinear system, it is more practical to go back to the vector Langevin equation in the integral form, which can be obtained by integrating (9.1), with an additional assumption that $g_{ij} = \delta_{ij}g_j$. This assumes that each dynamical variable, $x_i(t)$, is contaminated only by a single noise source.

To construct the algorithm, it is necessary to use a small time interval $\Delta t = h$ in the integral:

$$x_i(h) = x_i(0) + \int_0^h f_i(\boldsymbol{x}(t))dt + \int_0^h g_i(\boldsymbol{x}(t))dW_i(t). \qquad (9.31)$$

Note that it is necessary to carry out the expansion in terms of the powers of $h^{1/2}$ rather than in those of h because $\Delta W_{t+h,h} \propto h^{1/2}$. The expansion of $x_i(h)$ can be obtained by the iteration method [4] as

$$x_i(h) = x_i(0) + \delta x_i^{1/2} + \delta x_i^1 + \delta x_i^{3/2} + \delta x_i^2 + \cdots, \tag{9.32}$$

where

$$\delta x_i^{1/2} = g_i Z_{1,i}, \tag{9.33}$$

$$\delta x_i^1 = f_i h + \frac{1}{2} \sum_j g_{i,j} g_j Z_{1,i} Z_{1,j}, \tag{9.34}$$

$$\delta x_i^{3/2} = \sum_j (f_{i,j} g_j Z_{2,j} - g_{i,j} f_j Z_{2,i}) + \frac{1}{3!} \sum_{j,k} g_{i,j} g_{j,k} g_k Z_{1,i} Z_{1,j} Z_{1,k}$$

$$+ \sum_j g_{i,j} f_j h Z_{1,i} + \frac{1}{6} \sum_{j,k} g_{i,jk} g_j g_k Z_{1,i} Z_{1,j} Z_{1,k}, \tag{9.35}$$

$$\delta x_i^2 = \frac{1}{2} \sum_j f_{i,j} f_j h^2 + \frac{1}{2} \sum_{j,k} f_{i,j} g_{j,k} g_k Z_{3,jk} + \frac{1}{2} \sum_{j,k} f_{i,jk} g_j g_k Z_{3,jk}$$

$$+ \sum_{j,k} g_{i,j}\{[f_{j,k} g_k(Z_{1,i} Z_{2,k} - Z_{3,ik}) - g_{j,k} f_k(Z_{1,i} Z_{2,j} - Z_{3,ij})]$$

$$+ \frac{1}{2} g_{j,k} f_k(h Z_{1,i} Z_{1,j} - Z_{3,ij})\}$$

$$+ \frac{1}{24} \sum_{j,k,l} g_{i,j} g_{j,k} g_{k,l} g_l Z_{1,i} Z_{1,j} Z_{1,k} Z_{1,l}$$

$$+ \frac{1}{2} \sum_{j,k} g_{i,jk} g_k[\frac{1}{2} f_j(h Z_{1,i} Z_{1,k} - Z_{3,ik}) + \frac{1}{8} Z_{1,i} Z_{1,j} Z_{1,k} \sum_l g_{j,l} g_l Z_{1,l}]$$

$$+ \frac{1}{24} \sum_{j,k,l} g_{i,jkl} g_j g_k g_l Z_{1,i} Z_{1,j} Z_{1,k} Z_{1,l}$$

$$+ \frac{1}{24} \sum_{l,j,k} g_{i,l} g_{l,jk} g_j g_k Z_{1,i} Z_{1,j} Z_{1,k} Z_{1,l}, \tag{9.36}$$

where $f_i \equiv f_i(\boldsymbol{x}_0)$, $g_i \equiv g_i(\boldsymbol{x}_0)$, $g_{i,j} \equiv \dfrac{\partial}{\partial x_j} g_i(\boldsymbol{x}_0)$, $g_{i,jk} \equiv \dfrac{\partial^2}{\partial x_j \partial x_k} g_i(\boldsymbol{x}_0)$, $f_{i,j} \equiv \dfrac{\partial}{\partial x_j} f_i(\boldsymbol{x}_0)$, and $f_{i,jk} \equiv \dfrac{\partial^2}{\partial x_j \partial x_k} f_i(\boldsymbol{x}_0)$, and $Z_{1,i}$, $Z_{2,i}$, and $Z_{3,ij}$ are defined as follows,

$$Z_{1,i} \equiv \int_0^h dW_i(t), \tag{9.37}$$

$$Z_{2,i} \equiv \int_0^h \left[\int_0^t dW_i(s) \right] dt, \tag{9.38}$$

$$Z_{3,ij} \equiv \int_0^h \left[\int_0^t dW_i(s) \int_0^t dW_j(s) \right] dt. \tag{9.39}$$

As we can see above, the expansion is complicated. And it is still the integral in the Stratonovich sense rather than that in the Ito sense. However, when

$g_i(\boldsymbol{x}, t)$ is a constant, the terms in the expansion are reduced to the following simple expressions,

$$\delta x_i^{1/2} = g_i Z_{1,i}, \tag{9.40}$$

$$\delta x_i^1 = f_i h, \tag{9.41}$$

$$\delta x_i^{3/2} = \sum_j f_{i,j} g_j Z_{2,j}, \tag{9.42}$$

$$\delta x_i^2 = \frac{1}{2} \sum_j f_{i,j} f_j h^2 + \frac{1}{2} \sum_{j,k} f_{i,jk} g_j g_k Z_{3,jk}. \tag{9.43}$$

We thus have

$$x_i(h) = x_i(0) + h f_i + \frac{1}{2} h^2 \sum_j f_{i,j} f_j + \cdots + S_i, \tag{9.44}$$

$$S_i = g_i Z_{1,i} + \sum_j f_{i,j} g_j Z_{2,j} + \frac{1}{2} \sum_{j,k} f_{i,jk} g_j g_k Z_{3,jk} + \cdots. \tag{9.45}$$

Meanwhile, the stochastic integral in the Ito sense is the same as that in the Stratonovich sense when g_i is a constant. Therefore, it is most desirable that the differential equations modeling a system contain only constant diffusion terms with g_i being independent of both $\mathbf{x}$ and t. In order to carry out the algorithm numerically, a further step has to be established to avoid precalculation on the partial derivative of f_i in (9.41), (9.42), and (9.43). In general, Runge–Kutta method is used for this purpose. The detailed discussion can be found in [6].

9.2.4 Example: Dynamics of an Optically Injected Semiconductor Laser

We now use a single-mode semiconductor laser subject to optical injection as an example to demonstrate the simulation of the stochastic differential equations. Detailed discussions of the dynamics of this system can be found in Subsection 7.4.1 of Chapter 7. Here we concentrate on the mathematical aspects of formulating the differential equations for numerical simulation and the effect of noise on the dynamical characteristics of the system seen from the numerical results.

The dynamics of a semiconductor laser subject to the injection of an optical field is mathematically modeled by the following coupled equations [11],

$$\frac{dA}{dt} = -\frac{\gamma_c}{2} A + i(\omega_0 - \omega_c) A + \frac{\Gamma}{2}(1 - ib) g A + \eta A_i \exp(-i\Omega t) + F_{\mathrm{sp}}, \tag{9.46}$$

$$\frac{dN}{dt} = \frac{J}{ed} - \gamma_s N - \frac{2\epsilon_0 n^2}{\hbar \omega_0} g |A|^2, \tag{9.47}$$

where A is the total complex intracavity field amplitude of the laser at its free-oscillating frequency ω_0 and N is the carrier density. The injection field has a complex amplitude of A_i at an optical frequency of ω_i, which is detuned from the free-running frequency ω_0 of the injected laser by a detuning frequency of $\Omega = \omega_i - \omega_0 = 2\pi f$. Other variables in these coupled equations are defined in Chapter 10. The noise term, $F_{sp}(t)$, which originates from the spontaneous emission of the semiconductor laser, is described by a complex Langevin source term, $F_{sp}(t) = F_r(t) + iF_i(t)$. It is white Gaussian noise that is characterized by the following correlation relations [2,3],

$$\langle F_r(t)F_r(t')\rangle = \langle F_i(t)F_i(t')\rangle = \frac{R_{sp}}{2}\delta(t - t') , \tag{9.48}$$

$$\langle F_r(t)F_i(t')\rangle = 0, \tag{9.49}$$

where R_{sp} is the spontaneous emission rate. The focus of the discussions in the following is placed on the stochastic term of the spontaneous emission noise $F_{sp}(t)$. Other terms are all deterministic.

By writing A in terms of its magnitude and phase as $A = |A_0|(1 + a)e^{i\varphi}$, and N as $N = N_0(1+\tilde{n})$, these two coupled rate equations can be transformed into the following three coupled equations [12],

$$\frac{da}{dt} = \frac{1}{2}\left[\frac{\gamma_c\gamma_n}{\gamma_s\tilde{J}}\tilde{n} - \gamma_p(2a + a^2)\right](1 + a) + \xi\gamma_c\cos(2\pi ft + \varphi)$$
$$+ \frac{F_r\cos\varphi + F_i\sin\varphi}{|A_0|}, \tag{9.50}$$

$$\frac{d\varphi}{dt} = -\frac{b}{2}\left[\frac{\gamma_c\gamma_n}{\gamma_s\tilde{J}}\tilde{n} - \gamma_p(2a + a^2)\right] - \frac{\xi\gamma_c}{1 + a}\sin(2\pi ft + \varphi)$$
$$- \frac{1}{1 + a}\frac{F_r\sin\varphi - F_i\cos\varphi}{|A_0|}, \tag{9.51}$$

$$\frac{d\tilde{n}}{dt} = -\gamma_s\tilde{n} - \gamma_n\tilde{n}(1 + a)^2 - \gamma_s\tilde{J}(2a + a^2)$$
$$+ \frac{\gamma_s\gamma_p}{\gamma_c}\tilde{J}(2a + a^2)(1 + a)^2. \tag{9.52}$$

We can use an orthogonal transformation to further simplify the noise terms in (9.50) and (9.51) into $F_a = (F_r\cos\varphi + F_i\sin\varphi)/|A_0|$ and $F_\varphi = -(F_r\sin\varphi - F_i\cos\varphi)/|A_0|$, respectively, in which F_a and F_φ are still white Gaussian. However, the noise term in the phase equation (9.51) still contains a nonconstant factor of the form

$$\frac{1}{1 + a(t)} \tag{9.53}$$

that depends on the variable $a(t)$. Thus, recasting the coupled equations into the three equations given in (9.50)–(9.52) does not result in the most convenient form for numerical analysis when the laser noise is considered.

Instead of the amplitude a and the phase φ, the real and imaginary parts of $A/|A_0|$ defined by $A = |A_0|(a' + ia'')$ can be used together with $\tilde{n}$ defined by $N = N_0(1 + \tilde{n})$ to recast the coupled equations into the following form,

$$
\begin{aligned}
\frac{da'}{dt} &= \frac{1}{2}\left[\frac{\gamma_c\gamma_n}{\gamma_s\tilde{J}}\tilde{n} - \gamma_p(a'^2 + a''^2 - 1)\right](a' + ba'') + \xi\gamma_c\cos(2\pi ft) \\
&\quad + \frac{F_r}{|A_0|},
\end{aligned}
\tag{9.54}
$$

$$
\begin{aligned}
\frac{da''}{dt} &= \frac{1}{2}\left[\frac{\gamma_c\gamma_n}{\gamma_s\tilde{J}}\tilde{n} - \gamma_p(a'^2 + a''^2 - 1)\right](-ba' + a'') - \xi\gamma_c\sin(2\pi ft) \\
&\quad + \frac{F_i}{|A_0|},
\end{aligned}
\tag{9.55}
$$

$$
\begin{aligned}
\frac{d\tilde{n}}{dt} &= -\gamma_s\tilde{n} - \gamma_n\tilde{n}(a'^2 + a''^2) - \gamma_s\tilde{J}(a'^2 + a''^2 - 1) \\
&\quad + \frac{\gamma_s\gamma_p}{\gamma_c}\tilde{J}(a'^2 + a''^2 - 1)(a'^2 + a''^2).
\end{aligned}
\tag{9.56}
$$

Note that $a' = (1 + a)\cos\varphi$ and $a'' = (1 + a)\sin\varphi$, thus $a'^2 + a''^2 = (1 + a)^2$, when connecting (9.54) and (9.55) to (9.50) and (9.51). As F_r and F_i have the white Gaussian characteristics given in (9.48) and (9.49) and $|A_0|$ is a constant, we find that the noise terms in these equations both have a constant coefficient of $g_i = \sqrt{R_{sp}/2|A_0|^2}$ for $i = 1, 2$ because

$$
\frac{F_r}{|A_0|} = \sqrt{\frac{R_{sp}}{2|A_0|^2}}\, n_1(t) = g_1 n_1(t),
\tag{9.57}
$$

$$
\frac{F_i}{|A_0|} = \sqrt{\frac{R_{sp}}{2|A_0|^2}}\, n_2(t) = g_2 n_2(t),
\tag{9.58}
$$

where

$$
g_1 = g_2 = \sqrt{\frac{R_{sp}}{2|A_0|^2}},
\tag{9.59}
$$

$$
\langle n_i(s), n_j(t)\rangle = \delta_{ij}\delta(s - t),
\tag{9.60}
$$

$$
\langle n_i(s)\rangle = 0.
\tag{9.61}
$$

The constant coefficients g_1 and g_2 for the noise terms reduce the complexity of the computer algorithm. Therefore, these coupled equations are preferred for the numerical simulation of a semiconductor laser subject to optical injection when noise is considered.

We use the bifurcation diagrams, shown in Figures 9.1a and b, to give an overview of the nonlinear effect of noise on the different dynamical states of the system. This diagram is obtained by collecting the extrema of $a(t) = \sqrt{a'(t)^2 + a''(t)^2} - 1$ for each operating condition. The numerical simulation is

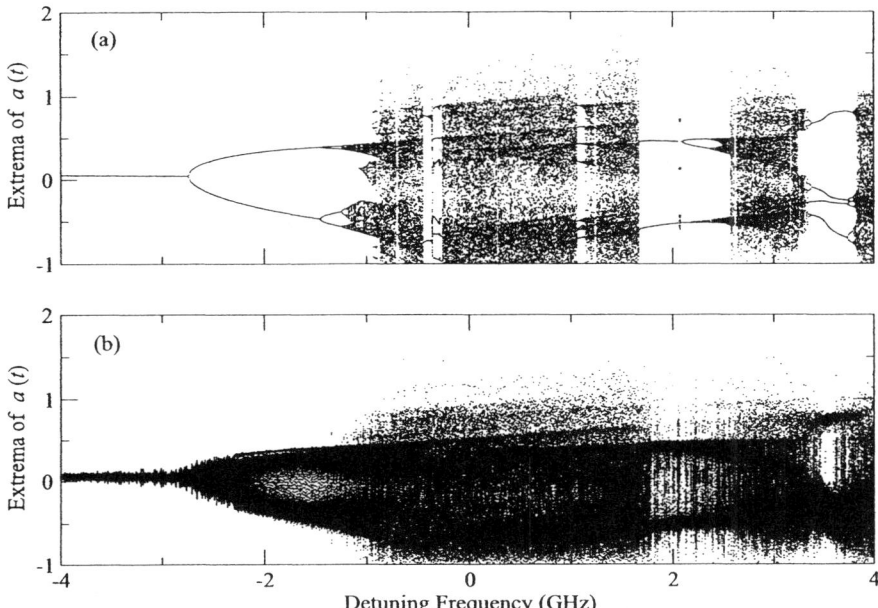

Fig. 9.1. Bifurcation diagrams (a) with $R_{\mathrm{sp}} = 0$ and (b) with $R_{\mathrm{sp}} = 4.7 \times 10^{18}$ $\mathrm{V^2 m^{-2} s^{-1}}$ of a semiconductor laser subject to optical injection with an injection parameter of $\xi = 0.03$ and a detuning frequency varying from -4 GHz to 4 GHz.

performed with $\xi = 0.03$ and f varying from -4 GHz to 4 GHz. The diagram without the consideration of the intrinsic laser noise is shown in Figure 9.1a, and that with the consideration of the noise is shown in Figure 9.1b. The dynamics of this system follows a period-doubling route to chaos [12]. It is seen in Figure 9.1a that the dynamics evolves from a stable locking state at $f = -4$ GHz through periodic states to chaotic states. When noise is present, it is observed in Figure 9.1b that all the states are blurred by the noise, but some states have dramatic changes in their dynamical characteristics by the noise. For example, the stable locking states and the chaotic states can still be recognized in Figure 9.1b, but some periodic states in windows within the chaos region cannot be easily distinguished from the nearby chaotic states.

Here we use the chaotic state obtained at $\xi = 0.03$ and $f = 0$ GHz and the periodic state obtained at $\xi = 0.03$ and $f = -0.4$ GHz in one of the windows within the chaos region as examples to demonstrate the nonlinear effect of the noise to different dynamical states. The effect of the noise on the chaotic state at $\xi = 0.03$ and $f = 0$ GHz is shown in Figure 9.2. The waveform, attractor,

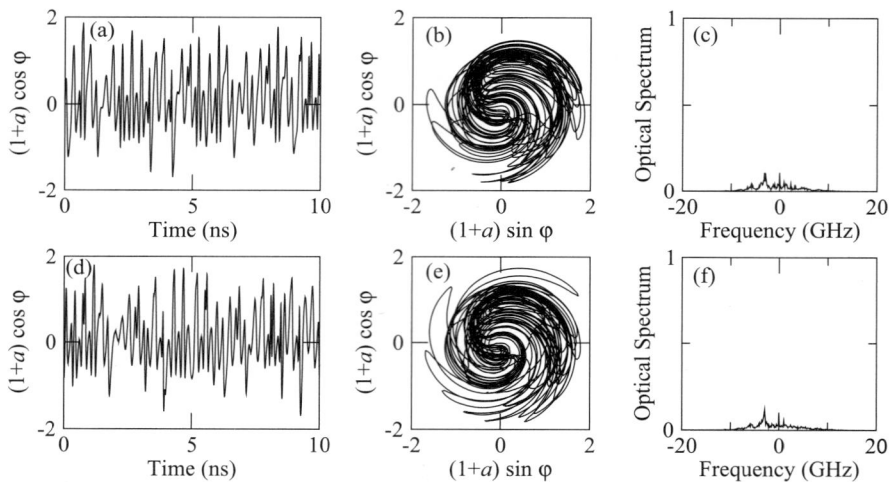

Fig. 9.2. Nonlinear effect of intrinsic laser noise on the chaotic state at $\xi = 0.03$ and $f = 0$ GHz: (a) Waveform, (b) attractor, and (c) optical spectrum are for $R_{\mathrm{sp}} = 0$. (d) Waveform, (e) attractor, and (f) optical spectrum are for $R_{\mathrm{sp}} = 4.7 \times 10^{18}$ V^{2}m^{-2}s^{-1}.

and optical spectrum in the absence of noise are shown in Figures 9.2a–c, respectively. The corresponding plots for the system in the same operating condition but in the presence of noise are shown in Figures 9.2d–f. By comparing Figures 9.2a–c with Figures 9.2d–f, the effect of the noise can barely be recognized because the power of the noise is spread in the bandwidth of the chaotic state. In reality, the noise only adds some fluctuations to this chaotic waveform and its attractor. Therefore, it increases the correlation dimension of the chaotic state by contaminating the attractor. This aspect is discussed in Subsection 9.3.4. This characteristic is quite general for the chaotic states in this system.

However, the effect of the noise on some dynamical states can be dramatic if noise-induced order or noise-induced chaos takes place [13]. An example of such effect is demonstrated in Figure 9.3 for $\xi = 0.03$ and $f = -0.4$ GHz. Each small plot in Figure 9.3 corresponds to that in Figure 9.2. In the absent of the intrinsic laser noise, the characteristics of this periodic state can be observed from its waveform, attractor, and optical spectrum, shown in Figures 9.3a–c, respectively. However, when noise is present, we find that the system is in a chaotic state with completely different characteristics for its waveform, attractor, and optical spectrum. The examples shown in Figures 9.2 and 9.3 demonstrate that the effect of the noise can be very different for different

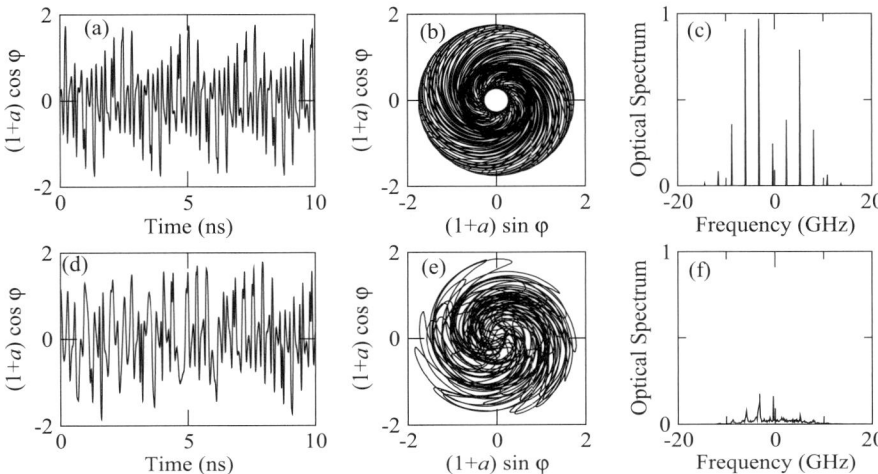

Fig. 9.3. Nonlinear effect of intrinsic laser noise on the state, which is periodic in the absence of noise but is chaotic in the presence of noise, at $\xi = 0.03$ and $f = -0.4$ GHz: (a) waveform, (b) attractor, and (c) optical spectrum for $R_{sp} = 0$; (d) waveform, (e) attractor, and (f) optical spectrum for $R_{sp} = 4.7 \times 10^{18}$ V^{2}m^{-2}s^{-1}.

dynamical states. They show that the effect of noise on a nonlinear system is not simply linear and additive but can be highly nonlinear.

9.3 Characterization of Chaos

The most fundamental characteristic of chaos is its sensitivity to the initial condition of the dynamical system. When there is a small amount of deviation between two initial conditions, this small deviation will be exponentially enlarged as time evolves. The dynamical system then evolves into different conditions represented by different points in its phase space. This phenomenon is called sensitivity to the initial condition. The corresponding divergence is measured by Lyapunov exponents. The entire trace of the dynamical system forms a chaotic attractor. Chaos occurring in nature is an attractor if the system is dissipative with a confined total energy.

Because of this characteristic, a chaotic attractor in its phase space does not repeat itself. Therefore, the geometric structure is usually fractal though some exception does exist in certain dynamical systems [10]. The divergence also reveals the characteristic of the nonperiodic motion of a chaotic dynamical system. The nonperiodic motion gives a chaotic system a broadband characteristic in the frequency domain. All of these signatures can be used to characterize chaos.

9.3.1 Divergence Characterized by Lyapunov Exponents

A dynamical system can be defined by an iterative equation

$$\mathbf{x}(k+1) = \mathbf{F}(\mathbf{x}(k)), \tag{9.62}$$

where $k = 1, 2, \ldots, n$, or by a differential equation

$$\frac{d\mathbf{x}}{dt} = \mathbf{F}(\mathbf{x}(t)), \tag{9.63}$$

where $\mathbf{x}(k)$ and $\mathbf{x}(t)$ are sets of variables describing the system dynamics. Such variables represented by $\boldsymbol{x}(k)$ or $\boldsymbol{x}(t)$ span the phase space of the corresponding dynamical system.

What we are concerned with here is the rate of divergence between two traces starting from two initial conditions that are close to each other. First, consider a system described by an iterative equation as given in (9.62). By defining the deviation, $\mathbf{e}$, of a neighboring trace away from the original trace as $\mathbf{e}(k) \equiv \mathbf{x}(k) - \mathbf{x}_0(k)$, where $\mathbf{x}_0$ indicates the original trace, we have

$$\mathbf{x}_0(k+1) + \mathbf{e}(k+1) = \mathbf{F}(\mathbf{x}_0(k) + \mathbf{e}(k))$$
$$= \mathbf{F}(\mathbf{x}_0(k)) + \mathbf{DF}(\mathbf{x}_0(k)) \cdot \mathbf{e}(k), \tag{9.64}$$

where the matrix $\mathbf{DF}$ is defined as

$$\mathbf{DF}(\mathbf{x}_0(k))_{ij} \equiv \frac{\partial F_i(\mathbf{x}_0(k))}{\partial x_j}. \tag{9.65}$$

Therefore, we obtain

$$\mathbf{e}(k+1) = \mathbf{DF}(\mathbf{x}_0(k)) \cdot \mathbf{e}(k) \tag{9.66}$$

for the iterative equation given in (9.62). Equation (9.66) is called the stability equation. It is important to emphasize that, in calculating the Lyapunov exponents for a system described by an iterative equation, this is the equation form that should be used.

For a system described by a differential equation as given in (9.63), we can obtain the following stability equation through a similar procedure,

$$\frac{d\mathbf{e}}{dt} = \mathbf{DF}(\mathbf{x}_0(t)) \cdot \mathbf{e}(t). \tag{9.67}$$

As can be seen here, the stability equation of the differential equation is different from that of the iterative equation. In order to numerically calculate Lyapunov exponents of a differential equation, one has to convert this stability equation into the form of the stability equation of the iterative equation. This can be achieved by sampling the attractor every τ_s time interval at $t = \tau_s, 2\tau_s, \ldots, n\tau_s$. In the following, we define $\mathbf{e}(k) = \mathbf{e}(t = k\tau_s)$ for convenience.

The stability equation of the differential equation can then be converted by realizing that (9.67) can be expanded as

$$\frac{\mathbf{e}(k+1) - \mathbf{e}(k)}{\tau_s} \simeq \frac{d\mathbf{e}}{dt} = \mathbf{DF}(\mathbf{x}_0(k)) \cdot \mathbf{e}(k). \tag{9.68}$$

Therefore, we can convert this stability equation into the following form,

$$\begin{aligned}
\mathbf{e}(k+1) &= \mathbf{DF}(\mathbf{x}_0(k)) \cdot \mathbf{e}(k)\tau_s + \mathbf{e}(k) \\
&= (\mathbf{I} + \mathbf{DF}(\mathbf{x}_0(k))\tau_s) \cdot \mathbf{e}(k) \\
&= \mathbf{G}(\mathbf{x}_0(k)) \cdot \mathbf{e}(k),
\end{aligned} \tag{9.69}$$

where the matrix $\mathbf{G}(\mathbf{x}_0(k)) \equiv \mathbf{I} + \mathbf{DF}(\mathbf{x}_0(k))\tau_s$ and $\mathbf{I}$ is a unity matrix.

Because most of the publications in the literature only discuss the calculation of the Lyapunov exponents from pure numerical data [16, 17], not from a differential equation, a brief discussion on the calculation of the Lyapunov exponents from a differential equation is provided here. In general, the complex eigenvalues of the Jacobian matrix $\mathbf{DF}(\mathbf{x}_0(k))$ provide the information of the dynamical stability of the system. The real part of each eigenvalue is a Lyapunov exponent that provides the information of divergence or convergence. The imaginary part provides the information of oscillation or cycling. In order to calculate the Lyapunov exponents of a dynamical system described by a differential equation, we do iteration on the matrix $\mathbf{G}(\mathbf{x}_0(k))$ instead of diagonalizing the Jacobian $\mathbf{DF}(\mathbf{x}_0(k))$. The Lyapunov exponents in the average sense, called the global Lyapunov exponents, can be obtained from the following procedure [16, 17],

$$\begin{aligned}
\mathbf{e}(k+L) &= \mathbf{G}(\mathbf{x}_0(k+L-1)) \cdot \mathbf{G}(\mathbf{x}_0(k+L-2)) \cdots \mathbf{G}(\mathbf{x}_0(k))\mathbf{e}(k). \\
&\equiv \mathbf{G}^L(\mathbf{x}_0(k)) \cdot \mathbf{e}(k),
\end{aligned} \tag{9.70}$$

where

$$\mathbf{G}^L(\mathbf{x}_0(k)) \equiv \mathbf{G}(\mathbf{x}_0(k+L-1)) \cdot \mathbf{G}(\mathbf{x}_0(k+L-2)) \cdots \mathbf{G}(\mathbf{x}_0(k)). \tag{9.71}$$

To calculate the real part of the complex eigenvalues of the Jacobian matrix, we actually calculate the eigenvalues of the following matrix:

$$|\mathbf{e}(k+L)|^2 = [\mathbf{G}^L(\mathbf{x}_0(k))]^\dagger \mathbf{G}^L(\mathbf{x}_0(k))|\mathbf{e}(k)|^2, \tag{9.72}$$

where the symbol $\dagger$ denotes the complex-conjugate transpose matrix. We then assign $[\mathbf{G}^L(\mathbf{x}_0(k))]^\dagger \mathbf{G}^L(\mathbf{x}_0(k))$ to a sequence of matrices:

$$\begin{aligned}
[\mathbf{G}^L(\mathbf{x}_0(k))]^\dagger \mathbf{G}^L(\mathbf{x}_0(k)) &= \mathbf{G}^\dagger(\mathbf{x}_0(k)) \cdot \mathbf{G}^\dagger(\mathbf{x}_0(k+1)) \cdots \\
&\quad \mathbf{G}^\dagger(\mathbf{x}_0(k+L-1)) \cdot \mathbf{G}(\mathbf{x}_0(k+L-1)) \cdots \\
&\quad \mathbf{G}(\mathbf{x}_0(k+1)) \cdot \mathbf{G}(\mathbf{x}_0(k)) \\
&= \mathbf{A}(2L) \cdot \mathbf{A}(2L-1) \cdot \mathbf{A}(2L-2) \cdots \\
&\quad \mathbf{A}(3) \cdot \mathbf{A}(2) \cdot \mathbf{A}(1),
\end{aligned} \tag{9.73} \\ \tag{9.74}$$

where

$$\mathbf{G}^\dagger(\mathbf{x}_0(k)) \equiv \mathbf{A}(2L), \tag{9.75}$$
$$\mathbf{G}^\dagger(\mathbf{x}_0(k+1)) \equiv \mathbf{A}(2L-1), \tag{9.76}$$
$$\mathbf{G}^\dagger(\mathbf{x}_0(k+2)) \equiv \mathbf{A}(2L-2), \tag{9.77}$$
$$\cdots$$
$$\mathbf{G}(\mathbf{x}_0(k+2)) \equiv \mathbf{A}(3), \tag{9.78}$$
$$\mathbf{G}(\mathbf{x}_0(k+1)) \equiv \mathbf{A}(2), \tag{9.79}$$
$$\mathbf{G}(\mathbf{x}_0(k)) \equiv \mathbf{A}(1). \tag{9.80}$$

In general, the multiplication of these matrices is ill-conditioned. Therefore, the following QR decomposition has to be used,

$$\mathbf{A}(j)\mathbf{Q}(j-1) = \mathbf{Q}(j)\mathbf{R}(j), \tag{9.81}$$

where $\mathbf{Q}(0) = \mathbf{I}$, an identity matrix. Following the tedious procedure of QR decomposition, we eventually obtain

$$\mathbf{A}(2L)\cdot\mathbf{A}(2L-1)\cdots\mathbf{A}(2)\cdot\mathbf{A}(1) = \mathbf{Q}(2L)\cdot\mathbf{R}(2L)\cdot\mathbf{R}(2L-1)\cdots\mathbf{R}(1). \tag{9.82}$$

Assume that

$$\mathbf{M}_1 = \mathbf{Q}(2L) \cdot \mathbf{R}(2L) \cdot \mathbf{R}(2L-1) \cdots \mathbf{R}(1)$$
$$\equiv \mathbf{Q}_1(2L) \cdot \mathbf{R}_1(2L) \cdot \mathbf{R}_1(2L-1) \cdots \mathbf{R}_1(1), \tag{9.83}$$
$$\mathbf{M}_2 = \mathbf{Q}_1^\dagger(2L) \cdot \mathbf{M}_1 \cdot \mathbf{Q}_1(2L), \tag{9.84}$$
$$= \mathbf{R}_1(2L) \cdot \mathbf{R}_1(2L-1) \cdots \mathbf{R}_1(1) \cdot \mathbf{Q}_1(2L). \tag{9.85}$$

By assigning $\mathbf{R}_1(j) \equiv \mathbf{A}_2(j)$ for $j > 1$ and $\mathbf{R}_1(1)\mathbf{Q}_1(2L) \equiv \mathbf{A}_2(1)$ and by applying (9.81) and (9.82) in (9.85), we can rewrite $\mathbf{M}_2$ as

$$\mathbf{M}_2 = \mathbf{A}_2(2L) \cdot \mathbf{A}_2(2L-1) \cdots \mathbf{A}_2(2) \cdot \mathbf{A}_2(1),$$
$$= \mathbf{Q}_2(2L) \cdot \mathbf{R}_2(2L) \cdot \mathbf{R}_2(2L-1) \cdots \mathbf{R}_2(1). \tag{9.86}$$

By repeating this tedious procedure K times until $\mathbf{Q}_K(2L)$ approaches an identity matrix with a desired accuracy, we eventually obtain

$$\mathbf{M}_K = \mathbf{Q}_K(2L) \cdot \mathbf{R}_K(2L) \cdot \mathbf{R}_K(2L-1) \cdots \mathbf{R}_K(1). \tag{9.87}$$

Then we have the Lyapunov exponents in average, the global Lyapunov exponents. These global Lyapunov exponents are denoted as $\overline{\lambda}_i$ and expressed as

$$\overline{\lambda}_i = \frac{1}{2L} \sum_{j=1}^{2L} \ln[(R_K(j))_{ii}], \tag{9.88}$$

where $(R_K(j))_{ii}$ is the ith diagonal element of the matrix $\mathbf{R}_K(j)$. Because the QR decomposition guarantees the order of $\overline{\lambda}_1 \geq \overline{\lambda}_2 \geq \cdots \geq \overline{\lambda}_n$, the largest global Lyapunov exponent is $\overline{\lambda}_1$. This procedure has been programmed to calculate the global Lyapunov exponents of a Lorentz system to check the accuracy. When a positive global Lyapunov exponent exists, the attractor is verified as a chaotic attractor. This procedure is also used to calculate the transverse Lyapunov exponents of synchronized systems, which are discussed in Subsection 9.4.1

When the dimension of the required phase space is too large to be handled by this matrix method in practice, another method can be used to calculate only the largest Lyapunov exponent. It is based on the concept that the largest Lyapunov exponent usually dominates the divergence [13, 18]. Therefore, the magnitude of the deviation $\mathbf{e}$ can be described as

$$|\mathbf{e}(k+L)| \approx |\mathbf{e}(k)| \exp(\overline{\lambda}_1 L). \tag{9.89}$$

Therefore, $\overline{\lambda}_1$ can be calculated as

$$\overline{\lambda}_1 \approx \frac{1}{L} \ln \left| \frac{\mathbf{e}(k+L)}{\mathbf{e}(k)} \right|. \tag{9.90}$$

In general, $\mathbf{e} = \mathbf{x} - \mathbf{x}_0$ in this method is calculated through the reconstruction of $\mathbf{x}_0$ from the time series of a single dynamical variable of the system. For example, the dynamical variable $x_i(t)$ as a function of time is chosen to construct the vector $\mathbf{x}_0$. This is achieved by assigning $x_i(t)$ as $y_1(t)$, $x_i(t + \tau_s)$ as $y_2(t)$, $x_i(t + 2\tau_s)$ as $y_3(t)$, and so forth until the whole set of $\{y_i(t), i = 1, \cdots, d\}$ can represent the dynamics of the attractor. The vectors $\mathbf{x}_0$ are then constructed as $\mathbf{x}_0(i) \equiv \mathbf{x}_0(t_i) = \{y_1(t_i), y_2(t_i), \ldots, y_{d-1}(t_i), y_d(t_i)\}$. The phase space expanded by $\{y_i(t), i = 1, \ldots, d\}$ is recognized as the embedding space of the attractor, and d is the dimension of the embedding space. Whether $\{y_i(t), i = 1, \ldots, d\}$ can represent the entire attractor or not depends on the proper choice of the time lag τ_s and the embedding dimension d. After the attractor is reconstructed in the embedding space, we can choose a point $\mathbf{x}_0(i)$ as a center, and a set of points $\mathbf{x}_0(j)$ within a small shell centered at $\mathbf{x}_0(i)$ as $\mathbf{x}$. Thus, the deviation $\mathbf{e} = \mathbf{x} - \mathbf{x}_0$ can be defined, and the magnitude of $\mathbf{e}$ is measured as the size of the shell. The set $\mathbf{x}$ is called the nearest neighboring points. Detailed discussions can be found in [13, 18].

The deviation $\mathbf{e}(k)$ can also be generated by a perturbation along the transmitter trace in the numerical simulation program. By calculating the average of the convergence or divergence of the perturbation, the largest Lyapunov exponent can be obtained. This method is very similar to the one discussed above for the calculation of the largest Lyapunov exponent. The difference is that this method generates the nearest neighboring points by the perturbation.

9.3.2 Geometric Structure Measured by Dimension of Chaotic Attractor

Another method to characterize a chaotic attractor is to characterize its geometric structure measured by dimensions. In general, a chaotic attractor is also a strange attractor, which means its geometric structure is fractal. There are several methods to characterize the geometric structure of a chaotic attractor. Here we discuss only the correlation dimension, which has been found to be a good characterization of a chaotic dynamical state [19, 20]. This correlation dimension is denoted by D_2. When D_2 is high, the complexity of the chaotic state is high.

To calculate D_2, we first define the correlation integral $C(N, r)$ of a dynamical state as [19]

$$C(N, r) = \frac{1}{N^2} \sum_{i,j=1}^{N} \theta(r - \|X_i - X_j\|), \qquad (9.91)$$

where $\theta(\cdot)$ is the Heaviside step function, X_i and X_j are vectors constructed in an embedding space from the time series of a single or several dynamical variables of the system to represent the attractor, N is the total number of the vectors, and r is a prescribed small distance.

The correlation dimension is obtained through the local slope of $\log C(N, r)$ calculated by

$$\nu(r_i) = \frac{\log C(N, r_{i-1}) - \log C(N, r_{i+1})}{\log r_{i-1} - \log r_{i+1}}. \qquad (9.92)$$

The value of $\nu(r_i)$ at the position where $\nu(r_i)$ shows a plateau provides the value of the correlation dimension D_2. This method has been proven to be efficient and has been implemented for different dynamical systems [13].

9.3.3 Other Signatures

Because most of the data collected for a dynamical system are contaminated by noise in a real world, it is usually difficult to verify from a single signature if the dynamics of a system is chaos or just a diffusion process. Several pieces of evidence have to be collected to make a correct judgment. Besides the Lyapunov exponents and the correlation dimension, a chaotic state can also be verified through the route to chaos. The common routes to chaos are the period-doubling route to chaos, the quasiperiodicity route to chaos, and the intermittency route to chaos. In addition, we can also check the frequency spectrum of a chaotic attractor to obtain another piece of evidence for confirming a chaotic state. Using all these signatures to verify a chaotic state is demonstrated in the following example.

9.3.4 Example: Chaos in an Optically Injected Semiconductor Laser

Here we use the chaotic state of the optical injection system operated at $\xi = 0.03$ and $f = 0$ GHz to demonstrate the use of the signatures. The route to this chaotic state by varying the detuning frequency is shown in the bifurcation diagram in Figure 9.1, which is shown in gray in Figure 9.4. This plot shows that the system dynamics evolves from a fixed point at $f = -4$ GHz through period-one, period-two, and period-four states, as the detuning frequency f is varied, and finally to the chaotic state at $f = 0$ GHz.

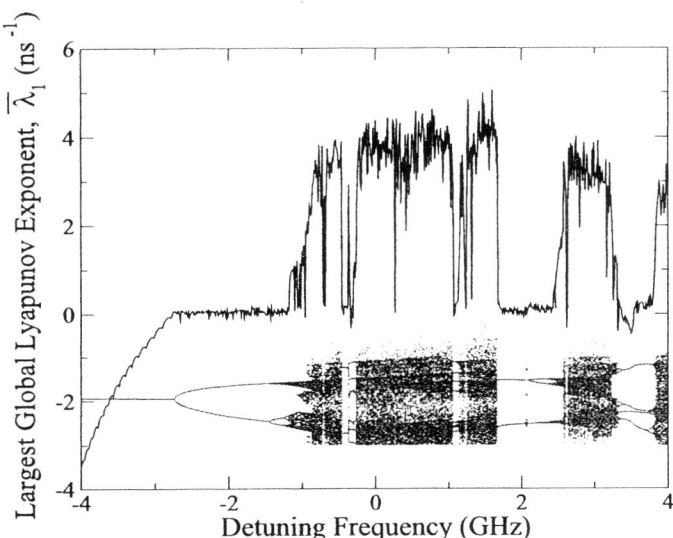

Fig. 9.4. Largest global Lyapunov exponent of the dynamical states as a function of detuning frequency ranging from $f = -4$ GHz to $f = 4$ GHz with the injection factor being $\xi = 0.03$. The gray line is the corresponding bifurcation diagram shown in Figure 9.1a.

Based on the procedure discussed in Subsection 9.3.1, the largest global Lyapunov exponent, $\overline{\lambda}_1$, is obtained and shown by the solid line in Figure 9.4. Ideally, fixed points should have global Lyapunov exponents being all negative, periodic states should have $\overline{\lambda}_1$ being zero, and chaotic states should have at least one positive Lyapunov exponent. However, it is usually very difficult to calculate the Lyapunov exponents into the desired accuracy, say exact $\overline{\lambda}_1 = 0$ for periodic states. Besides, Lyapunov exponents are not dimensionless, but are inversely proportional to the time scale. The numerical value of a Lyapunov exponent changes with time unit though the characteristic time of the dynamics is usually a good choice. Thus, it is usually not conclusive to announce a state being chaos based solely on the existence of a positive largest Lyapunov exponent, especially when the time scale is around the characteristic time of the dynamics, and the numerical value is closed to zero but is positive. As is shown in Figure 9.4, the fixed points have negative $\overline{\lambda}_1$ when they are not next to the periodic states. The periodic states judged from the bifurcation diagram have $\overline{\lambda}_1$ not exact zero but around zero, which depend on the time unit. This judgment can be further confirmed by the fact that $\overline{\lambda}_1$ is nearly a constant for all the periodic states based on the bifurcation diagram. Therefore, by combining the bifurcation diagram and the Lyapunov exponents, we can say that the dynamical states with $\overline{\lambda}_1$ significantly larger than the constant part of the solid curve shown in Figure 9.4 are chaotic, including the state with $\xi = 0.03$ and $f = 0$ GHz. The chaotic characteristic of this state can also be further examined by its optical spectrum shown in Figure 9.2c, which shows the broadband characteristic of chaos. This provides an additional piece of evidence to verify this chaotic state.

We can also examine the geometric structure of this chaotic state by calculating its correlation dimension D_2. As is mentioned in Section 9.3.2, the correlation dimension is determined by the plateau of $\nu(r_i)$. The $\nu(r_i)$ curves marked with different embedding dimensions d are shown in Figure 9.5a. When d is smaller than D_2, the plateau parts of the curves saturate to d, as is seen in the curves with $d = 1$ and $d = 2$ in Figure 9.5a. When d is larger than D_2, the plateau parts of the curves approximately stay as a constant and do not increase with d, as is seen in the curves with $d \geq 3$ in Figure 9.5a. Therefore, the correlation dimension of the chaotic attractor can be obtained from the common flat part of all the slope curves with $d \geq 3$ in Figure 9.5a. From Figure 9.5a, the correlation dimension of the dynamical state for $R_{\mathrm{sp}} = 0$ in the absent of noise is measured to be around $D_2 \simeq 2.4$.

The plateau is a very important characteristic for the determination of the correlation dimension from these curves. Because the structure of a chaotic attractor is generally fractal, the structure should remain unchanged when one zooms in or zooms out. This is the reason to have the flat part, where the slope of the correlation integral is independent of the correlation scale. However, when the scale being considered is so small that it reaches the small-scale uniformity limited by the device accuracy, the fractal structure disappears. The local slope of each curve then increases due to the contribution of all the noise, which includes the laser noise and the data truncation noise. On the other ends of all the curves, where the scale being considered is so large that the measured range for calculating the correlation dimension does not only cover the local structure but also covers remote structures, the local slope becomes irregular. It can then either increase or decrease, depending on the specific global property of the attractor. Further increasing the scale will make the chaotic attractor indistinguishable from a single point. At this large scale, we obtain the data showing that the dimension of the chaotic attractor is zero for a single point. All these characteristics can be observed in Figure 9.5a. Therefore, the correlation dimension is obtained only from the flat parts of all curves.

When there is noise in the system, the attractor is contaminated by the noise. The $\nu(r_i)$ curves contaminated by the laser noise measured by $R_{\mathrm{sp}} = 4.7 \times 10^{18}$ $\mathrm{V^2 m^{-2} s^{-1}}$ and $R_{\mathrm{sp}} = 2 \times 4.7 \times 10^{18}$ $\mathrm{V^2 m^{-2} s^{-1}}$ are shown in Figures 9.5b and c, respectively. As is seen, the noise increases the plateau level of each $\nu(r_i)$ curves with $d \geq 3$ because the dimension of the noise is infinity.

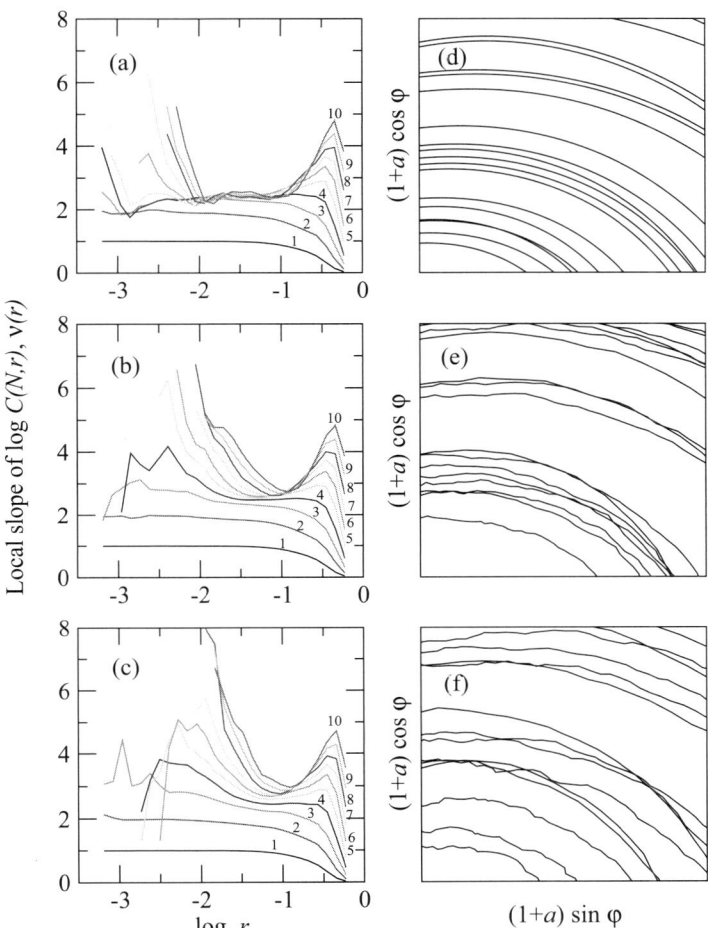

Fig. 9.5. Correlation dimension as a function of embedding dimension for the chaotic attractor at $\xi = 0.03$ and $f = 0$ GHz with different levels of noise: (a) $\nu(r_i)$ curves with $R_{sp} = 0$; (b) $\nu(r_i)$ curves with $R_{sp} = 4.7 \times 10^{18}$ V^{2}m^{-2}s^{-1}; (c) $\nu(r_i)$ curves with $R_{sp} = 2 \times 4.7 \times 10^{18}$ V^{2}m^{-2}s^{-1}; (d), (e), and (f) are the portions of the corresponding attractors, respectively. Each $\nu(r_i)$ curve is marked by the value of its embedding dimension.

Therefore, the common plateau of the $\nu(r_i)$ curves with $d \geq 3$ disappears, and it becomes difficult to evaluate D_2. The local enlargement of the attractors for $R_{\mathrm{sp}} = 4.7 \times 10^{18}$ $\mathrm{V}^2\mathrm{m}^{-2}\mathrm{s}^{-1}$ and $R_{\mathrm{sp}} = 2 \times 4.7 \times 10^{18}$ $\mathrm{V}^2\mathrm{m}^{-2}\mathrm{s}^{-1}$ is shown in Figures 9.5e and f, and that of the attractor for $R_{\mathrm{sp}} = 0$ is shown in Figure 9.5d for comparison. The whole attractors with $R_{\mathrm{sp}} = 0$ and $R_{\mathrm{sp}} = 4.7 \times 10^{18}$ $\mathrm{V}^2\mathrm{m}^{-2}\mathrm{s}^{-1}$ are shown in Figure 9.2b and e, respectively. The effect of the noise with different strength can be observed from the fluctuating traces of the attractor shown in Figures 9.5e and f.

When a dynamical system is contaminated by noise, the verification of its dynamical state becomes more complicated and more difficult. The concept of using several pieces of evidence to verify a chaotic state becomes even more important. The route to chaos provides the unique characteristic of a chaotic system. The bifurcation diagram provides the characteristic of the waveform. The Lyapunov exponents provide the aspect of the sensitivity to the initial condition. The optical spectrum provides the characteristic of the broadband. The correlation dimension D_2 provides the characteristic of the geometric structure. All of these signatures are demonstrated in this example.

9.4 Robustness of Chaos Synchronization

The approach to synchronizing chaos considered here is based on the concept described by Kocarev and Parlitz [23]. For two chaotic systems physically connected by a signal $\mathbf{s}(t)$, each system can be rewritten mathematically as if the system were driven by a common driving signal $\mathbf{D}(\mathbf{s}(t))$, and can be expressed as the following,

$$\frac{d\mathbf{x}}{dt} = \mathbf{F}(\mathbf{x}, \mathbf{D}(\mathbf{s}(t))) \ , \tag{9.93}$$

$$\frac{d\mathbf{y}}{dt} = \mathbf{G}(\mathbf{y}, \mathbf{D}(\mathbf{s}(t))) \ , \tag{9.94}$$

where $\mathbf{D}(\mathbf{s}(t))$ is a function of the signal $\mathbf{s}(t)$. With the vector function $\mathbf{G}$ being equal to the vector function $\mathbf{F}$, these two chaotic systems can be synchronized to each other if the difference $\mathbf{e} = \mathbf{x} - \mathbf{y}$ possesses a fixed point with zero value. This fixed point exists when the average local Lyapunov exponents of the difference $\mathbf{e}$ are all negative. These average local Lyapunov exponents are also called the average local transverse Lyapunov exponents of the synchronized attractors.

9.4.1 Transverse Lyapunov Exponents in the Case of Perfect Parameter Match

Now, we examine the robustness of chaos synchronization by considering the transverse Lyapunov exponents. For illustration purposes, only the additive driving signal is considered here. In this subsection, only chaos synchronization with perfectly matched parameters is considered, for which $\mathbf{G} = \mathbf{F}$. This is an ideal situation that serves the purpose of examining the existence of chaos synchronization.

For synchronization achieved by coupling through an additive driving signal, the general equations (9.93) and (9.94) of the synchronization theory can be rewritten as

$$\frac{d\mathbf{x}}{dt} = \mathbf{F}(\mathbf{x}) + \alpha\mathbf{D}(\mathbf{x}), \tag{9.95}$$

$$\frac{d\mathbf{y}}{dt} = \mathbf{G}(\mathbf{y}) + \alpha\mathbf{D}(\mathbf{x}), \tag{9.96}$$

where $\alpha\mathbf{D}(\mathbf{x})$ has replaced the driving signal $\mathbf{s}(t)$ because this special type of driving force is more suitable when the chaos synchronization system is designed for the communication purpose, and the coupling strength α has been separated from $\mathbf{s}(t)$. By defining $\mathbf{f}(\mathbf{x}) \equiv \mathbf{F}(\mathbf{x}) + \alpha\mathbf{D}(\mathbf{x})$ and $\mathbf{g}(\mathbf{y}) \equiv \mathbf{G}(\mathbf{y}) + \alpha\mathbf{D}(\mathbf{y})$, we obtain

$$\frac{d\mathbf{x}}{dt} = \mathbf{F}(\mathbf{x}) + \alpha\mathbf{D}(\mathbf{x}) = \mathbf{f}(\mathbf{x}) \ , \tag{9.97}$$

$$\begin{aligned}\frac{d\mathbf{y}}{dt} &= \mathbf{G}(\mathbf{y}) + \alpha\mathbf{D}(\mathbf{x}) \\ &= \mathbf{G}(\mathbf{y}) + \alpha\mathbf{D}(\mathbf{y}) + \alpha\mathbf{D}(\mathbf{x}) - \alpha\mathbf{D}(\mathbf{y}) \\ &= \mathbf{g}(\mathbf{y}) + \alpha\mathbf{D}(\mathbf{x}) - \alpha\mathbf{D}(\mathbf{y}) \ . \end{aligned} \tag{9.98}$$

When all the parameters are matched, $\mathbf{G} = \mathbf{F}$ and $\mathbf{g} = \mathbf{f}$.

The equations given in (9.97) and (9.98) are the general equations for most of the proposed setups utilizing an additive driving signal to achieve chaos synchronization for the communication purpose. The first equation describes the chaotic dynamics of the transmitter. The second equation describes the synchronization dynamics of the receiver. Now the separation between the traces $\mathbf{x}(t)$ and $\mathbf{y}(t)$ is defined as $\mathbf{e} = \mathbf{y} - \mathbf{x}$. Then the equation describing the synchronization dynamics can be written as the following,

$$\begin{aligned}\frac{d\mathbf{e}}{dt} &= \mathbf{f}(\mathbf{x} + \mathbf{e}) - \mathbf{f}(\mathbf{x}) - \alpha[\mathbf{D}(\mathbf{x} + \mathbf{e}) - \mathbf{D}(\mathbf{x})] \\ &= \left[\frac{\partial\mathbf{f}}{\partial\mathbf{x}}(\mathbf{x}) - \alpha\frac{\partial\mathbf{D}}{\partial\mathbf{x}}\right] \cdot \mathbf{e} \ . \end{aligned} \tag{9.99}$$

By defining μ as the eigenvalues of the Jacobian $\left[\dfrac{\partial\mathbf{f}}{\partial\mathbf{x}}(\mathbf{x}) - \alpha\dfrac{\partial\mathbf{D}}{\partial\mathbf{x}}\right]$, we obtain the transverse Lyapunov exponents, denoted as λ_{T}, as the real part of μ:

$$\lambda_{\mathrm{T}} = \mathrm{Re}(\mu). \tag{9.100}$$

The transverse Lyapunov exponents can be obtained through the method discussed in Section 9.3. The transverse Lyapunov exponents so obtained are called the average transverse Lyapunov exponents. One can also use the perturbation method to calculate the largest average Lyapunov exponents.

9.4.2 Transverse Lyapunov Exponents in the Presence of Parameter Mismatch

In practice with real physical systems, it is not possible to achieve chaos synchronization without any parameter mismatch. Therefore, it is important to discuss the method of calculating the transverse Lyapunov exponents when the system parameters are mismatched. Instead of using (9.97) and (9.98), we use another set of equations, in which $\mathbf{g} \neq \mathbf{f}$, to describe the synchronization when the parameters of the transmitter and the receiver are not perfectly matched:

$$\frac{d\mathbf{x}}{dt} = \mathbf{f}(\mathbf{x}), \tag{9.101}$$

$$\frac{d\mathbf{y}}{dt} = \mathbf{g}(\mathbf{y}) + \alpha \mathbf{D}(\mathbf{x}) - \alpha \mathbf{D}(\mathbf{y}). \tag{9.102}$$

In this case, because the function describing the dynamics of the transmitter is not the same as that describing the dynamics of the receiver, the stability of synchronization is no longer described by the deviation between the transmitter trace and the receiver trace in the coupled phase space. A new concept has to be invented.

We propose the following concept to deal with this situation. Instead of considering the stability of synchronization as the deviation between the transmitter dynamics and the receiver dynamics, we consider the synchronization stability as the deviation between the receiver trace and its nearby traces, using the receiver trace attracted by the transmitter trace in their phase space of the synchronized dynamics as the reference. Therefore, the deviation vector, $\mathbf{e}(t)$, is defined in another manner as

$$\mathbf{e}(t) = \mathbf{y}(t) - \mathbf{y}_0(t), \tag{9.103}$$

where $\mathbf{y}_0(t)$ is the original trace of the receiver synchronized to the transmitter, and $\mathbf{y}(t)$ is the perturbed trace of the receiver. Therefore, the proposed concept can be quantified to describe the synchronization stability in the presence of parameter mismatch:

$$\frac{d\mathbf{e}}{dt} = \left[\frac{\partial \mathbf{g}}{\partial \mathbf{y}}(\mathbf{y}_0) - \alpha \frac{\partial \mathbf{D}}{\partial \mathbf{y}}(\mathbf{y}_0) \right] \cdot \mathbf{e}. \tag{9.104}$$

If the function $\mathbf{g}(\mathbf{y})$ is replaced by the function $\mathbf{f}(\mathbf{y})$, this equation is identical to (9.99).

We can use the same procedure described in Subsection 9.3.1 to implement this concept to find the transverse Lyapunov exponents. Alternatively, we can generate a small perturbation in the numerical program on the receiver trace and then calculate the time evolution of the perturbation. However, we can only obtain the largest transverse Lyapunov exponent from the latter method, which is actually adequate for the stability analysis.

The concept proposed here for analyzing the synchronization stability when the system parameters are mismatched can be considered as a special case of the generalized chaos synchronization proposed by Kocarev and Parlitz [24].

9.4.3 Evaluation of the Quality of Synchronization

The simplest and most intuitive method to quantitatively measure the quality of synchronization is to calculate the difference between the traces of the transmitter output and the receiver output as $\mathbf{e}(t) = \mathbf{y}(t) - \mathbf{x}(t)$. For the practical purpose of chaotic communications, this synchronization error is normalized to the size of its chaotic attractor. Although there has not been a standard definition of synchronization error, a common one is defined as the following,

$$\zeta = \frac{\langle |X(t) - Y(t)| \rangle}{\langle |X(t)| \rangle}. \tag{9.105}$$

When the synchronization error ζ is small, the quality of synchronization is high. Due to its simplicity, this direct measurement of the quality of synchronization is widely used in the numerical simulation of chaos synchronization systems.

In an experiment, however, the nonsimultaneous digitization of experimental data does not allow such direct and precise comparison of the transmitter and receiver outputs to provide a convincing result. A measurement method that is not very sensitive to the digitization error is then more desirable. A common concept known as correlation coefficient is thus usually used for estimating the quality of synchronization of real systems. The correlation coefficient is defined as the following [15],

$$\rho = \frac{\langle [X(t) - \langle X(t) \rangle][Y(t) - \langle Y(t) \rangle] \rangle}{\langle |X(t) - \langle X(t) \rangle|^2 \rangle^{1/2} \langle |Y(t) - \langle Y(t) \rangle|^2 \rangle^{1/2}}, \tag{9.106}$$

where $X(t)$ and $Y(t)$ are the outputs of the transmitter and the receiver, respectively, and $\langle \cdot \rangle$ denotes the time average. The correlation coefficient is bounded as $-1 \leq \rho \leq 1$. A larger value for $|\rho|$ means a higher quality of synchronization. Instead of quantifying the quality of synchronization using the synchronization error, this correlation coefficient measures the similarity of the two attractors.

9.4.4 Example: Synchronization of Optically Injected Semiconductor Lasers

We now use a single-mode semiconductor laser subject to optical injection as an example to demonstrate the analysis of the robustness and the quality of the chaos synchronization. Detailed discussions of the chaos synchronization of this system can be found in Subsection 10.4.1 of Chapter 10, and the configuration can be found in Figure 10.15. Here we concentrate on the mathematical analysis of the robustness and quality of synchronization.

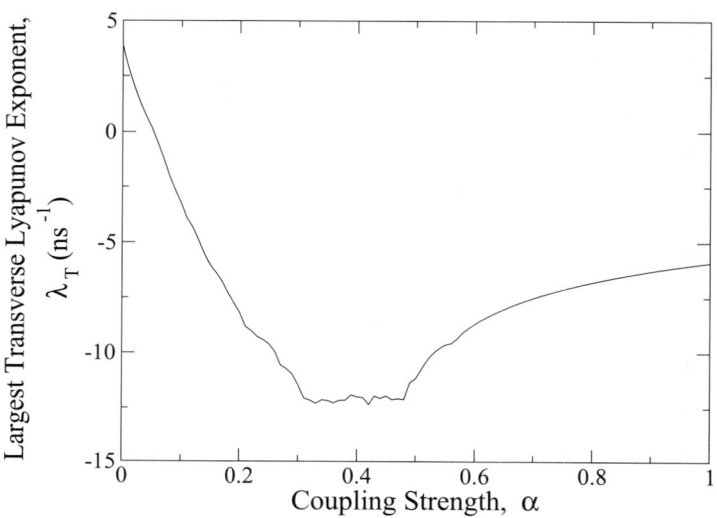

Fig. 9.6. Largest global transverse Lyapunov exponent of chaos synchronization as a function of the coupling strength. The chaotic state is generated at $\xi = 0.03$ and $f = 0$ GHz. (Reprinted with permission from [22], ©2000 IEEE.)

The dynamics of a semiconductor laser subject to the injection of an optical field is mathematically modeled by the following coupled equations [11],

$$\frac{dA^{\mathrm{T}}}{dt} = -\frac{\gamma_{\mathrm{c}}^{\mathrm{T}}}{2}A + i(\omega_0^{\mathrm{T}} - \omega_{\mathrm{c}}^{\mathrm{T}})A^{\mathrm{T}} + \frac{\Gamma}{2}(1 - ib^{\mathrm{T}})gA^{\mathrm{T}}$$
$$+ \eta A_{\mathrm{i}}\exp(-i\Omega t), \tag{9.107}$$

$$\frac{dN^{\mathrm{T}}}{dt} = \frac{J}{ed} - \gamma_{\mathrm{s}}N^{\mathrm{T}} - \frac{2\epsilon_0 n^2}{\hbar\omega_0}g|A^{\mathrm{T}}|^2, \tag{9.108}$$

as is expressed in (9.46) and (9.47). The superscript T labels the variables of the transmitter. The receiver, driven by the transmitted signal $\alpha A^{\mathrm{T}}(t) + A_{\mathrm{i}}\exp(-i\Omega t + \theta)$ with $A_{\mathrm{i}}\exp(-i\Omega t + \theta)$ being the optical injection signal, is modeled by the following equations,

$$\frac{dA^{\mathrm{R}}}{dt} = -\frac{\gamma_{\mathrm{c}}^{\mathrm{R}} - 2\alpha\eta}{2}A^{\mathrm{R}} + i(\omega_0^{\mathrm{T}} - \omega_{\mathrm{c}}^{\mathrm{T}} + \Delta\omega_{\mathrm{c}})A^{\mathrm{R}} + \frac{\Gamma}{2}(1 - ib^{\mathrm{R}})g^{\mathrm{R}}A^{\mathrm{R}}$$

$$+\eta A_i \exp(i\Omega t + i\theta) + \alpha\eta(A^T - A^R), \tag{9.109}$$

$$\frac{dN^R}{dt} = \frac{J^R}{ed} - \gamma_s^R N^R - \frac{2\epsilon_0 n^2}{\hbar\omega_0^T} g^R |A^R|^2, \tag{9.110}$$

where the superscript R labels the variables of the receiver, ω_c is the longitudinal mode frequency of the cold laser cavity, $\Delta\omega_c = \omega_c^T - \omega_c^R$ is the difference between the cold-cavity frequencies of the transmitter and the receiver, α is the coupling strength of the transmitter output to the receiver, and θ is the relative optical phase difference [22]. The definitions of all other parameters can be found in Subsection 7.7.4. Based on the synchronization concept proposed by Kocarev and Parlitz [23], the existence of the perfect synchronization solution, $A^R = A^T$, requires $\theta = 0$, $\Delta\omega_c = 0$, and that the two lasers be identical except that $\gamma_c^R = \gamma_c^T + 2\eta\alpha$ [22].

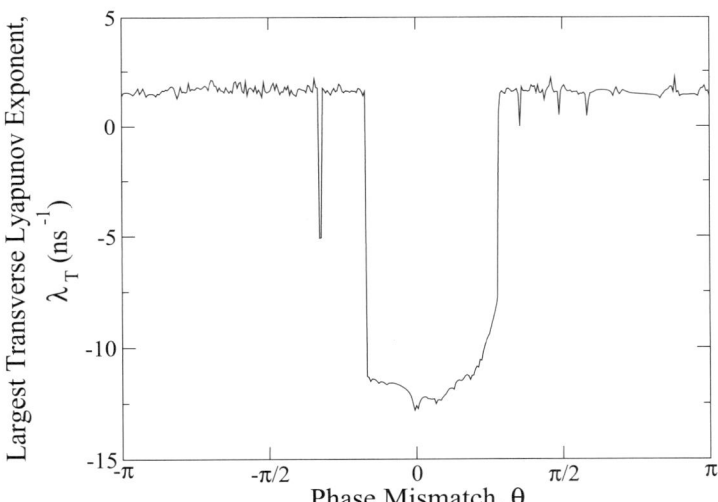

Fig. 9.7. Largest global transverse Lyapunov exponent of chaos synchronization as a function of the phase mismatch. The chaotic state is generated at $\xi = 0.03$ and $f = 0$ GHz.

The largest global transverse Lyapunov exponent, λ_T, of the coupled system with different coupling strength, α, is shown in Figure 9.6. The value for each α is calculated by assigning the matrix **DF** in (9.67) as

$$\mathbf{DF} = \left[\frac{\partial \mathbf{f}}{\partial \mathbf{x}}(\mathbf{x}) - \alpha\frac{\partial \mathbf{D}}{\partial \mathbf{x}}\right] \tag{9.111}$$

and following the procedure described in Section 9.3.1. It is important to know that λ_T being negative does not guarantee that the local transverse Lyapunov exponent at each segment of the synchronized trace is negative. Therefore, the Lyapunov exponent begins to be negative around $\alpha = 0.06$, but robust synchronization will not occur until α is larger than 0.15 [22]. We can see that an increase in the coupling strength does not necessarily result in an increase in the robustness of synchronization. This phenomenon has been observed in other systems [25]. The most robust synchronization occurs when the value of the coupling strength index is around $\alpha = 0.4$. This coupling strength is used when we calculate λ_T for the system operated under the condition of the phase mismatch.

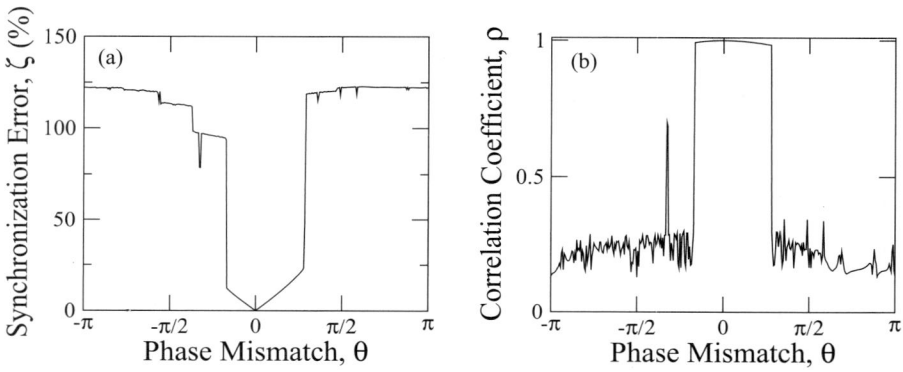

Fig. 9.8. Quality of chaos synchronization as a function of the phase mismatch θ measured by (a) the synchronization error η and (b) the correlation coefficient ρ. The chaotic state is generated at $\xi = 0.03$ and $f = 0$ GHz.

The effect of the mismatch in many parameters on the robustness and quality of synchronization has been studied for this system [22]. Here we consider only the phase mismatch θ for the demonstration of the numerical methods discussed above. The dependence of λ_T on the phase mismatch θ is shown in Figure 9.7. The value of λ_T is calculated in this situation by assigning the matrix $\mathbf{DF}$ in (9.67) as

$$\mathbf{DF} = \left[\frac{\partial \mathbf{g}}{\partial \mathbf{y}}(\mathbf{y}) - \alpha \frac{\partial \mathbf{D}}{\partial \mathbf{y}} \right] \tag{9.112}$$

and then following the procedure described in Section 9.3.1. As shown, the tolerance of the chaos synchronization to the phase mismatch is in a range of above $\pi/2$ around $\theta = 0$. It is important to check if this characteristic is consistent with the quality of synchronization. The quality measured by ζ and

ρ is shown in Figures 9.8a and b, respectively. As is shown, this tolerance can also be observed by examining the quality of the synchronization. A comparison between the curves of λ_T, ζ, and ρ demonstrates the consistency of these measures.

9.5 Chaotic Communications

The standard performance measure of a communication system is the BER for the decoded message as a function of the channel signal-to-noise ratio (SNR) in the transmission channel. The channel SNR is defined as

$$\text{SNR} = \frac{P_{\text{m}}}{\sigma_{\mathbf{X}}^2}, \tag{9.113}$$

where P_{m} is the power of the transmitted message, and $\sigma_{\mathbf{X}}^2$ is the variance of the channel noise $\mathbf{X}(t)$. The channel SNR is a function of the channel noise, which is taken to be additive white Gaussian noise, and the bit energy of the transmitted message, which depends on the modulation index of the message.

The BER of the decoded message is a function of the channel noise and the intrinsic noise of the devices in the system, both of which cause synchronization error between the receiver and the transmitter. In a chaotic communication system utilizing chaos synchronization, the synchronization error is caused mainly by the noise, including the channel noise and the transmitter and receiver noise, the message encoding process, and the parameter mismatch between the transmitter and the receiver. The synchronization error is contributed by two forms of error: synchronization deviation, associated with the accuracy of synchronization, and desynchronization bursts, associated with the robustness of synchronization. Synchronization deviation is simply the synchronization error when the system is synchronized, but not perfectly and precisely. Desynchronization bursts are characterized by sudden desynchronization between the transmitter and the receiver. A desynchronization burst can cause a large, abrupt difference between the waveforms of the transmitter and the receiver. Because a system takes some finite time to resynchronize after a desynchronization burst, the bits that follow a desynchronization burst within the resynchronization time are destroyed.

9.5.1 Numerical Algorithm for Treating Channel Noise

In the example of simulating the laser dynamics, only the intrinsic noise of the semiconductor laser system is considered. To extend the numerical analysis to the system performance, the numerical method for treating white Gaussian channel noise is necessary and nontrivial. In order to integrate the channel noise into the dynamics of the receiver, the randomness of the channel noise has to be considered. As is discussed in the simulation of a stochastic differential equation, it is not correct to calculate the channel signal first and then apply the channel signal as a deterministic function of time to the differential equation that describes the dynamics of the receiver. Besides, the channel signal has to be evaluated before it is injected into the receiver in order to include the linear effect of the channel noise. This requires that the effect of the channel noise on system performance be the same as that in a traditional communication system when the nonlinear effect of the channel noise is not considered. The consistency between the linear and the nonlinear effects of channel noise must be satisfied.

The numerical method discussed here is based on the assumption that the channel signal is directly coupled into the receiver at a constant coupling rate. Based on this assumption, the equation of the receiver can be obtained by adding white Gaussian channel noise in (9.102) as the following,

$$\frac{d\boldsymbol{y}}{dt} = \mathbf{g}(\boldsymbol{y}) + \alpha \boldsymbol{D}(\boldsymbol{x}) - \alpha \boldsymbol{D}(\boldsymbol{y}) + \boldsymbol{N}(\boldsymbol{X}(t), \boldsymbol{y}; \eta), \qquad (9.114)$$

where $\boldsymbol{N}(\cdot)$ is a general function regarding the effect of channel noise on the receiver, η is the coupling rate, and $\boldsymbol{X}(t) = (X_1(t), \ldots, X_k(t))$ is a generalized k-dimensional white Gaussian channel noise. Each component of $\boldsymbol{X}(t)$ can be expressed as

$$X_i(t) = \sqrt{\frac{N_i}{2}} n_i(t), \qquad (9.115)$$

where $N_i/2$ is the power spectral density of $X_i(t)$, and $n_i(t)$ is a normalized white Gaussian variable with a zero mean, which satisfies (9.2) and (9.3). The variance of $X_i(t)$ is indicated by σ_i^2. Because all of $X_i(t)$ are independent of each other, the variance σ_X^2 of the entire channel noise $\boldsymbol{X}(t)$ with $\{\sigma_i^2, \ i = 1, \ldots, k\}$ has the following relationship,

$$\sigma_X^2 = \sum_{i=1}^{k} \sigma_i^2, \qquad (9.116)$$

and the power spectral density $N_{\boldsymbol{X}}/2$ of $\boldsymbol{X}(t)$ with $\{N_i/2, \ i = 1, \ldots, k\}$ has the following relationship,

$$\frac{N_{\boldsymbol{X}}}{2} = \sum_{i=1}^{k} \frac{N_i}{2}. \qquad (9.117)$$

Because the dynamics of the transmitter is not affected by channel noise, it is not important in this discussion. Therefore, we only focus on the dynamics of the receiver when the white Gaussian channel noise is considered.

In order to simplify the simulation of stochastic differential equations as is discussed in Subsection 9.2.3, it is preferred that $\boldsymbol{N}(\boldsymbol{X}(t), \boldsymbol{y}; \eta)$ is not a function of $\boldsymbol{y}$. Thus, the dynamical variables of the receiver should be chosen so that $\boldsymbol{N}(\boldsymbol{X}(t), \boldsymbol{y}; \eta) = C\eta \boldsymbol{X}(t)$, where C is a constant generated from the choice of $\boldsymbol{y}$. When this objective is accomplished, each dynamical variable of the receiver is governed by the equation expressed as:

$$\frac{dy_i}{dt} = g_i(\boldsymbol{y}) + \alpha D_i(\boldsymbol{x}) - \alpha D_i(\boldsymbol{y}) + C\tilde{\eta}\sqrt{\frac{N_i}{2}}n_i(t). \tag{9.118}$$

Therefore, the dynamics of the receiver affected by the white Gaussian channel noise can be simulated by following the discussion in Section 9.2 once the value of $N_i/2$ is known. This equation calculates the nonlinear effect of channel noise on the quality of synchronization and thus message decoding.

However, the channel noise has to be evaluated before it is injected into the receiver because its linear effect has to match that in the traditional communication systems. To generate $\boldsymbol{X}(t)$, its components $X_i(t)$ are paired and generated through the Box−Muller method [26] as

$$X_{2j-1} = \sqrt{-2\sigma_{2j-1}^2 \ln a_j} \cdot \cos(2\pi b_j), \tag{9.119}$$

$$X_{2j} = \sqrt{-2\sigma_{2j}^2 \ln a_j} \cdot \sin(2\pi b_j), \tag{9.120}$$

where $j = 1, \ldots, [k/2]_{\text{int}}$ with $[\cdot]_{\text{int}}$ defined as the nearest integer smaller than the value in the bracket, and the pair of a_j and b_j are independent random variables evenly distributed in the interval $(0, 1]$. The a_j and b_j variables for a given j value are independent of each other, and each pair is also independent of other pairs of different j values. If the dimension k of $\boldsymbol{X}(t)$ is an odd number, the last unpaired $X_k(t)$ can be generated by (9.119) with $2j - 1 = k$. This evaluation calculates the linear effect of channel noise on message decoding.

It is important to ensure that the nonlinear effect of the channel noise is consistent to the linear effect. Because the nonlinear effect of the channel noise on the dynamics of the receiver is simulated by knowing the value of $N_i/2$ and the linear effect of the channel noise is evaluated by knowing the value of σ_i^2, a connection between $N_i/2$ and σ_i^2 has to be established. The consistent connection is built based on the twofold effect of this channel noise on the system performance in communication: When the nonlinear effect of the channel noise on the dynamics of the receiver is considered, the receiver will provide natural filtering on the bandwidth of the channel noise. In this situation, the bandwidth of the channel noise does not have to be predefined. When the effect of the channel noise on the dynamics of the receiver is not considered, the effective bandwidth of the channel noise is defined by the bandwidth of the encoding signal. This assumption is the same as the one widely used in a traditional communication system.

Based on this fact, the relationship between σ_i^2 and $N_i/2$ can be established as the following,

$$\sigma_i^2 = \frac{N_i}{2}(2W)$$

$$= \frac{N_i}{2}(f_{\text{m}})$$

$$= \frac{N_i}{2T_{\mathrm{b}}}, \tag{9.121}$$

where W is the resolving power and is equal to the half bit rate, and the bit rate f_{m} is equal to $1/T_{\mathrm{b}}$ with T_{b} being the bit duration. Thus, the SNR defined in (9.113) can be expressed as

$$\begin{aligned}
\mathrm{SNR} &= \frac{P_{\mathrm{m}}}{\sigma_{\mathbf{X}}^2} \\
&= \frac{E_{\mathrm{b}}/T_{\mathrm{b}}}{\sum_{i=1}^{k} \sigma_i^2} \\
&= \frac{E_{\mathrm{b}}}{\sum_{i=1}^{k} N_i/2},
\end{aligned} \tag{9.122}$$

where E_{b} is the energy per bit. Therefore, once we choose the values of E_{b}, T_{b} (or f_{m}), and SNR, the channel noise as a time series can be evaluated by knowing the variance $\{\sigma_i^2;\ i = 1, \ldots, k\}$, and the effect of the channel noise on the dynamics of the receiver can be simulated by knowing the values of $\{N_i;\ i = 1, \ldots, k\}$ and η.

9.5.2 Example: Chaotic Optical Communication Using Optically Injected Semiconductor Lasers

Here we use the optical injection system as an example to show the BER as a function of SNR obtained by this method [14]. The rate equation describing the dynamics of the transmitter is given by (9.107) and (9.108), which is not our main interest in this subsection. The rate equation of the receiver after receiving the channel signal including the white Gaussian channel noise is described as the following,

$$\frac{dA^{\mathrm{R}}}{dt} = -\frac{\gamma_{\mathrm{c}}}{2}A^{\mathrm{R}} + i(\omega_0^T - \omega_{\mathrm{c}}^{\mathrm{T}})A^{\mathrm{R}} + \frac{\Gamma}{2}(1 - ib^{\mathrm{R}})gA^{\mathrm{R}} + F_{\mathrm{sp}}^{\mathrm{R}}$$
$$+[1 + m(t)]\eta A_{\mathrm{i}}\exp(-i\Omega t) + \eta\alpha A^{\mathrm{T}}(t) - \eta\alpha A^{\mathrm{R}}(t) + \eta X(t) \tag{9.123}$$
$$\frac{dN^{\mathrm{R}}}{dt} = \frac{J}{ed} - \gamma_{\mathrm{s}}N^{\mathrm{R}} - \frac{2\epsilon_0 n^2}{\hbar\omega_0}g|A^{\mathrm{R}}|^2, \tag{9.124}$$

where $m(t)$ is the message and $X(t)$ is the optical white Gaussian channel noise. In this system, $X(t)$ is a complex noise term that can be considered as a two-dimensional white Gaussian variable $X(t) = (X_{\mathrm{r}}(t), X_{\mathrm{i}}(t)) \equiv (X_1(t), X_2(t))$, where $X_{\mathrm{r}}(t)$ is the real part and $X_{\mathrm{i}}(t)$ is the imaginary part of $X(t)$. The encoding message $m(t)$ used in the simulation has non-return-to-zero (NRZ) random digital bits "1" and "0", with the amplitude of "1" indicated by $m(t) = \varepsilon$ and that of "0" as $m(t) = 0$. The strength of the message can be varied by adjusting the parameter ε. The condition of perfect parameter matching is assumed.

As is discussed in Subsection 9.2.4, A^R is a complex field, which is a two-dimensional variable. In order to ensure that the channel noise term $N(X(t), y; \eta)$ in (9.114) is decomposed into the form in (9.118), the complex field A^R is decomposed into its real and imaginary parts as $A^R = |A_0|(a'^R + ia''^R)$. Together with $\tilde{n}^R$ defined by $N^R = N_0(1 + \tilde{n}^R)$, the coupled equations (9.123) and (9.124) are recast into the following form,

$$
\frac{da'^R}{dt} = \frac{1}{2}\left\{ \frac{\gamma_c \gamma_n}{\gamma_s \tilde{J}} \tilde{n}^R - \gamma_p\left[(a'^R)^2 + (a''^R)^2 - 1\right]\right\}(a'^R + ba''^R)
$$
$$
+ [1 + m(t)]\xi\gamma_c \cos(2\pi ft) + \frac{F_r}{|A_0|}
$$
$$
+ \alpha\eta a'^T - \alpha\eta a'^R + \eta\frac{X_1(t)}{|A_0|}, \tag{9.125}
$$

$$
\frac{da''^R}{dt} = \frac{1}{2}\left\{ \frac{\gamma_c \gamma_n}{\gamma_s \tilde{J}} \tilde{n} - \gamma_p\left[(a'^R)^2 + (a''^R)^2 - 1\right]\right\}(-ba'^R + a''^R)
$$
$$
- [1 + m(t)]\xi\gamma_c \sin(2\pi ft) + \frac{F_i}{|A_0|}
$$
$$
+ \alpha\eta a''^T - \alpha\eta a''^R + \eta\frac{X_2(t)}{|A_0|}, \tag{9.126}
$$

$$
\frac{d\tilde{n}^R}{dt} = -\gamma_s\tilde{n}^R - \gamma_n\tilde{n}^R\left[(a'^R)^2 + (a''^R)^2\right] - \gamma_s\tilde{J}\left[(a'^R)^2 + (a''^R)^2 - 1\right]
$$
$$
+ \frac{\gamma_s\gamma_p}{\gamma_c}\tilde{J}\left[(a'^R)^2 + (a''^R)^2 - 1\right]\left[(a'^R)^2 + (a''^R)^2\right]. \tag{9.127}
$$

As the result, the channel noise in (9.125) and (9.126) above have the same form as that in (9.118).

Because $X_1(t)$ and $X_2(t)$ have the same magnitude, $\sigma_X^2 = 2\sigma_1^2 = 2\sigma_2^2$ and $N_X = 2N_1 = 2N_2$. Thus, the SNR can be expressed as

$$
\text{SNR} = \frac{E_b/T_b}{2\sigma_1^2} = \frac{E_b}{N_1}
$$
$$
= \frac{E_b/T_b}{2\sigma_2^2} = \frac{E_b}{N_2}. \tag{9.128}
$$

Once the values of ε, T_b (or f_m), and SNR are determined, the values of E_b, σ_1^2, σ_2^2, N_1, and N_2 can be obtained through (9.128). Therefore, the dynamics of the receiver can be numerically simulated through (9.125)–(9.127) with the white Gaussian channel noise evaluated through (9.119) and (9.120).

To examine this method by using this example, it is important to know that the nonlinear effect of channel noise does not contribute to the BER of this laser system when the term $\eta X(t)$ in (9.126) is set to zero. This assumes that the channel noise does not inject into the receiver laser; thus the quality of synchronization is not affected by the channel noise. Therefore, when $\eta X(t) = 0$, the performance of this system can be directly compared to that

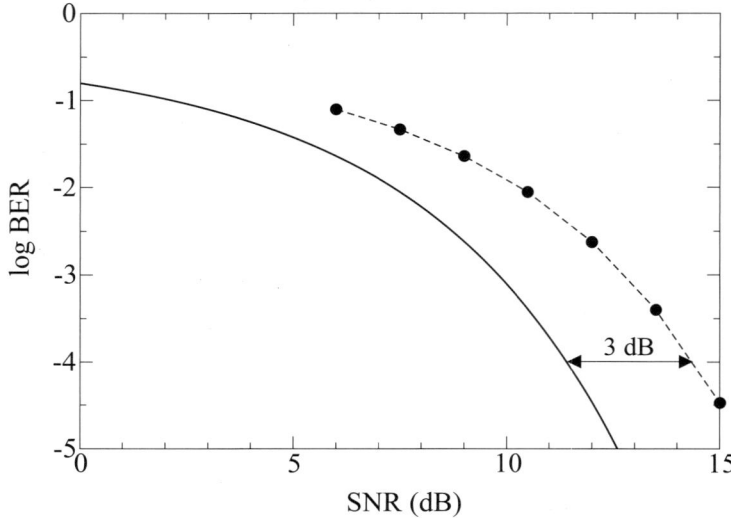

Fig. 9.9. Comparison between the BER as a function of SNR for BPSK and that for the optically injected semiconductor laser system when the channel noise is not injected into the receiver. The solid curve indicates the BER of BPSK, and the dashed curve marked with solid circles indicates the BER of the optical injection system when the nonlinear effect of the channel noise on the receiver is turned off numerically. The transmitter is operated at $\xi = 0.03$ and $f = 0$ GHz to generate the chaotic waveform as the message carrier. Perfect chaos synchronization is assumed.

of a traditional communication system. As an examination of this method, the BER of this laser system with $\eta X(t) = 0$ is compared to that of a traditional binary phase shift keying (BPSK) system. In Figure 9.9, the BER as a function of the SNR for BPSK is shown as the solid curve, and that for the optical injection system with $\eta X(t) = 0$ is shown as the dashed curve marked by solid circles. The solid circles show the value of BER obtained through the numerical simulation using (9.125)−(9.127), and the dashed curve just provides the visual aid. Because the encoding message used in BPSK has NRZ random digital binary bits "0" and "1" with $m(t) = -\varepsilon/2$ as the bit "0" and $m(t) = \varepsilon/2$ as the bit "1" [15], the E_b of BPSK is only half that of the encoding method used in this optical injection system. Therefore, a 3-dB difference in the SNR for the same BER should be expected. This 3-dB difference is marked in Figure 9.9.

The BER as a function of the SNR from the numerical simulation of (9.125)−(9.127) with nonzero $\eta X(t)$ is shown in Figure 9.10 [14]. The transmitter is operated at $\xi = 0.03$ and $f = 0$ GHz to generate the chaotic waveform as the message carrier. The intrinsic laser noise of the transmitter and that of the receiver are both ignored in this simulation to see clearly the effect of the channel noise. Each curve is obtained by fixing the message amplitude while

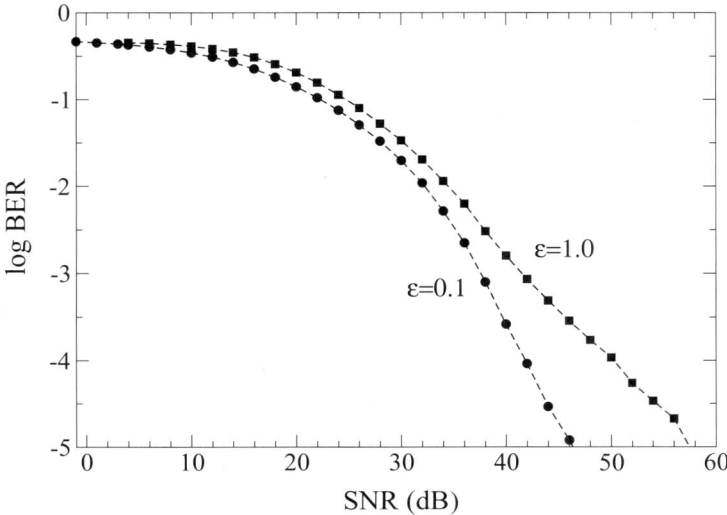

Fig. 9.10. BER as a function of SNR(dB) for the optically injected semiconductor laser system when the system performs chaos synchronization. The dashed curve marked with solid circles is obtained when $\varepsilon = 0.1$, and the dashed curve marked with solid squares is obtained when $\varepsilon = 1.0$. The transmitter is operated at $\xi = 0.03$ and $f = 0$ GHz to generate the chaotic waveform as the message carrier. (Reprinted with permission from [14], ©2002 IEEE.)

changing the strength of the channel noise. The dashed curve marked with solid circles is obtained when $\varepsilon = 0.1$, and the dashed curve marked with solid squares is obtained when $\varepsilon = 1.0$. As is shown, the performance analysis of the system with the same SNR but different ε, or equivalently different E_b, results in different values of the BER. Because the channel noise increases with E_b for a given SNR, this dependence of the BER on E_b at a fixed SNR demonstrates the nonlinear effect of the channel noise on the system performance.

9.6 Conclusions

This chapter provides a detailed discussion of the most important numerical tools needed in the analysis of chaotic systems performing chaos synchronization and chaotic communications. Basic concepts, theoretical framework, and computer algorithms are reviewed. The subjects covered include the concepts and numerical simulations of stochastic nonlinear systems, the complexity of a chaotic attractor measured by Lyapunov exponents and correlation dimension, the robustness of synchronization measured by the transverse Lyapunov exponents in parameter-matched systems and parameter-mismatched systems, the quality of synchronization measured by the correlation coefficient and the synchronization error, and the treatment of channel noise for quantifying the performance of a chaotic communication system. Optically injected single-mode semiconductor lasers are used as examples to demonstrate the use of these numerical tools.

The discussion on a dynamical system described by a stochastic differential equation shows that the integral of a stochastic term in the equation is very different from that of a deterministic term. When the first order of the integral is considered, the integral of a stochastic term is proportional to the square root of the infinitesimal time interval. In comparison, that of the deterministic term is proportional to the infinitesimal time interval. There are two types of integrals for solving a stochastic differential equation, namely, the Ito integral and the Stratonovich integral. Which method has to be used depends on the characteristics of the noise. It is revealed that the calculation in the Stratonovich sense follows the rules of the Riemann–Stieltjes integral, but the white noise term in the equation is no longer white noise. The calculation in terms of the Ito sense assures that the white noise treated in the integral is still white noise, but the calculation does not follow the rules of the Riemann–Stieltjes integral. A connection between these two methods is given in this chapter.

Two quantitative measures, namely, the Lyapunov exponents and the correlation dimension, for a chaotic attractor are discussed. The Lyapunov exponents measure the sensitivity of a chaotic system to initial conditions and perturbations. The correlation dimension increases with the complexity of a chaotic attractor. Numerical methods for calculating these parameters are outlined. When the dimension of a chaotic attractor is very high, the calculation of the largest Lyapunov exponent is preferred because it greatly simplifies the numerical computation.

The robustness of synchronization is measured by the transverse Lyapunov exponents. Either the whole set of transverse Lyapunov exponents or the largest transverse Lyapunov exponent can be used to measure the robustness. Usually only the largest transverse Lyapunov exponent is important. Thus, the calculation of the largest transverse Lyapunov exponent is preferred when the system is very complex. Because perfect parameter matching between a transmitter and a receiver is generally not possible in a real system, a new concept of robustness of synchronization is introduced for a system with parameter mismatch. For a perfectly parameter-matched synchronization system, the robustness of synchronization is measured between the attractors of the transmitter and the receiver. However, when the parameters of the system are not perfectly matched, this comparison between the transmitter and the receiver is impossible. Therefore, the comparison is made between the unperturbed and perturbed receiver attractors when synchronization is achieved. For the examination of the quality of synchronization, the correlation coefficient and the synchronization error obtained by the direct comparison between the transmitter and the receiver waveforms are used. Because it is very difficult to obtain both chaotic waveforms with exactly simultaneous digitization when they are measured in experiment, a direct comparison between the transmitter and the receiver waveforms is usually not accurate. Therefore, the measurement using the correlation coefficient is preferred for the experimental data because it is less sensitive to digitization timing errors.

The performance of a communication system is commonly measured by the BER as a function of SNR. In addition to the noise in the transmitter and the receiver, the noise of the communication channel has to be considered in evaluating the BER and SNR of the system. The white channel noise in a synchronized chaotic communication system interacts nonlinearly with the nonlinear system. For example, the channel noise can cause the receiver to lose synchronization with the transmitter, thus resulting in a large bit error. Therefore, both the linear effect of the channel noise as its role in conventional communication systems and the nonlinear effects of the channel noise have to be considered in a chaotic communication system. An approach to integrating the linear and nonlinear effects of the channel noise into the system consistently is addressed.

Acknowledgments

This work was supported by the U.S. Army Research Office under MURI grant DAAG55-98-1-0269.

References

1. C. C. Chen and K. Yao, Stochastic-calculus-based numerical evaluation and performance analysis of chaotic communication systems, *IEEE Trans. Circuits Syst. I*, vol. 47, pp. 1663−1672, 2000.
2. T. B. Simpson and J. M. Liu, Spontaneous emission, nonlinear optical coupling, and noise in laser diodes, *Opt. Commun.*, vol. 112, pp. 43−47, 1994.
3. J. M. Liu, C. Chang, T. B. Simpson, Amplitude noise enhancement caused by nonlinear interaction of spontaneous emission field in laser diodes, *Opt. Commun.*, vol. 120, pp. 282−286, 1995.
4. R. Mannella and V. Pallesche, Fast and precise algorithm for computer simulation of stochastic differential equations, *Phys. Rev. A*, vol. 40, pp. 3381−3386, 1989.
5. R. L. Honeycutt, Stochastic Runge−Kutta algorithms I. White noise, *Phys. Rev. A*, vol. 45, pp. 600−603, 1992.
6. E. Helfang, Numerical integration of stochastic differential equations, *The Bell Syst. Tech. J.*, vol.58, pp. 2289−2298, 1979.
7. R. F. Fox, I. R. Gatland, R. Roy, and G. Vemuri, Fast, accurate algorithm for numerical simulation of exponentially corrected colored noise, *Phys. Rev. A*, vol. 38, pp. 5938−5940, 1988.
8. C. W. Gardiner, *Handbook of Stochastic Methods for Physics, Chemistry, and the Nature Sciences*, 2nd ed. (Springer, New York, 2001).

9. S. Cyganowski, P. Koleden, and J. Ombach, *From Elementary Probability to Stochastic Differential Equations with MAPLE* (Springer, New York, 2001).

10. V. S. Anishchenko, V. V. Astakhov, A. B. Neiman, T. E. Vadivasova, and L. Schimansky-Geier, *Nonlinear Dynamics of Chaotic and Stochastic Systems* (Springer, New York, 2001).

11. J. M. Liu, H. F. Chen, X. J. Meng, and T. B. Simpson, Modulation bandwidth, noise, and stability of a semiconductor laser subject to strong injection locking, *IEEE Photon. Techno. Lett.*, vol. 9, pp. 1325−1327, 1997.

12. T. B. Simpson, J. M. Liu, A. Gavrielides, V. Kovanis, and P. M. Alsing, Period-doubling route to chaos in a semiconductor laser subject to optical injection, *Appl. Phys. Lett.*, vol. 64, pp. 3539−3541, 1994.

13. S. K. Hwang, J. B. Gao, and J. M. Liu, Noise-induced chaos in an optically injected semiconductor laser model,*Phys. Rev. E*, vol. 61, pp. 5162−5170, 2000.

14. S. Tang, H. F. Chen, S. K. Hwang and J. M. Liu, Message encoding and decoding through chaos modulation in chaotic optical communications, *IEEE Trans. on Circuits Syst. I*, vol. 49, pp. 163−169, 2002.

15. S. Haykin, *Communication Systems*, 3rd ed. (John Wiley & Sons, New York, 1994).

16. H. D. I. Abarbanel, R. Brown, and M. B. Kennel, Lyapunov exponents in chaotic systems: their importance and their evaluation using observed data, *Int. J. of Modern Phys.*, vol. 5, pp. 1347−1375, 1991.

17. R. Brown, P. Bryant, and H. D. I. Abarbanel, Computing the Lyapunov spectrum of a dynamical system from an observed time series, *Phys. Rev. A*, vol. 43, pp. 2787−2806, 1991.

18. J. B. Gao, S. K. Hwang, and J. M. Liu, Effects of intrinsic spontaneous-emission noise on the nonlinear dynamics of an optically injected semiconductor laser, *Phys. Rev. A*, vol. 59, pp. 1582−1585, 1999.

19. P. Grassberger and I. Procaccia, Characterization of strange attractors, *Phys. Rev. Lett.*, vol. 50, pp. 346−349, 1983.

20. J. Theiler, Estimating fractal dimension, *J. Opt. Soc. Am. A*, vol. 7, pp. 1055−1073, 1990.

21. E. V. Grigorieva, H. Haken, and S. A. Kaschenko, Theory of quasiperiodicity in model of lasers with delayed optoelectronic feedback, *Opt. Commun.*, vol. 165, pp. 279−292, 1999.

22. H. F. Chen and J. M. Liu, Open-loop chaotic synchronization of injection-locked semiconductor lasers with gigahertz range modulation, *IEEE J. Quantum Electron.*, vol. 36, pp. 27−34, 2000.

23. L. Kocarev and U. Parlitz, General approach for chaotic synchronization with applications to communication, *Phys. Rev. Lett.*, vol. 74, pp. 5028−5031, 1995.

24. L. Kocarev and U. Parlitz, Generalized synchronization, predictability, and equivalence of unidirectionally coupled dynamical systems, *Phys. Rev. Lett.*, vol. 76, pp. 1816−1911, 1996.

25. L. M. Pecora, T. L. Carroll, G. A. Johnson, D. J. Mar, and J. F. Heagy, Fundamentals of synchronization in chaotic systems, concepts, and applications, *Chaos*, vol. 7, pp. 520−543, 1997.

26. D. E. Knuth, *Seminumerical Algorithms*, vol. 2 of *The Art of Computer Programming*, 3rd ed. (Addison−Wesley, Reading, MA), p. 122, 1997.

10

Dynamics and Synchronization of Semiconductor Lasers for Chaotic Optical Communications

Jia-Ming Liu, How-Foo Chen, and Shuo Tang

Summary. The objective of this chapter is to provide a complete picture of the nonlinear dynamics and chaos synchronization of single-mode semiconductor lasers for chaotic optical communications. Basic concepts and theoretical framework are reviewed. Experimental results are presented to demonstrate the fundamental concepts. Numerical computations are employed for mapping the dynamical states and for illustrating certain detailed characteristics of the chaotic states. Three different semiconductor laser systems, namely, the optical injection system, the optical feedback system, and the optoelectronic feedback system, that are of most interest for high-bit-rate chaotic optical communications are considered. The optical injection system is a nonautonomous system that follows a period-doubling route to chaos. The optical feedback system is a phase-sensitive delayed-feedback autonomous system for which all three known routes, namely, period-doubling, quasiperiodicity, and intermittency, to chaos can be found. The optical feedback system is a phase-insensitive delayed-feedback autonomous system that follows a quasiperiodicity route to chaotic pulsing. Identical synchronization in unidirectionally coupled configurations is the focus of discussions for chaotic communications. For optical injection and optical feedback systems, the frequency, phase, and amplitude of the optical fields of both transmitter and receiver lasers are all locked in synchronism when complete synchronization is accomplished. For the optoelectronic feedback system, chaos synchronization involves neither the locking of the optical frequency nor the synchronization of the optical phase. For both optical feedback and optoelectronic feedback systems, where the transmitter is configured with a delayed feedback loop, anticipated and retarded synchronization can be observed as the difference between the feedback delay time and the propagation time from the transmitter laser to the receiver laser is varied. For a synchronized chaotic communication system, the message encoding process can have a significant impact on the quality of synchronization and thus on the message recoverability at the receiver end. It is shown that high-quality synchronization can be maintained when a proper encoding scheme that maintains the symmetry between the transmitter and the receiver is employed.

10.1 Introduction

In a synchronized chaotic communication system, a chaos generator is used to generate a chaotic waveform, which has the characteristics of a noiselike time series and a broadband spectrum. The message to be transmitted is encoded in the time domain on the chaotic waveform through a certain chaotic encryption scheme. An identical chaos generator at the receiver end regenerates the chaotic waveform. Message decoding is then accomplished by comparing the received signal with this reproduced chaotic waveform. This basic concept is illustrated in Figure 10.1. Recent theoretical and experimental progresses on the control and synchronization of nonlinear dynamical systems have allowed the demonstration of chaotic communication systems functioning in both radio and optical frequency regions [1, 2].

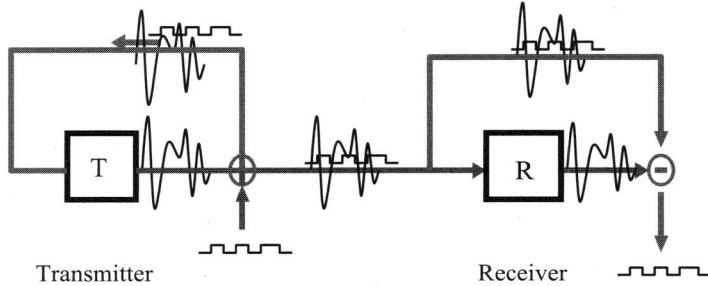

Fig. 10.1. Basic concept of synchronized chaotic communication with unidirectionally coupled transmitter and receiver. Ideally, the transmitter and the receiver are identical chaotic oscillators that are driven by the same signal.

To develop a synchronized chaotic communication system, the nonlinear dynamics of the devices used as the transmitter and receiver need to be thoroughly studied and understood first so that the desired chaotic states can be generated and controlled. Then, various issues regarding chaos synchronization between the transmitter and the receiver have to be investigated. For the implementation of a synchronized chaotic communication system, one of the most important such issues is the stability and quality of synchronization between the receiver and the transmitter in the presence of an encoded message because a communication system at work needs to carry a message. This is a profound issue that depends on the nonlinear dynamics of the transmitter and receiver and on the encryption scheme used to encode and decode the message. It has important implications on the performance of a chaotic communication system.

Chaotic states exist in many nonlinear dynamical systems. For chaotic optical communications, one is concerned with lasers that can be used as nonlinear dynamical devices to generate chaotic optical waveforms. Both fiber lasers [3–6] and semiconductor lasers [7–14] have been considered for this purpose. Most research in this area has focused on semiconductor lasers because of their dominant position in optical communication systems, their ability to support high-bit-rate messages, and the rich nonlinear dynamics they can be induced to display.

Of most interest for chaotic optical communications are three different semiconductor laser systems: the optical injection system, the optical feedback system, and the optoelectronic feedback system. These three systems all enter chaotic states under proper operating conditions, but they have very different nonlinear dynamics. They are modeled differently and require different numerical and experimental tools for analysis and investigation. The conditions and characteristics of chaos synchronization are also very different for these three systems. For example, both the optical injection system and the optical feedback system are sensitive to optical frequency and phase, but the optoelectronic feedback system is not. Both optical feedback system and optoelectronic feedback system are delayed feedback systems that have increased complexity as the delay time increases, whereas the optical injection system has no feedback mechanism. Chaotic optical communications implemented for various message encoding and decoding schemes using these three systems also have different performance characteristics [15]. They also have different degrees of susceptibility to internal and external perturbations from noise.

In this chapter, we review, based on our research results, the dynamics and synchronization of the three semiconductor laser systems that are considered for chaotic optical communications. Numerical methods for the simulation and analysis of these chaotic systems are discussed in Chapter 6. The performance characteristics of synchronized chaotic optical communication systems based on these laser systems for various message encoding and decoding schemes are addressed in Chapter 8.

10.2 Basic Concepts of Laser Dynamics

The dynamics of a laser is governed by the coupling of three macroscopic physical quantities: the intracavity laser field, $\mathbf{E}$, the material polarization, $\mathbf{P}$, and the population inversion density, N. The population inversion provides the laser gain; it is coupled to the laser field through the material polarization. On the quantum-mechanical level, $\mathbf{P}$ and N are respectively determined by the off-diagonal and diagonal density-matrix elements associated with the laser transition levels. Each of the three physical quantities has its characteristic relaxation time: the photon lifetime, τ_c, also known as the cavity decay time, for the intracavity laser field $\mathbf{E}$; the phase relaxation time, T_2, for the polarization $\mathbf{P}$; and the population relaxation time, T_1, for the population inversion density N.

When only the temporal characteristics of a laser are of interest, as is the case in our consideration for chaotic optical communications at the present stage, the spatial dependencies of $\mathbf{E}$, $\mathbf{P}$, and N can be integrated out to result in purely time-dependent coupled differential equations that determine the temporal dynamical behavior of the laser. The equation for $\mathbf{E}$, which originates from the wave equation, is originally a second-order differential equation. It can be reduced to a first-order differential equation by taking the slowly varying amplitude approximation, which is always valid for a laser because $\omega \gg \tau_c^{-1}$ for an optical frequency at ω. The equation for $\mathbf{P}$ can be similarly reduced to the first order. The equation for N is originally a first-order differential equation because it originates from the first-order equation of motion of the density matrix.

Thus, the temporal dynamics of a laser is mathematically governed by coupled first-order differential equations of the three quantities $\mathbf{E}$, $\mathbf{P}$, and N with characteristic relaxation time constants τ_c, T_2, and T_1, respectively. For a single-mode laser, the intracavity laser field is simply the field of the single oscillating mode. Then, its dynamics is governed by three coupled first-order differential equations of the following general form,

$$\frac{d\mathbf{E}}{dt} = F(\mathbf{E}, \mathbf{P}, N) \quad \text{with relaxation time constant } \tau_{\mathrm{c}}, \qquad (10.1)$$

$$\frac{d\mathbf{P}}{dt} = G(\mathbf{E}, \mathbf{P}, N) \quad \text{with relaxation time constant } T_2, \qquad (10.2)$$

$$\frac{dN}{dt} = H(\mathbf{E}, \mathbf{P}, N) \quad \text{with relaxation time constant } T_1, \qquad (10.3)$$

where $F(\mathbf{E}, \mathbf{P}, N)$, $G(\mathbf{E}, \mathbf{P}, N)$, and $H(\mathbf{E}, \mathbf{P}, N)$ are functions of $\mathbf{E}$, $\mathbf{P}$, and N that are characteristic of a particular laser. For a multimode laser, however, the total intracavity laser field is the combination of all oscillating laser modes. These mode fields can be coupled, but each of them has to be described by a separate equation. Thus, the dynamics of a multimode laser is governed by more than three coupled first-order differential equations of the following general form,

$$\frac{d\mathbf{E}_q}{dt} = F(\ldots \mathbf{E}_q \ldots, \mathbf{P}, N) \quad \text{with relaxation time constant } \tau_{\mathrm{c}}, \quad (10.4)$$

$$\frac{d\mathbf{P}}{dt} = G(\ldots \mathbf{E}_q \ldots, \mathbf{P}, N) \quad \text{with relaxation time constant } T_2, \quad (10.5)$$

$$\frac{dN}{dt} = H(\ldots \mathbf{E}_q \ldots, \mathbf{P}, N) \quad \text{with relaxation time constant } T_1, \quad (10.6)$$

where q is the laser mode index that takes on as many different values as the number of oscillating laser modes.

For complex nonlinear dynamics such as chaos to be possible for a mathematically continuous dynamical system such as a laser, whose dynamics is described by differential equations, two basic conditions must be satisfied. The first condition is clearly that there exists a certain nonlinear physical mechanism to make the system nonlinear for its dynamics to be described by coupled nonlinear differential equations. The second condition is that the nonlinear system must have more than two dimensions because any mathematically continuous one- or two-dimensional system cannot have chaotic dynamics. These are universal requirements. In the following, we examine these two issues for laser systems.

10.2.1 Nonlinear Mechanism

A laser is inherently nonlinear for the simple fact that a laser has a threshold above which the population inversion in its gain medium is clamped. The gain medium is the bare minimum component of a laser, and it functions nonlinearly when a laser oscillates above threshold. Thus, the basic nonlinear mechanism that always exists in a laser is the nonlinearity of its gain medium. The response of the gain medium to the laser field is characterized by a resonant susceptibility $\chi_{\mathrm{res}}(\omega)$ associated with atomic transitions at the oscillating laser frequency. The imaginary part, $\chi_{\mathrm{res}}''(\omega)$, of this resonant susceptibility contributes to the optical gain, $g(\mathbf{E}, N)$, that is a function of both the laser field and the population inversion density in the gain medium, whereas the real part, $\chi_{\mathrm{res}}'(\omega)$, of the resonant susceptibility modifies the refractive index, $n(\mathbf{E}, N)$, of the gain medium, making it also a function of both the laser field and the population inversion density in the gain medium. The real and imaginary parts of the susceptibility of a given medium are not independent of each other but are related through the Kramers–Kronig relations imposed by the causality requirement for any physical system. Thus $g(\mathbf{E}, N)$ and $n(\mathbf{E}, N)$ of a gain medium are directly related to each other. Their relationship is in general a function of the optical frequency ω, the optical field $\mathbf{E}$, and the population density N, as well as other parameters such as temperature. For practical purposes, however, their relationship at the laser oscillating frequency ω is usually characterized by a linewidth enhancement factor, also known as antiguidance factor, b, defined as

$$b = -\frac{2\omega}{c}\frac{\partial n/\partial N}{\partial g/\partial N} = \frac{\chi_{\mathrm{res}}'}{\chi_{\mathrm{res}}''}. \tag{10.7}$$

This factor is often treated as a constant for a given laser though it actually changes as the operating condition of the laser is varied. A positive value for b indicates that the refractive index of the gain medium decreases when the gain increases. This factor causes the laser frequency to change in response to changes in laser pumping condition and laser intensity. Thus, for a laser with a nonvanishing b value, nonlinearity in the optical gain also leads to nonlinearity in the refractive index of the gain medium; any gain nonlinearity that affects the amplitude of the laser field also affects its phase and frequency.

Besides the intrinsic nonlinearity of the gain medium, additional nonlinear mechanisms can be introduced through other nonlinear elements, incorporated either intentionally or unintentionally, in a laser. Though there exists a wide range of possibilities, such nonlinear elements can generally be categorized as active or passive in terms of their operation and as absorptive or reactive in terms of their basic physical mechanism. An active nonlinear element, such as an electro-optic modulator, is one that is operated by an externally applied signal, whereas a passive one, such as an all-optical nonlinear component, is simply operated by its response to the laser field without any actively applied signal. An absorptive nonlinear element, such as a saturable absorbing material, has its nonlinearity in the imaginary part of its susceptibility; thus, it changes its absorption coefficient in response to changes in the laser field parameters. A reactive nonlinear element, such as a nonlinear optical phase modulator, has its nonlinearity in the real part of its susceptibility; thus, it changes its refractive index in response to changes in the laser field. An active nonlinear element can have either absorptive or reactive mechanism, or a combination of both, and so can a passive nonlinear element.

10.2.2 Laser Classification

It seems that any laser has at least three dynamical degrees of freedom because it has three dynamical variables of $\mathbf{E}$, $\mathbf{P}$, and N and both $\mathbf{E}$ and $\mathbf{P}$ are complex vectorial quantities consisting of amplitudes and phases. Indeed, even a free-running, single-mode laser can have three degrees of freedom, thus with its dynamics described in a three-dimensional phase space, if all of its three dynamical variables $\mathbf{E}$, $\mathbf{P}$, and N are independent of, but coupled to, one another. However, the reality is that most free-running, single-mode lasers are stable because they actually have only one or two degrees of freedom. Chaotic dynamics is not possible for such lasers.

For most lasers, not all three dynamical variables are independent of one another because of the differences in their characteristic relaxation time constants. Furthermore, for a free-running, single-mode laser, both the directions and the phases of $\mathbf{E}$ and $\mathbf{P}$ are not part of the laser dynamics because the field vectors of a laser mode are well-defined and the phases of a free-running laser are decoupled from all other variables. Thus the degrees of freedom for many lasers can be significantly reduced. Arecchi et al. [16,17] classified homogeneously broadened single-mode lasers into three different classes based on their differences in the characteristic relaxation time constants of the dynamical variables.

For class A lasers, the photon relaxation time is much larger than both phase and population relaxation times of the gain medium: $\tau_c \gg T_1, T_2$. Physically, both polarization $\mathbf{P}$ and population N can respond very quickly to any variations in the laser field $\mathbf{E}$. Thus, $\mathbf{P}$ and N both follow the variations of $\mathbf{E}$. Mathematically, both variables $\mathbf{P}$ and N can be adiabatically eliminated by expressing them as a function of $\mathbf{E}$, reducing the independent dynamical variables to only $\mathbf{E}$. Only the differential equation of $\mathbf{E}$ remains to define the dynamics of a class A laser. For a free-running, single-mode class A laser, the vector and phase of the complex vectorial field variable $\mathbf{E}$ are irrelevant to the laser dynamics. Then the only remaining differential equation for $\mathbf{E}$ can be reduced to a scalar differential equation in terms of the intracavity photon density, S, which is a real variable. Thus, the only possible solution for the laser is a constant stable output representing a fixed point in the phase space. No unstable dynamics can be expected.

For class B lasers, both photon relaxation time and population relaxation time are much larger than the phase relaxation time of the material polarization: $\tau_c, T_1 \gg T_2$. In this situation, the polarization $\mathbf{P}$ follows the variations of both $\mathbf{E}$ and N. The variable $\mathbf{P}$ can be adiabatically eliminated by expressing it as a function of $\mathbf{E}$ and N. Thus, a class B laser is dynamically described by the coupled differential equations of $\mathbf{E}$ and N. For a free-running, single-mode class B laser, the vector and phase of the field variable $\mathbf{E}$ are also irrelevant to the laser dynamics, and the differential equation of $\mathbf{E}$ can be reduced to a scalar one of the real variable S. The dynamics of the laser is then described by two coupled first-order differential equations. The only dynamical behavior possible for such a system is periodic oscillation. Therefore, class B lasers can show relaxation oscillations but not instabilities and chaos.

For class C lasers, all three relaxation time constants are of comparable magnitudes: $\tau_c \approx T_1 \approx T_2$. None of the three dynamical variables can be eliminated. All three differential equations are required to describe the dynamics of a free-running, single-mode class C laser. Therefore, for certain laser parameters and operating conditions, complex dynamical behaviors including instabilities and chaos are possible for a single-mode class C laser oscillating in a free-running condition.

Most lasers, particularly those that have well-developed commercial applications, belong to either class A or class B. Instabilities and chaotic dynamics do not naturally occur in these lasers when they are free running because they have at most two independent dynamical variables. To induce instabilities and complex dynamics in such lasers, it is necessary to increase their dynamical dimensions. This objective can be accomplished by many different approaches. For a single-mode laser, it is necessary to introduce some form of external perturbation or feedback to increase the dimension of the laser. For a multimode laser, the dimension of the laser is increased by the multiple oscillating modes if the phases or magnitudes, or both, of the fields of the multiple oscillating laser modes are nonlinearly coupled, either directly or indirectly. This is the reason why instabilities and complex dynamics are often observed in multimode lasers but not in single-mode lasers unless a single-mode laser is somehow perturbed.

10.3 Single-Mode Semiconductor Lasers

In considering the temporal nonlinear dynamics of a semiconductor laser, here we take a lumped-circuit approach by integrating the spatial variations of the laser variables over the entire laser so that $\mathbf{E}$, $\mathbf{P}$, and N are macroscopic time-dependent dynamical variables of the laser. This approach is sufficiently accurate for describing the temporal dynamics of most semiconductor lasers, particularly single-mode semiconductor lasers, because the dynamical variations of the laser field are generally on a time scale larger than the time it takes a photon to make a trip through the laser cavity.

For a semiconductor laser, the population density N that provides the laser gain is the carrier density determined by that of the injected electron-hole pairs in the active region of the laser. The gain coefficient g of a semiconductor laser gain medium is a function of both the carrier density N and the laser field $\mathbf{E}$. As is discussed above, this is the source of intrinsic nonlinearity for a semiconductor laser. For a semiconductor laser, the dependence of the gain coefficient on the laser field appears as a function of $|\mathbf{E}|^2$, which is proportional to the laser intensity and the intracavity photon density S. Thus, the gain can be expressed as a function of N and S: $g(N, S)$. The resonant material susceptibility associated with the laser transition and responsible for the laser gain is in general a second-rank tensor, but we take it to be a scalar χ_{res} for a semiconductor laser when laser modes of different polarization directions are not considered at the same time. The resonant susceptibility of a semiconductor laser gain medium in a laser oscillating at a frequency ω is related to $g(N, S)$ as

$$\chi_{\mathrm{res}} = -\frac{nc}{\omega}(b + i)g(N, S), \tag{10.8}$$

where n is the refractive index of the gain medium and b is the linewidth enhancement factor defined in (10.7). For a semiconductor laser, b is positive. It typically has a value in the range of $2 < b < 7$, but it can be as small as 1 or as large as 10 for some lasers. Because of this factor, the gain is coupled to the refractive index; thus nonlinearity in the laser gain also leads to nonlinearity in the refractive index of the laser medium. This coupling can significantly enrich the nonlinear dynamics of a semiconductor laser for which the phase of the laser field is a dynamical variable [18], such as the optical injection system or the optical feedback system. It is irrelevant to the dynamics of a semiconductor laser for which the phase of the laser field is not a dynamical variable, such as the optoelectronic feedback system.

The population relaxation time T_1 for a semiconductor laser is the spontaneous carrier recombination lifetime, τ_{s}. For most semiconductor lasers, including the most commonly used single-mode lasers such as the distributed feedback (DFB) lasers and the vertical cavity surface-emitting lasers, the photon lifetime τ_{c} is on the order of a few picoseconds, the carrier recombination lifetime τ_{s} is on the order of a few hundred picoseconds to a few nanoseconds, and the polarization phase relaxation time T_2 is on the order of about 100 femtoseconds. Thus, a semiconductor laser is a class B laser because $\tau_{\mathrm{s}} > \tau_{\mathrm{c}} \gg T_2$. The resonant material polarization in a semiconductor laser gain medium responds almost instantaneously to the variations in the laser field; thus $\mathbf{P}$ is related to the laser field $\mathbf{E}$ as follows,

$$\mathbf{P}(t) = \epsilon_0 \chi_{\mathrm{res}} \mathbf{E}(t). \tag{10.9}$$

The polarization equation is then not needed for describing the semiconductor laser dynamics because the polarization of the gain medium is now completely defined by the laser field $\mathbf{E}$ and, through the dependence of χ_{res} on N, by the carrier density N. Consequently, the dynamics of a single-mode semiconductor laser can be completely described by two coupled first-order differential equations for $\mathbf{E}$ and N as in (10.1) and (10.3), respectively, and the equation for $\mathbf{P}$ in (10.2) is eliminated.

10.3.1 Rate Equations

The condition $\tau_{\mathrm{s}}, \tau_{\mathrm{c}} \gg T_2$ that makes a semiconductor laser a class B laser allows the dynamical equations of a single-mode semiconductor laser to be reduced to two coupled rate equations in the photon density S and the carrier density N. This procedure is known as the rate equation approximation.

In the lumped-circuit approach used here to describe the temporal laser dynamics, the intracavity field of a single-mode semiconductor laser with a free-running mode frequency ω_0 can be expressed as

$$\mathbf{E}(t) = \hat{e} A(t) e^{-i\omega_0 t}, \tag{10.10}$$

where $\hat{e}$ is the unit vector for the laser mode field and $A(t)$ is the total complex intracavity field amplitude at the free-oscillating frequency ω_0. The intracavity photon density is related to the laser field amplitude A by

$$S(t) = \frac{2n^2 \epsilon_0}{\hbar \omega_0} |A(t)|^2. \tag{10.11}$$

For a laser that has a photon lifetime of τ_{c}, the intracavity photon density decays at a photon decay rate, also called cavity decay rate, of $\gamma_{\mathrm{c}} = \tau_{\mathrm{c}}^{-1}$ through internal losses and output-coupling losses of the laser. This decay of photon density has to be counterbalanced by a growth of photon density for the laser to continue oscillating. The growth of the photon density is contributed by the amplification of the laser field through the optical gain in the laser medium. Thus, we can define a gain parameter g that quantifies the concept of photon growth rate as

$$g(N, S) = \frac{c}{n} g(N, S). \tag{10.12}$$

Note that the gain coefficient g has the unit of m^{-1}, whereas the gain parameter g has the unit of s^{-1}.

With the relationship between $\mathbf{E}$ and A defined in (10.10), the dynamical equation for $\mathbf{E}$ can be expressed as one in A. Then, the two coupled first-order differential equations needed to describe the dynamics of a single-mode semiconductor laser can be expressed in A and N as [8]:

$$\frac{dA}{dt} = -\frac{\gamma_c}{2}A + i(\omega_0 - \omega_c)A + \frac{\Gamma}{2}(1 - ib)gA + F_{sp}, \tag{10.13}$$

$$\frac{dN}{dt} = \frac{J}{ed} - \gamma_s N - \frac{2\epsilon_0 n^2}{\omega_0}g|A|^2, \tag{10.14}$$

where ω_c is the longitudinal mode frequency of the cold laser cavity, Γ is the confinement factor determined by the overlap of the laser mode field distribution with the active gain region, J is the injection current density, e is the electronic charge, d is the active layer thickness of the laser, $\gamma_s = \tau_s^{-1}$ is the spontaneous carrier decay rate, and n is the refractive index of the semiconductor medium. The spontaneous emission noise source $F_{sp} = F_r + iF_i$ has uncorrelated real and imaginary components and an effective delta-function self-correlation in time [19]:

$$\langle F_r(t)F_r(t')\rangle = \langle F_i(t)F_i(t')\rangle = \frac{R_{sp}}{2}\delta(t - t'), \tag{10.15}$$

$$\langle F_r(t)F_i(t')\rangle = 0, \tag{10.16}$$

where R_{sp} gives the rate of spontaneous emission into the mode. Without perturbation, no complex dynamics will develop from this single-mode laser. The laser has a free-running steady state that is characterized by a gain parameter saturated at the threshold gain parameter value of

$$g_{th} = \frac{\gamma_c}{\Gamma}, \tag{10.17}$$

a carrier density of N_{th} clamped at that reached at the laser threshold, and a field amplitude of A_0 with a corresponding photon density of S_0.

Note that the equation for the field amplitude A given in (10.13) is a complex differential equation because A is a complex quantity consisting of magnitude and phase. Because the phase of the laser field of a free-running, single-mode semiconductor laser is not coupled to the magnitude of the field or the carrier density, it can be eliminated from the rate equations by converting (10.13) into a real differential equation in terms of the photon density S. Therefore, the dynamics of a free-running, single-mode semiconductor laser operating above threshold is sufficiently described by two real, first-order rate equations, one for S and another for N:

$$\frac{dS}{dt} = -\gamma_c S + \Gamma gS + 2\sqrt{S_0 S}F_s, \tag{10.18}$$

$$\frac{dN}{dt} = \frac{J}{ed} - \gamma_s N - gS, \tag{10.19}$$

where F_s is a stochastic noise term derived from F_{sp}. Clearly, there are only two degrees of freedom in the two variables S and N for a free-running, single-mode semiconductor laser. As is discussed in the preceding section, such a two-dimensional continuous dynamical system cannot have any interesting dynamics except for relaxation oscillations between the photon and carrier densities.

10.3.2 Dynamical Parameters

We see from the coupled equations given in (10.13) and (10.14) that there are three parameters, γ_c, γ_s, and b, that appear in these dynamical equations of a semiconductor laser. We also see that the b parameter disappears from the coupled equations given in (10.18) and (10.19) after the phase of the laser field is eliminated. As is discussed above, the linewidth enhancement factor b has an effect on the laser dynamics only when the phase of the laser field participates in the laser dynamics.

Besides these three dynamical parameters, there are two more dynamical parameters that specifically make the laser a nonlinear system. As we can see from (10.13) and (10.14) or from (10.18) and (10.19), the laser would be a linear system characterized by linear differential equations if the gain parameter g were a constant that is independent of N and S. The system becomes nonlinear only because g is a function of N and S. A relation of g to either N or S in any form would already make the system nonlinear, but the fact is that g is a function of both N and S. There are different ways of expressing g as a function of N and S from theoretical and experimental viewpoints. Here we take a macroscopic approach to define the dynamical parameters that characterize the nonlinearities of the laser gain. This approach is consistent with our lumped-circuit model and proves to be very convenient and accurate for both experimental measurement and theoretical modeling [20].

When a single-mode semiconductor laser is in a free-running steady state, its gain and carrier density are clamped at their respective threshold values of g_{th} and N_{th}, but the photon density reaches a constant value of S_0 determined by the operating level above the laser threshold. Under any dynamical perturbation, the gain can deviate from g_{th} due to the variations in the carrier and photon densities caused by the perturbation. The dependence of the gain parameter on the carrier and photon densities can be expressed as

$$g = g_{th} + g_n(N - N_{th}) + g_p(S - S_0), \qquad (10.20)$$

where g_n is the differential gain parameter characterizing the dependence of the gain on the carrier density and g_p is the nonlinear gain parameter characterizing the effect of gain compression due to the saturation of gain by intracavity photons. It has been found empirically that both g_n and g_p stay quite constant over large ranges of carrier density and photon density in a given laser. For most practical purposes, they can be treated as constants over the operating range of a laser. These parameters are normally measured experimentally though they can also be calculated theoretically. Note that $g_n > 0$ but $g_p < 0$. It is convenient to define a differential carrier relaxation rate, γ_n, and a nonlinear carrier relaxation rate, γ_p, as

$$\gamma_n = g_n S_0, \qquad \gamma_p = -\Gamma g_p S_0. \qquad (10.21)$$

Both γ_n and γ_p are rates that have the same unit of s^{-1} as γ_c and γ_s, and they all have positive values. Because g_n and g_p are quite constant, both γ_n and γ_p vary linearly with the laser power.

Table 10.1. Measured Parameters of a Semiconductor Laser

Parameter	Symbol	Value[*]
Laser output power	P	4 to 15 mW
Intracavity photon density	S	$6 \times 10^{13} P$ cm^{-3}
Confinement factor	Γ	0.4
Injection coupling rate	η	1×10^{11} s^{-1}
Linewidth enhancement factor	b	3 ± 1
Relaxation resonance frequency	f_{r}	$(0.954P)^{1/2}$ GHz
Relaxation resonance angular frequency	Ω_{r}	$(37.66P)^{1/2}$ GHz
Total carrier relaxation rate	γ_{r}	$(1.458 + 0.435P) \times 10^{9}$ s^{-1}
Spontaneous carrier relaxation rate	γ_{s}	1.458×10^{9} s^{-1}
Spontaneous carrier lifetime	τ_{s}	686 ps
Differential carrier relaxation rate	γ_{n}	$0.155P \times 10^{9}$ s^{-1}
Nonlinear carrier relaxation rate	γ_{p}	$0.28P \times 10^{9}$ s^{-1}
Cavity decay rate	γ_{c}	2.4×10^{11} s^{-1}
Photon lifetime	τ_{c}	4.2 ps
Differential gain parameter	g_{n}	2.6×10^{-6} cm^{3}s^{-1}
Nonlinear gain parameter	g_{p}	-1.18×10^{-5} cm^{3}s^{-1}
Spontaneous emission rate	R_{sp}	4.7×10^{18} V^{2}m^{-2}s^{-1}

Source: This table is adapted from [20] and [21].
[*]The laser power, P, in the values of the power-dependent parameters is in mW.

The dynamics of a single-mode semiconductor laser is determined by five intrinsic dynamical parameters: γ_{c}, γ_{s}, γ_{n}, γ_{p}, and b. For a given laser, γ_{c} and γ_{s} are constants that are independent of the laser power, but γ_{n} and γ_{p} are linearly proportional to the laser power. The b parameter is often treated as a constant, but it can vary with the operating condition of the laser. These five intrinsic dynamical parameters of a given laser can be experimentally measured using a four-wave mixing technique [20]. The spectral characteristics and the strength of the noise source F_{sp} can also be experimentally measured [21]. For a single-mode semiconductor laser without any additional arrangement to increase its system dimension, the only dynamical behavior that can be observed is relaxation oscillation. The relaxation oscillation of a free-running, single-mode semiconductor laser has a relaxation rate of [20]

$$\gamma_{\mathrm{r}} = \gamma_{\mathrm{s}} + \gamma_{\mathrm{n}} + \gamma_{\mathrm{p}}, \tag{10.22}$$

and a relaxation resonance frequency of [20]

$$\Omega_{\mathrm{r}} = 2\pi f_{\mathrm{r}} = (\gamma_{\mathrm{c}}\gamma_{\mathrm{n}} + \gamma_{\mathrm{s}}\gamma_{\mathrm{p}})^{1/2}. \tag{10.23}$$

It can be seen from these relations that γ_{r} varies linearly with, but not proportional to, the laser power, whereas Ω_{r} and f_{r} are proportional to the square root of the laser power. Table 10.1 shows, as an example, the dynamical parameters of a semiconductor laser that are measured using a four-wave mixing technique with optical injection [20].

The operating condition of a solitary semiconductor laser is solely control by the injection current density J. It is thus convenient to define an operational parameter: $\tilde{J} = (J/ed - \gamma_s N_0)/\gamma_s N_0$, which is the normalized injection parameter indicating how high the laser is biased above its threshold. The intrinsic dynamical parameters and the noise sources of a given laser are not externally controllable parameters and therefore cannot be directly varied at will, though γ_n and γ_p are linearly proportional to the laser power and can be indirectly varied by varying the operating point of the laser. Therefore, for a solitary semiconductor laser, the only parameter that can be varied at will is the operational parameter $\tilde{J}$.

10.3.3 Reformulated Dynamical Equations

Because the field amplitude A is a complex variable, the coupled dynamical equations in terms of A and N given in (10.13) and (10.14), respectively, can be tranformed into a set of three real equations in a form that shows explicitly the dependence on the dynamical laser parameters. This transformation is done by separating the magnitude and phase of the field amplitude and by defining normalized dimensionless dynamical variables that measure the variations with respect to the free-running condition of the laser in much the same manner as normalizing J to $\tilde{J}$. The field amplitude can be expressed in terms of its magnitude and phase as

$$A = |A|e^{i\varphi} = |A_0|(1 + a)e^{i\varphi}, \tag{10.24}$$

and the carrier density can be expressed as $N = N_0(1 + \tilde{n})$. Thus we have three real variables: $a = (|A| - |A_0|)/|A_0|$, φ, and $\tilde{n} = (N - N_0)/N_0$. The coupled dynamical equations in (10.13) and (10.14) can then be tranformed into the following three real equations,

$$\frac{da}{dt} = \frac{1}{2}\left[\frac{\gamma_c\gamma_n}{\gamma_s\tilde{J}}\tilde{n} - \gamma_p(2a + a^2)\right](1 + a) + F_a, \tag{10.25}$$

$$\frac{d\varphi}{dt} = -\frac{b}{2}\left[\frac{\gamma_c\gamma_n}{\gamma_s\tilde{J}}\tilde{n} - \gamma_p(2a + a^2)\right] + \frac{F_\varphi}{1 + a}, \tag{10.26}$$

$$\frac{d\tilde{n}}{dt} = -\gamma_s\tilde{n} - \gamma_n(1 + a)^2\tilde{n} - \gamma_s\tilde{J}(2a + a^2)$$
$$+ \frac{\gamma_s\gamma_p}{\gamma_c}\tilde{J}(2a + a^2)(1 + a)^2. \tag{10.27}$$

The noise sources are related to F_{sp} as $F_a = (F_r\cos\varphi - F_i\sin\varphi)/|A_0|$, $F_\varphi = (F_r\sin\varphi + F_i\cos\varphi)/|A_0|$. Thus,

$$\langle F_a(t)F_a(t')\rangle = \langle F_\varphi(t)F_\varphi(t')\rangle = \frac{R_{sp}}{2|A_0|^2}\delta(t - t'), \tag{10.28}$$

$$\langle F_a(t) F_\varphi(t') \rangle = 0. \tag{10.29}$$

Although there are three dynamical equations in terms of the three real variables a, φ, and $\tilde{n}$, we see from these equations that the phase variable φ does not appear in either the equation for a or that for $\tilde{n}$. Clearly, the dynamical variations of a and $\tilde{n}$ are not affected by the value of the phase φ though they determine the phase variations. Therefore, the coupled equations in (10.25) and (10.27) for a and $\tilde{n}$ completely determine the dynamics of the laser. The phase equation given in (10.26) can be ignored in this situation. As a consequence, this laser is a two-dimensional dynamical system, as is discussed above.

A similar transformation can be carried out for the coupled equations of S and N in (10.18) and (10.19) by expressing $S = S_0(1 + \tilde{s})$ to define $\tilde{s} = (S - S_0)/S_0$ and by defining $\tilde{n}$ as above. Then, (10.18) and (10.19) are transformed into the following form in terms of the normalized dimensionless variables $\tilde{s}$ and $\tilde{n}$,

$$\frac{d\tilde{s}}{dt} = \frac{\gamma_c \gamma_n}{\tilde{J} \gamma_s} \tilde{n}(\tilde{s} + 1) - \gamma_p \tilde{s}(\tilde{s} + 1) + \tilde{F}_s, \tag{10.30}$$

$$\frac{d\tilde{n}}{dt} = -\gamma_s \tilde{n} - \gamma_s \tilde{J} \tilde{s} - \gamma_n \tilde{n}(1 + \tilde{s}) + \frac{\gamma_s \gamma_p}{\gamma_c} \tilde{J} \tilde{s}(1 + \tilde{s}), \tag{10.31}$$

where $\tilde{F}_s = 2\sqrt{1 + \tilde{s}} F_a$. These two coupled equations can also be obtained directly from (10.25) and (10.27) by using the relation that $\tilde{s} = 2a + a^2$.

10.4
Nonlinear Dynamics of Single-Mode Semiconductor Lasers

Complex dynamics, such as sustained periodic oscillations or chaos, can be generated in a single-mode semiconductor laser by a certain external perturbation that provides additional degrees of freedom, thus increasing the dynamical dimension of the laser. The complexity of the dynamics increases as the dimension of the perturbed laser system is increased. Many different schemes for perturbing a single-mode semiconductor laser into chaotic oscillations are possible. What have been considered and intensively studied in recent years for chaotic optical communications are the following three different semiconductor laser systems: the optical injection system, the optical feedback system, and the optoelectronic feedback system. These three systems all enter chaotic states under proper operating conditions, but they have very different nonlinear dynamics. They are modeled differently and require different numerical and experimental tools for analysis and investigation. The conditions and characteristics of chaos synchronization are also very different for these three systems. In this section, we review the theoretical model, the numerical results, and the experimentally observed phenomena regarding the dynamics of these three systems.

10.4.1 Single-Mode Semiconductor Laser with Optical Injection

Optical Injection

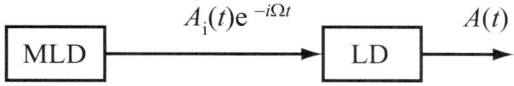

Fig. 10.2. Single-mode semiconductor laser subject to optical injection from a master laser. LD: laser diode; MLD: master laser diode.

Figure 10.2 shows schematically a single-mode semiconductor laser subject to optical injection from a master laser. The injection field has a complex amplitude of A_i at an optical frequency of ω_i, which is detuned from the free-running frequency ω_0 of the slave laser by a detuning frequency of $\Omega = \omega_i - \omega_0$. The following coupled equations are used to describe the dynamics of this optically injected single-mode semiconductor laser [20]:

$$\frac{dA}{dt} = -\frac{\gamma_c}{2}A + i(\omega_0 - \omega_c)A + \frac{\Gamma}{2}(1 - ib)gA + F_{\text{sp}}$$
$$+\eta A_i \exp(-i\Omega t), \tag{10.32}$$

$$\frac{dN}{dt} = \frac{J}{ed} - \gamma_s N - \frac{2\epsilon_0 n^2}{\hbar\omega_0}g|A|^2, \tag{10.33}$$

where η is the injection coupling rate. Compared to the coupled equations in (10.13) and (10.14) for the solitary laser, an additional term representing the injection field is added to the field equation. This modification completely changes the dynamics of the system by adding one more dimension to the system.

To see clearly how the injection field changes the dimension of the system, we transform, by following the procedure outlined in the preceding section, the coupled field and carrier equations into the following three coupled equations in terms of a, φ, and $\tilde{n}$,

$$\frac{da}{dt} = \frac{1}{2}\left[\frac{\gamma_c\gamma_n}{\gamma_s \tilde{J}}\tilde{n} - \gamma_p(2a + a^2)\right](1 + a) + F_a + \xi\gamma_c\cos(2\pi ft + \varphi) \tag{10.34}$$

$$\frac{d\varphi}{dt} = -\frac{b}{2}\left[\frac{\gamma_c\gamma_n}{\gamma_s \tilde{J}}\tilde{n} - \gamma_p(2a + a^2)\right] + \frac{F_\varphi}{1 + a} - \frac{\xi\gamma_c}{1 + a}\sin(2\pi ft + \varphi) \tag{10.35}$$

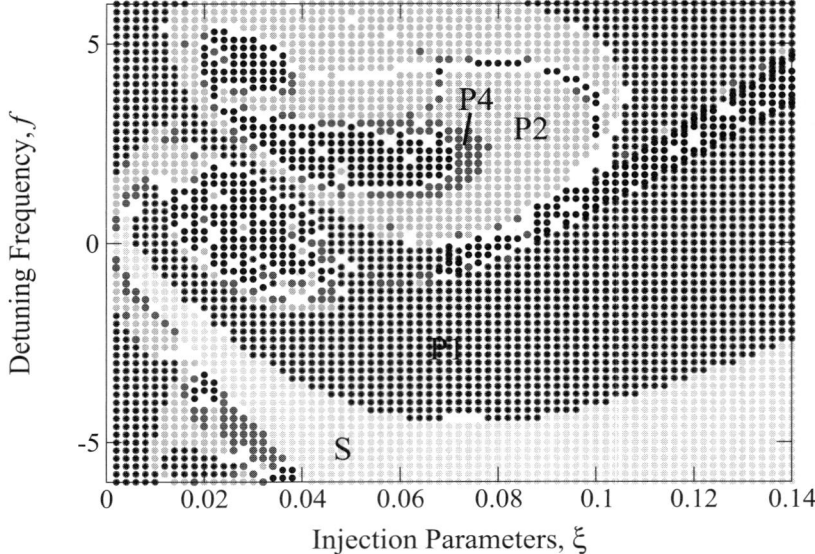

Fig. 10.3. Numerally calculated mapping of the dynamical states of an optically injected semiconductor laser operating at a constant current level of $\tilde{J} = 2/3$ as a function of the injection parameter ξ and the detuning frequency f. In this map, S, P1, P2, and P4 indicate the regions of the stable injection-locked states, the period-one states, the period-two states, and the period-four states, respectively, and the chaotic states are indicated by the solid-black region. Laser parameters listed in Table 10.1 are used to generate this map. (Reprinted with permission from [7], ©2000 IEEE.)

$$\frac{d\tilde{n}}{dt} = -\gamma_{\mathrm{s}}\tilde{n} - \gamma_{\mathrm{n}}(1+a)^2\tilde{n} - \gamma_{\mathrm{s}}\tilde{J}(2a+a^2)$$

$$+\frac{\gamma_{\mathrm{s}}\gamma_{\mathrm{p}}}{\gamma_{\mathrm{c}}}\tilde{J}(2a+a^2)(1+a)^2, \tag{10.36}$$

where $\xi = \eta|A_{\mathrm{i}}|/(\gamma_{\mathrm{c}}|A_0|)$ is a normalized injection parameter, and $f = \Omega/2\pi$ is the detuning frequency. Compared to the equations in (10.25)-(10.27) for a solitary laser where both a and $\tilde{n}$ are independent of φ, we find that a and φ are now mutually coupled through the additional terms in (10.34) and (10.35) caused by the injection field. Because a and $\tilde{n}$ are still coupled, all three equations in this system are now mutually coupled, meaning that all three variables a, φ, and $\tilde{n}$ are relevant to the dynamics of this system. Therefore, an optically injected single-mode semiconductor laser is a three-dimensional nonautonomous system that is capable of generating complex nonlinear dynamics. As the phase φ of the laser field is one of the three coupled dynamical variables, the dynamics of this system is phase-sensitive. We also see from (10.35) the effect of the b parameter on the dynamics of the laser through its coupling of the variations in the field amplitude and carrier density to changes in the phase of the laser field.

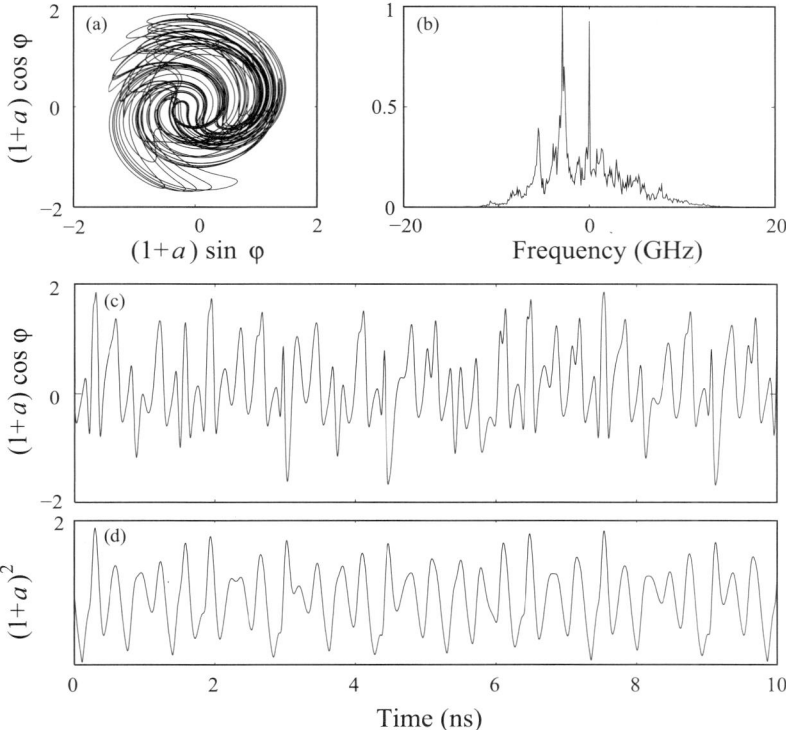

Fig. 10.4. Numerically calculated (a) attractor, (b) optical spectrum, (c) optical field, and (d) optical intensity of the chaotic state at the operating condition of $\tilde{J} = 2/3$, $\xi = 0.03$, and $f = 0$ for an optical injected single-mode semiconductor laser that has the parameters listed in Table 10.1.

The dynamics of an optically injected single-mode semiconductor laser is determined by the five intrinsic parameters, γ_c, γ_s, γ_n, γ_p, and b, discussed in the preceding section. In addition, there are now three operational parameters, the current injection level $\tilde{J}$, the optical injection parameter ξ, and the detuning frequency $f = \Omega/2\pi$, that can be externally varied to control the dynamics of the laser in operation. The dependence of the dynamics of an optically injected laser on these intrinsic and operational parameters has been thoroughly studied [18]. Figure 10.3 shows as an example a numerally calculated mapping of the dynamical states of an optically injected laser operating at a constant current level of $\tilde{J} = 2/3$ as a function of the injection parameter ξ and the detuning frequency f. Laser parameters listed in Table 10.1 are used to generate this map. It can be seen from Figure 10.3 that an optically injected semiconductor laser follows a period-doubling route to chaos when one of its operational parameters is varied and the other two are fixed [22].

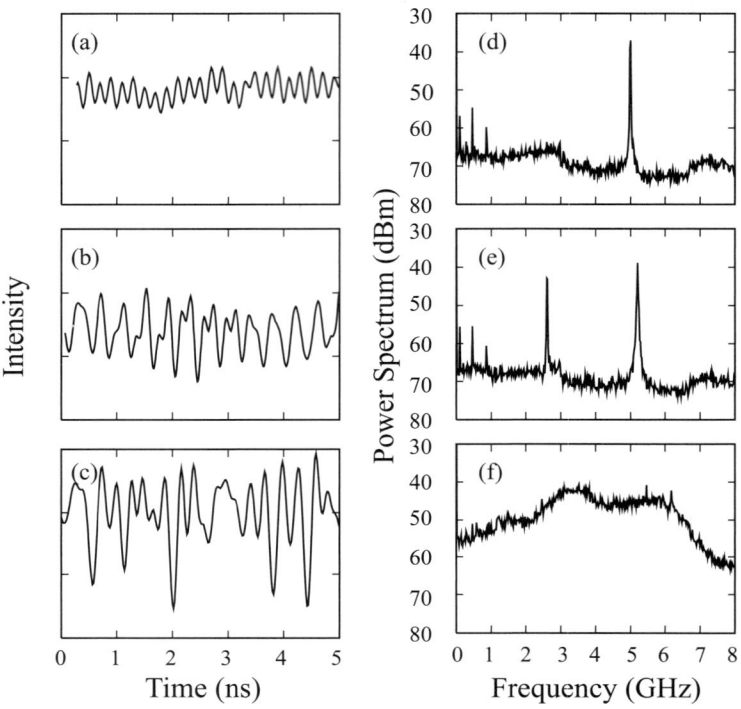

Fig. 10.5. Experimentally measured temporal waveforms and corresponding power spectra of period-one, period-two, and chaotic states of an optically injected single-mode semiconductor laser.

Figure 10.4 shows the numerically calculated attractor, optical spectrum, optical field, and optical intensity of a representative chaotic state of the system at an operating condition of $\tilde{J} = 2/3$, $\xi = 0.03$, and $f = 0$. Experimentally measured temporal waveforms and power spectra of different dynamical states of an optically injected InGaAsP single-mode DFB laser at 1.3 μm wavelength are presented in Figure 10.5. Because of the laser noise, the period-four states seen in the numerical mapping are very difficult to observe experimentally.

10.4.2 Single-Mode Semiconductor Laser with Optical Feedback

Figures 10.6a and b show two possible schemes of delayed optical feedback to a single-mode semiconductor laser. In the single-pass scheme, the feedback loop consists of an optical isolator so that a fraction of the laser output is fed back to the laser in only one delay time of τ defined by the length of the feedback

Optical Feedback

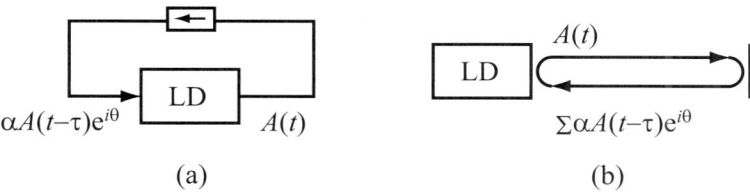

$\alpha A(t-\tau)e^{i\theta}$

$A(t)$

$\Sigma \alpha A(t-\tau)e^{i\theta}$

(a)   (b)

Fig. 10.6. Schematics of single-mode semiconductor lasers with optical feedback in (a) single-pass configuration and (b) multiple-pass configuration.

loop. In the multiple-pass scheme, which often consists of an external mirror to partially reflect the laser output back to the laser, a diminishing fraction of the laser output continues to be fed back to the laser after every integral multiple of delay time τ because the external mirror forms an external cavity with the output laser mirror. The following coupled equations are used to describe the dynamics of this optically injected single-mode semiconductor laser:

$$\frac{dA}{dt} = -\frac{\gamma_c}{2}A + i(\omega_0 - \omega_c)A + \frac{\Gamma}{2}(1 - ib)gA + F_{sp} + \alpha\eta A(t - \tau)e^{i\theta} \quad (10.37)$$

$$\frac{dN}{dt} = \frac{J}{ed} - \gamma_s N - \frac{2\epsilon_0 n^2}{\hbar\omega_0}g|A|^2, \quad (10.38)$$

where α is the fraction of the laser output field that is fed back, η is the coupling rate for injecting the feedback laser field back into the cavity, τ is the feedback delay time, and θ is the phase difference between the feedback field and the intracavity field at the feedback injection point. The field equation in (10.37) for this system is equivalent to the well-known Lang–Kobayashi equation [23]. Strictly speaking, this equation is valid only for the single-pass feedback scheme shown in Figure 10.6a because it does not account for the multiple reflections between the output laser mirror and the external mirror of the multiple-pass scheme shown in Figure 10.6b. Thus, significant errors can occur when this model is used for a multiple-pass system at a high feedback strength. Compared to the coupled equations in (10.13) and (10.14) for the solitary laser, the feedback delay field term added to the field equation for this system can add more than one dimension to the system. Thus, this system can have even more complicated dynamics than the optical injection system.

The coupled field and carrier equations in (10.37) and (10.38) can be transformed into the following three coupled equations in terms of a, φ, and $\tilde{n}$:

$$\frac{da}{dt} = \frac{1}{2}\left[\frac{\gamma_c\gamma_n}{\gamma_s\tilde{J}}\tilde{n} - \gamma_p(2a + a^2)\right](1 + a) + F_a$$
$$+\xi\gamma_c[1 + a(t - \tau)]\cos[\varphi(t - \tau) - \varphi(t) + \theta], \quad (10.39)$$

$$\frac{d\varphi}{dt} = -\frac{b}{2}\left[\frac{\gamma_{\mathrm{c}}\gamma_{\mathrm{n}}}{\gamma_{\mathrm{s}}\tilde{J}}\tilde{n} - \gamma_{\mathrm{p}}(2a + a^2)\right] + F_\varphi$$

$$+\xi\gamma_{\mathrm{c}}\frac{1 + a(t - \tau)}{1 + a(t)}\sin[\varphi(t - \tau) - \varphi(t) + \theta], \qquad (10.40)$$

$$\frac{d\tilde{n}}{dt} = -\gamma_{\mathrm{s}}\tilde{n} - \gamma_{\mathrm{n}}(1 + a)^2\tilde{n} - \gamma_{\mathrm{s}}\tilde{J}(2a + a^2)$$

$$+\frac{\gamma_{\mathrm{s}}\gamma_{\mathrm{p}}}{\gamma_{\mathrm{c}}}\tilde{J}(2a + a^2)(1 + a)^2, \qquad (10.41)$$

where $\xi = \alpha\eta/\gamma_{\mathrm{c}}$ is a normalized feedback parameter measuring the feedback strength. We see that, like those for the optical injection system, a, φ, and $\tilde{n}$ are all relevant dynamical variables for the optical feedback system because all three differential equations for them are mutually coupled. However, different from the optically injected laser, a single-mode semiconductor laser with optical feedback is an autonomous system that can have a dimension larger than three because two of its three dynamical equations are delay-differential equations. The dimension of the system increases with the delay time. The dynamics of this system is sensitive to the phase θ.

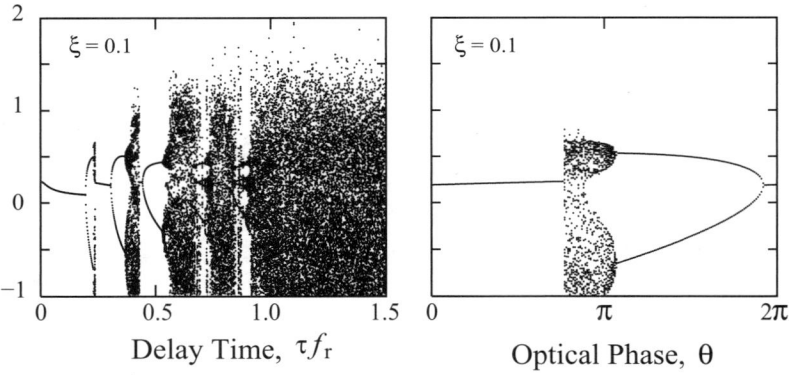

Fig. 10.7. Numerically calculated bifurcation diagrams for the dynamical states of a semiconductor laser with optical feedback (a) as a function of the normalized delay time τf_{r} for a fixed phase of $\theta = 0$ and (b) as a function of the phase θ for a fixed normalized delay time of $\tau f_{\mathrm{r}} = 0.3$. For both cases, the laser parameters listed in Table 10.1 are used, and the system is operated at $\tilde{J} = 2/3$ and $\xi = 0.1$.

The dynamics of a single-mode semiconductor laser with optical feedback is determined by the five intrinsic parameters, γ_{c}, γ_{s}, γ_{n}, γ_{p}, and b, as well as by three operational parameters, the current injection level $\tilde{J}$, the optical feedback parameter ξ, and the feedback delay time τ. The operational parameters can be externally varied to control the dynamics of the laser in operation. A semiconductor laser with optical feedback has much more complicated dynamics than one with optical injection because of its increasing dimension with the feedback delay time and because of the additional phase factor of θ. The phase factor θ accounts for the microscopic variations of the feedback length on the scale of a fraction of the optical wavelength. Thus, the dynamics of this system depends on the feedback delay time at two levels: one is the macroscopic level at a fixed value of θ, and the other is the microscopic level at a fixed delay time with θ varying within a 2π range. Figures 10.7a and b show examples of the bifurcation diagrams for these two cases.

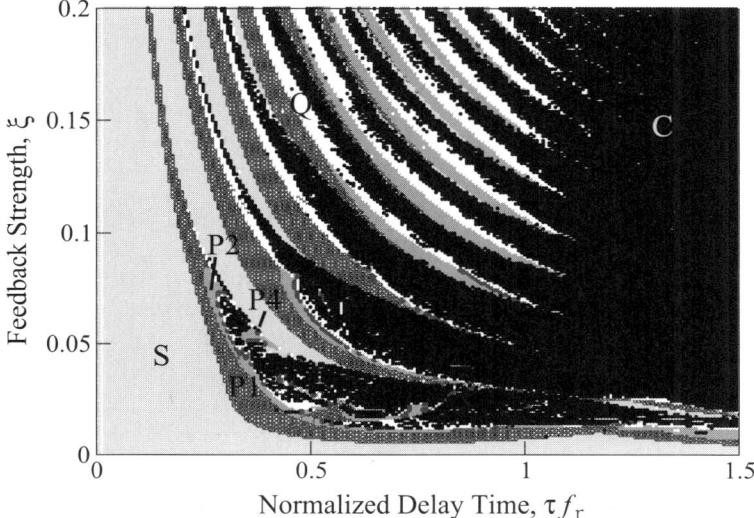

Fig. 10.8. Numerically calculated mapping of the high-frequency dynamical states of a semiconductor laser with single-pass optical feedback operating at a constant current level of $\tilde{J} = 2/3$ as a function of the feedback parameter ξ and the normalized delay time $\hat{\tau} = \tau f_{\mathrm{r}}$. In this map, S, P1, P2, P4, and Q indicate the regions of the stable states, the period-one states, the period-two states, the period-four states, and the quasiperiodic states, respectively, and the chaotic states are indicated by the solid-black region. Laser parameters listed in Table 10.1 are used to generate this map.

Besides its increased dimensionality and phase sensitivity, the dynamics of a semiconductor laser with optical feedback has other complications. There are basically two types of chaotic dynamics for this system: (1) low-frequency fluctuations (LFFs), which are often observed near the laser threshold [24–26] but can also occur at high injection levels [27], and (2) broadband chaos. The LFFs are characterized by sudden power dropouts of the laser in the megahertz range. Dynamical states in the LFF regime follow an intermittency route to chaos [28]. The broadband chaos, by comparison, has high-frequency continuous waveforms. Dynamical states in this regime are characterized by high-frequency fluctuations in the gigahertz range. In this regime, this system can follow either a period-doubling route or a quasiperiodicity route to chaos though the period-doubling route dominates. LFF chaos is not useful for high-bit-rate chaotic communications. Thus only dynamical states relevant to broadband high-frequency chaos are considered here.

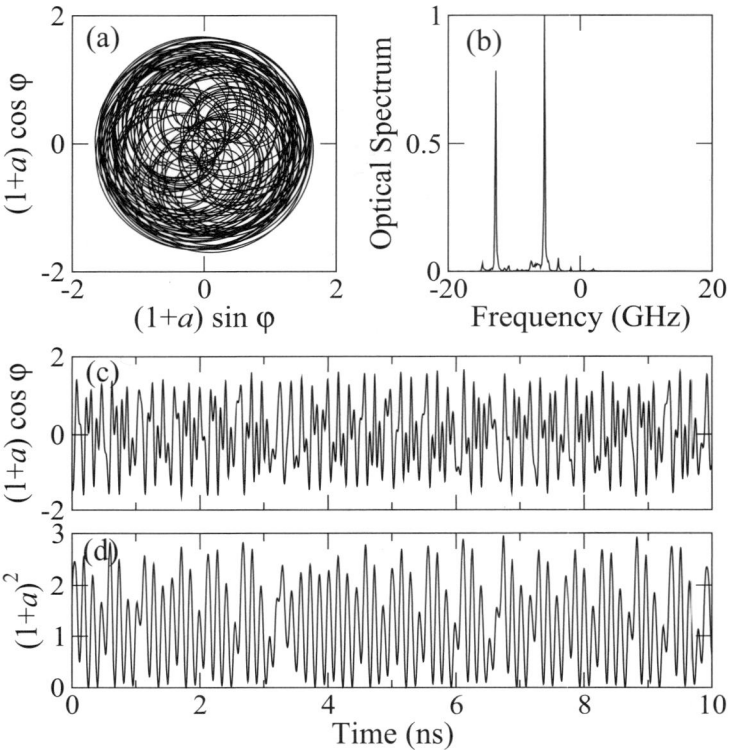

Fig. 10.9. Numerically calculated (a) attractor, (b) optical spectrum, (c) optical field, and (d) optical intensity of the chaotic state at the operating condition of $\tilde{J} = 2/3$, $\xi = 0.1$, and $\tau f_r = 0.3$ for a single-mode semiconductor laser with optical feedback that has the parameters listed in Table 10.1.

Figure 10.8 shows as an example a numerically calculated mapping of the high-frequency dynamical states of a semiconductor laser with single-pass optical feedback operating at a constant current level of $\tilde{J} = 2/3$ as a function of the feedback parameter ξ and the normalized delay time $\hat{\tau} = \tau f_r$, where f_r is the free-running relaxation frequency of the laser defined in the preceding section. Laser parameters listed in Table 10.1 are used to generate this map. It can be seen from Figure 10.8 that the high-frequency dynamics of this system can follow either a period-doubling route or a quasiperiodicity route to chaos when one of its operational parameters is varied while the other two are fixed.

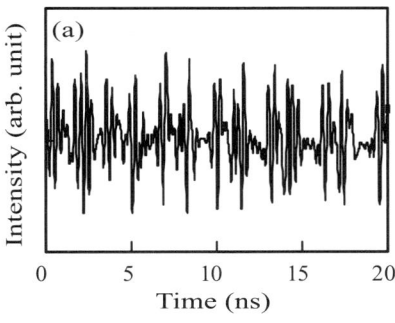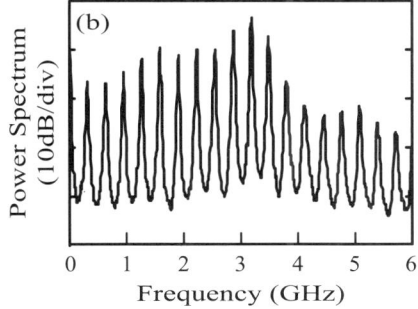

Fig. 10.10. Experimentally measured (a) temporal waveform and (b) power spectrum of a chaotic state of a single-mode semiconductor laser with optical feedback. (Reprinted with permission from [29], ©2002 AIP.)

Figure 10.9 shows the numerically calculated attractor, optical spectrum, optical field, and optical intensity of a representative chaotic state of the system at an operating condition of $\tilde{J} = 2/3$, $\xi = 0.1$, and $\hat{\tau} = \tau f_r = 0.3$. Experimentally measured temporal waveform and power spectrum of a chaotic state of an InGaAsP single-mode DFB laser with optical feedback are presented in Figures 10.10a and b, respectively [29].

10.4.3 Single-Mode Semiconductor Laser with Optoelectronic Feedback

Optoelectronic Feedback

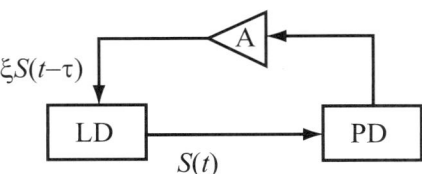

Fig. 10.11. Schematic diagram of a semiconductor laser with delayed optoelectronic feedback.

A semiconductor laser with delayed optoelectronic feedback is schematically shown in Figure 10.11. In this configuration, a combination of photodetector and amplifier is used to convert the optical output of the laser into an electrical signal that is fed back to the laser by adding it to the injection current. Because the photodetector responds only to the intensity of the laser output, the feedback signal contains the information on the variations of the laser intensity, which is proportional to the photon density in the laser cavity. Therefore, the phase of the laser field is not part of the dynamics of this system. The dynamics of this laser can be described by the following coupled equations of the photon and carrier densities,

$$\frac{dS}{dt} = -\gamma_c S + \Gamma g S + 2\sqrt{S_0 S} F_s, \tag{10.42}$$

$$\frac{dN}{dt} = \frac{J}{ed}[1 + \xi y(t - \tau)] - \gamma_s N - g S, \tag{10.43}$$

$$y(t) = \int_{-\infty}^{t} d\eta f(t - \eta) S(\eta) / S_0, \tag{10.44}$$

where ξ is the feedback strength, τ is the feedback delay time, and $f(t)$ is the normalized response function of the optoelectronic feedback loop which accounts for the finite bandwidths of the photodetector, the amplifier, and the electrical parasitic effects of the laser. In the ideal situation where the feedback loop has a flat, infinite bandwidth, then $f(t) = \delta(t)$ and $y(t) = S(t)/S_0$. When $f(t)$ is not simply $\delta(t)$, as is the case in a realistic situation, the system is characterized by coupled delay-differential-integral equations.

The coupled equations in (10.42) and (10.43) can be transformed into the following three coupled equations in terms of $\tilde{s}$ and $\tilde{n}$,

$$\frac{d\tilde{s}}{dt} = \frac{\gamma_c \gamma_n}{\tilde{J}\gamma_s}\tilde{n}(\tilde{s} + 1) - \gamma_p \tilde{s}(\tilde{s} + 1) + \tilde{F}_s, \tag{10.45}$$

$$\frac{d\tilde{n}}{dt} = \gamma_s \xi(1 + \tilde{J})y(t - \tau) - \gamma_s \tilde{n} - \gamma_s \tilde{J}\tilde{s} - \gamma_n \tilde{n}(1 + \tilde{s})$$
$$+ \frac{\gamma_s \gamma_p}{\gamma_c}\tilde{J}\tilde{s}(1 + \tilde{s}), \tag{10.46}$$

$$y(t) = \int_{-\infty}^{t} d\eta f(t - \eta)[1 + \tilde{s}(\eta)]. \tag{10.47}$$

Though there are only two dynamical variables $\tilde{s}$ and $\tilde{n}$, both of which are real scalar quantities, this system is a delayed-feedback autonomous system that can have a high dimension when the delay time is sufficiently long. Therefore, it can have chaotic dynamics. The dynamics of this system is not phase-sensitive.

It can be seen that the dynamics of a single-mode semiconductor laser with optoelectronic feedback is again determined by the five intrinsic parameters: γ_c, γ_s, γ_n, γ_p, and b. The parameter b does not have a direct effect on the dynamics of this system but only has an indirect effect through the coupling between the amplitude and phase of noise fluctuations not explicitly shown in the above model. The dynamics of this system can be controlled by three operational parameters: $\tilde{J}$, ξ, and τ that can be externally varied in operation. In addition, the response function $f(t)$ of the feedback loop also plays an important role in the dynamics of the system [30]. The significance of the

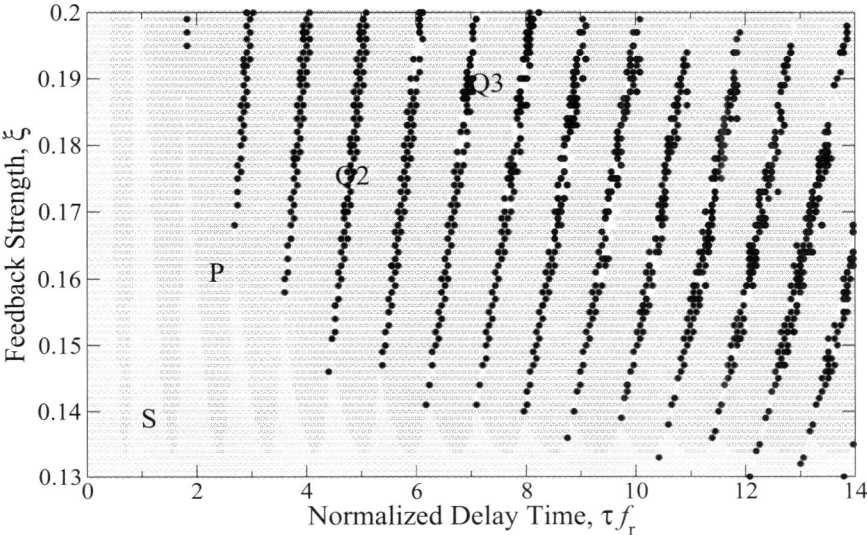

Fig. 10.12. Numerically calculated mapping of the dynamical states of a single-mode semiconductor laser with optoelectronic feedback operating at a constant current level of $\tilde{J} = 2/3$ as a function of the feedback parameter ξ and the normalized delay time $\hat{\tau} = \tau f_r$. In this map, S, P, Q2, and Q3 indicate the regions of the stable states, the periodic pulsing states, the two-frequency quasiperiodic pulsing states, and the three-frequency quasiperiodic pulsing states, respectively, and the chaotic pulsing states are indicated by the solid black region. Laser parameters listed in Table 10.1 are used to generate this map.

function $f(t)$ on the dynamics of the system depends on its form. In the ideal situation of an infinite bandwidth, $f(t) = \delta(t)$, the dynamics of the system is not affected by the bandwidth limitation. On the other extreme when $f(t)$ represents that of a narrow bandpass filter, the chaotic dynamics can be completely eliminated. In our experimental and numerical studies, we have found that a practical broad, but finite, bandpass filter has the effect of changing the pulsing frequency but not the general dynamics of the system [30]. The optoelectronic feedback can be either positive, with $\xi > 0$ [30], or negative, with $\xi < 0$ [31]. The dynamical phenomena for both cases have been studied in detail [30, 31].

Figure 10.12 shows as an example a numerically calculated mapping of the dynamical states of a single-mode semiconductor laser with positive optoelectronic feedback operating at a constant current level of $\tilde{J} = 2/3$ as a function of the feedback parameter ξ and the normalized delay time $\hat{\tau} = \tau f_r$, where f_r is the free-running relaxation frequency of the laser. The dynamics

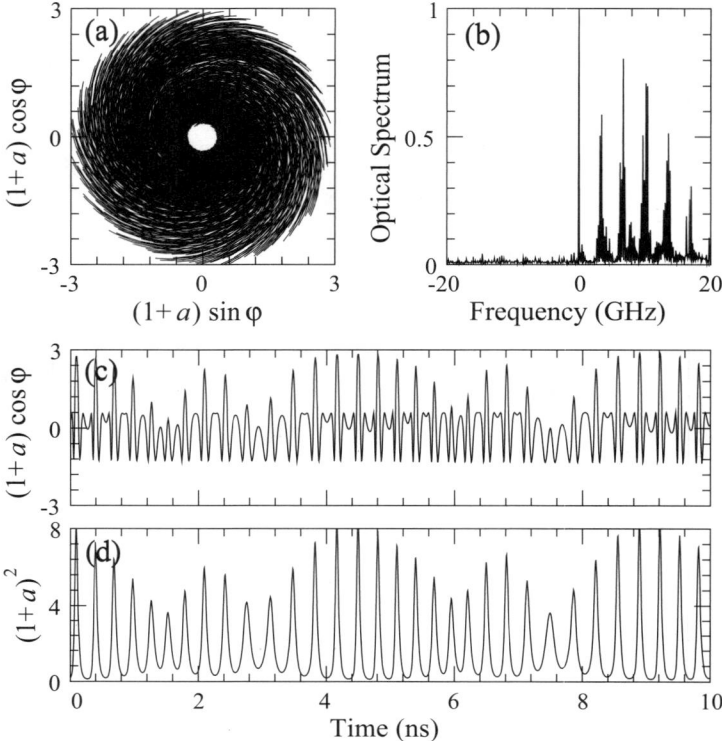

Fig. 10.13. Numerically calculated (a) attractor, (b) optical spectrum, (c) optical field, and (d) intensity of the chaotic state at the operating condition of $\tilde{J} = 2/3$, $\xi = 0.18$, and $\hat{\tau} = 12.5$ for a single-mode semiconductor laser with optoelectronic feedback that has the parameters listed in Table 10.1.

of this system is very different from that of the optical injection system and that of the optical feedback system. The instability induced by the optoelectronic feedback causes the laser to pulse, starting from a regular pulsing state and then following a quasiperiodic pulsing route to a chaotic pulsing state as one of the operational parameters is varied while the other two are fixed. It is found that the pulse spacing correlates very well with the pulse peak. In a regular pulsing state, both the pulse spacing and peak are constant. In a quasiperiodic pulsing state, both vary quasiperiodically. In a chaotic pulsing state, both vary chaotically with the same signature.

Figure 10.13 shows the numerically calculated attractor, optical spectrum, optical field, and optical intensity of a representative chaotic state of the system at an operating condition of $\tilde{J} = 2/3$, $\xi = 0.18$, and $\hat{\tau} = 12.5$. Experimentally measured temporal waveforms and power spectra of different dynamical states of an InGaAsP single-mode DFB laser with optoelectronic feedback

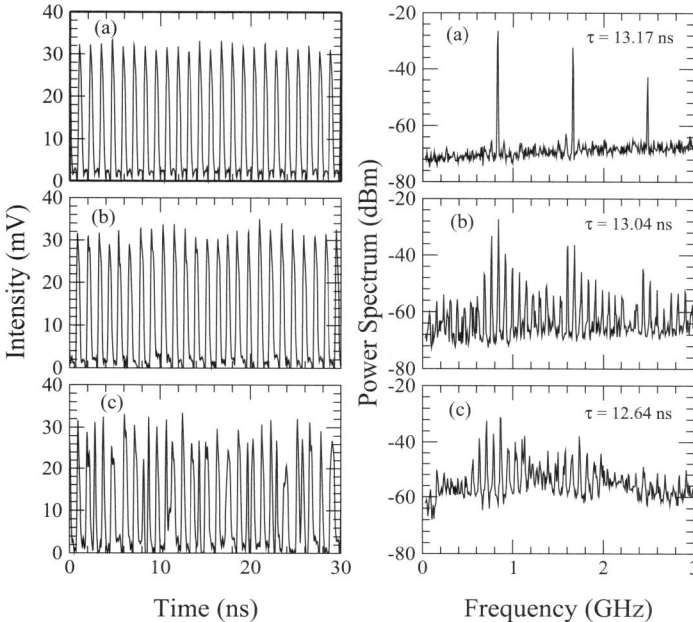

Fig. 10.14. Experimentally measured temporal waveforms and corresponding power spectra of (a) periodic pulsing, (b) two-frequency quasiperiodic pulsing, and (c) chaotic pulsing states of a single-mode semiconductor laser with positive optoelectronic feedback.

at different delay times are presented in Figure 10.14. Because of the laser noise, the three-frequency quasiperiodic pulsing states seen in the numerical mapping are very difficult to observe experimentally.

10.5 Basic Concept of Chaos Synchronization

A chaotic system is a dynamical system that is intrinsically sensitive to initial conditions by definition. For a chaotic system described in its phase space, two nearby initial conditions as two points in the phase space exponentially separate from each other when its dynamics evolves forward in time. Because of this exponential divergence on different initial conditions, it seems not possible to operate two chaotic systems to perform synchronized dynamics even when they are identical systems. Indeed, it is not possible to synchronize two isolated identical chaotic systems because of their sensitivity to initial conditions and to any small perturbations. However, when there is a proper driving signal to couple two chaotic systems, synchronization is possible. The feasibility of synchronizing two chaotic systems was first realized based on the theory proposed by Pecora and Carroll [32], and later generalized to the concept of passive-active synchronization proposed by Kocarev and Parlitz [33]. Based on these theories, chaos synchronization is possible if the difference between the outputs of the two chaotic systems possesses a stable fixed point with zero value when a driving signal is used to couple these two systems. Because chaos is usually noiselike in terms of its complexity and long-term unpredictability, chaos synchronization has generated great interest for its potential application in private communications. Because the discussions here focus on chaos synchronization for communication purposes, we consider only *identical chaos synchronization* in which two synchronized chaotic oscillators generate identical outputs. Generalized chaos synchronization, achieved when the outputs of two synchronized chaotic oscillators are related by any function but not an identity function, is not addressed in the following discussions.

10.5.1 General Review

There are two methods, one proposed by Pecora and Carroll [32] and another proposed by Kocarev and Parlitz [33], to achieve chaos synchronization. The difference between these two methods is significant when the concept of chaos synchronization is implemented in real systems.

To demonstrate the theory of chaos synchronization proposed by Pecora and Carroll [32], we first consider two chaotic systems described by $\mathbf{x}$ and $\mathbf{y}$, and their respective equations. We assume that these two dynamical systems are identical. Each of these two systems performing synchronization can be decomposed into two subsystems, $\mathbf{x} = (\mathbf{x_1}, \mathbf{x_2})$ and $\mathbf{y} = (\mathbf{y_1}, \mathbf{y_2})$. One of the subsystems, say $\mathbf{x_1}$, decomposed from the first chaotic system, is used as the driving signal to couple the two chaotic systems and, at the same time, to replace the corresponding dynamical variables, represented by $\mathbf{y_1}$, in the second chaotic system. Synchronization is accomplished when the remaining subsystems represented by $\mathbf{x_2}$ and $\mathbf{y_2}$ are synchronized. This method can be mathematically expressed as the following,

$$\mathbf{x} = (\mathbf{x_1}, \mathbf{x_2}) \ , \tag{10.48}$$

$$\frac{d\mathbf{x_1}}{dt} = \mathbf{F_1}(\mathbf{x_1}, \mathbf{x_2}) \ , \tag{10.49}$$

$$\frac{d\mathbf{x_2}}{dt} = \mathbf{F_2}(\mathbf{x_1}, \mathbf{x_2}) \ , \tag{10.50}$$

$$\frac{d\mathbf{y_2}}{dt} = \mathbf{G_2}(\mathbf{x_1}, \mathbf{y_2}) \ , \tag{10.51}$$

where $\mathbf{G_2} = \mathbf{F_2}$ is the required condition for the existence of perfect synchronization. When synchronization takes place, the deviation, $\mathbf{x_2} - \mathbf{y_2}$, between these two subsystems asymptotically approaches zero. The achievement of chaos synchronization in this method depends on the proper choice of the subsystem which works as the driving signal [32]. The subsystem used to provide the driving signal has to be extractable and separable from the first chaotic system. When the coupling signal is absent, the second system is actually identical to one of the subsystems decomposed from the first chaotic system. Therefore, the dimension of the second system is smaller than that of the first chaotic system, and it is equal to one of the subsystems of the first chaotic system. For a three-dimensional chaotic system described by differential equations, the second system cannot be a chaotic system because its maximum possible embedding dimension is two. This method is generally referred to as the complete replacement method [34].

The method later proposed by Kocarev and Parlitz is known as the active−passive method [33]. In this method, the driving signal, $\mathbf{s}(t) = \mathbf{h}(\mathbf{x})$, is a function of the dynamical variables $\mathbf{x}$ of the first chaotic system. For two chaotic systems that are physically coupled by a signal $\mathbf{s}(t)$, each system can be rewritten mathematically as if the system is driven by a common driving signal $\mathbf{D}(\mathbf{s}(t))$ so that they can be mathematically expressed as the following,

$$\frac{d\mathbf{x}}{dt} = \mathbf{F}(\mathbf{x}, \mathbf{D}(\mathbf{s}(t))) \ , \tag{10.52}$$

$$\frac{d\mathbf{y}}{dt} = \mathbf{G}(\mathbf{y}, \mathbf{D}(\mathbf{s}(t))) \ , \tag{10.53}$$

where $\mathbf{D}(\mathbf{s}(t))$ is a function of the signal $\mathbf{s}(t)$. With the vector function $\mathbf{G}$ being equal to the vector function $\mathbf{F}$, these two chaotic systems can be synchronized to each other if the difference $\mathbf{e} = \mathbf{x} - \mathbf{y}$ possesses a stably fixed point with a zero value. The fixed point exists when the average local Lyapunov exponents of the difference $\mathbf{e}$ are all negative.

As is shown in the equations in (10.52) and (10.53), the driving signal does not replace any dynamical variable of the second system. Instead, the variables of the second system that correspond to those contained in the driving signal asymptotically approach those of the first system. When they are synchronized, the remaining variables of the second system also synchronize with the remaining variables of the first system. In this method, the number of the dynamical variables of the second system is the same as that of the first system because the dynamical variables describing the second system are not partially replaced by the driving signal. Whether the two systems have the same dynamical dimension when they are decoupled depends on the configuration of the setups to generate chaos and to synchronize both systems.

10.5.2 Synchronization of Semiconductor Lasers

In chaos synchronization utilizing semiconductor lasers as chaos generators, the dynamical variables used for the driving signal are not always separable from others and some are simply not extractable from a laser. When the output laser field of the first laser system is transmitted and coupled, both its

magnitude and phase are transmitted and coupled to the second laser. It is
not possible to only transmit and couple the magnitude but not the phase,
or only the phase but not the magnitude. Therefore, unless the phase is not
part of the dynamics of the lasers, such as in the case of lasers with optoelectronic feedback, the synchronization between two laser systems depends on
the coupling of the two variables, the magnitude and phase of the laser field,
at the same time. Furthermore, the carrier density is not directly accessible
externally. Therefore, it cannot be used as the driving signal to couple the
first and the second lasers. Because a semiconductor laser is an integrated
entity, all the dynamical variables are definitely not replaceable. Therefore,
chaos synchronization of semiconductor laser systems cannot be implemented
with the complete replacement method proposed by Pecora and Carroll, but
it is feasible if the active-passive method proposed by Kocarev and Parlitz is
used.

In semiconductor laser systems, the coupling between two laser systems is
usually achieved through the entire optical field or the field intensity of the
first laser to additively connect the two systems. For such systems using an
additive driving signal to connect two dynamical systems, the general equations in (10.52) and (10.53) of the synchronization theory can be rewritten
as

$$\frac{d\mathbf{x}}{dt} = \mathbf{F}(\mathbf{x}) + \alpha \mathbf{D}(\mathbf{x}) , \tag{10.54}$$

$$\frac{d\mathbf{y}}{dt} = \mathbf{G}(\mathbf{y}) + \alpha \mathbf{D}(\mathbf{x}) , \tag{10.55}$$

where $\alpha \mathbf{D}(\mathbf{x})$ has replaced the driving signal $\mathbf{D}(\mathbf{s}(t))$ because this special type
of driving force is more suitable when the chaos synchronization system is de-
signed for communication purposes, and the coupling strength α has been
separated from $\mathbf{D}(\mathbf{s}(t))$. As is mentioned in the theory of chaos synchronization, in order to achieve perfect chaos synchronization, the dynamics of two
laser systems have to be described by identity equations. Therefore, if the
first term, $\mathbf{F}(\mathbf{x})$, on the right-hand side of (10.54) describes the dynamics of a
free-running semiconductor laser, the signal $\alpha \mathbf{D}(\mathbf{x})$ is also the signal to drive
the first semiconductor laser into chaotic states. If $\mathbf{F}(\mathbf{x})$ already contains what
drives the laser into chaos, then the signal $\alpha \mathbf{D}(\mathbf{x})$ can further influence the
laser dynamics, sometimes increasing its complexity and dimensionality [35].

These general equations of chaos synchronization through an additive cou-
pling can be written in another form. By defining $\mathbf{f}(\mathbf{x}) \equiv \mathbf{F}(\mathbf{x}) + \alpha \mathbf{D}(\mathbf{x})$ and
$\mathbf{g}(\mathbf{y}) \equiv \mathbf{G}(\mathbf{y}) + \alpha \mathbf{D}(\mathbf{y})$, we can rewrite the equations as the following,

$$\frac{d\mathbf{x}}{dt} = \mathbf{F}(\mathbf{x}) + \alpha \mathbf{D}(\mathbf{x}) = \mathbf{f}(\mathbf{x}) , \tag{10.56}$$

$$\frac{d\mathbf{y}}{dt} = \mathbf{G}(\mathbf{y}) + \alpha \mathbf{D}(\mathbf{x})$$

$$= \mathbf{G}(\mathbf{y}) + \alpha \mathbf{D}(\mathbf{y}) + \alpha \mathbf{D}(\mathbf{x}) - \alpha \mathbf{D}(\mathbf{y})$$

$$= \mathbf{g}(\mathbf{y}) + \alpha (\mathbf{D}(\mathbf{x}) - \mathbf{D}(\mathbf{y})) . \tag{10.57}$$

This expression makes a clear connection between the method proposed by Pecora and Carroll and that proposed by Kocarev and Parlitz: Because $\mathbf{D}(\mathbf{x})$ or $\mathbf{D}(\mathbf{y})$ contains only partial variables describing laser dynamics, the achievement of chaos synchronization requiring $\mathbf{D}(\mathbf{x}) - \mathbf{D}(\mathbf{y}) = 0$ effectively replaces these partial variables in the second equation by those in the first equation. This expression also provides an intuitive picture about the occurrence of synchronization though the dynamics of each system is sensitive to initial conditions. This issue is addressed in more detail in Chapter 9.6.

No matter which expression is used to describe the synchronization, we have $\mathbf{G} = \mathbf{F}$, thus $\mathbf{g} = \mathbf{f}$ when all the parameters are matched. Thus, one can see from (10.54) and (10.55) or from (10.56) and (10.57) that perfect synchronization is possible only when $\mathbf{G} = \mathbf{F}$ and $\mathbf{g} = \mathbf{f}$. Any mismatch between the equations describing the two systems can either deteriorate the quality of synchronization or make the synchronization impossible.

10.6 Chaos Synchronization of Single-Mode Semiconductor Lasers

Chaos synchronization theory requires that the two laser systems to be identically synchronized be described by identical equations, as is discussed in Section 10.5. A single-mode semiconductor laser is a class B laser that does not have complex dynamics in its free-running condition in the absence of an external perturbation. Thus an external perturbation is necessary to drive the lasers into chaos. When this condition is combined with the requirement for identical chaos synchronization, the external perturbation that drives the transmitter laser into chaos has to be contained in the driving signal, $\alpha \mathbf{D}(\mathbf{x})$, that is coupled to the receiver laser. Therefore, the external perturbation chosen to drive the transmitter laser into chaos usually limits the design of the system to perform chaos synchronization. For this reason, the criteria used to categorize different semiconductor laser systems based on different external perturbations can be used to categorize the systems devised to achieve chaos synchronization. In this section, we discuss chaos synchronization of the three semiconductor laser systems categorized according to the external perturbation used to drive the transmitter into chaos as discussed in Section 10.4. They are the optical injection system [7], the optical feedback system [9–12], and the optoelectronic feedback system [13,14]. All three systems considered here for chaos synchronization for the purpose of chaotic optical communication are unidirectionally coupled systems.

Different systems have different characteristics regarding chaos synchronization. For optical injection and optical feedback systems, the frequency, phase, and amplitude of the optical fields of both transmitter and receiver lasers are all locked in synchronism when synchronization is accomplished [7–29] because both phase and amplitude of the laser field participate in the dynamics of these systems. Besides the complete synchronization of the entire optical field, the optical injection system exhibits synchronization sensitivity to the optical phase difference between the injection field and the transmitter laser field. The optical feedback system exhibits phase sensitivity only when the receiver has a closed-loop configuration. For the optoelectronic feedback system, chaos synchronization does not involve the locking of the optical frequency or the synchronization of the optical phase because the phase of the laser field does not participate in the nonlinear dynamics of this system. Synchronization of this system is not phase sensitive [13]. For both optical feedback and optoelectronic feedback systems, where the transmitter is configured with a delayed feedback loop, anticipated and retarded synchronization can be observed as the difference between the feedback delay time and the propagation time from the transmitter to the receiver laser is varied [36].

Although these three systems have different synchronization characteristics, all of them exhibit multiple synchronous scenarios regardless of the differences in their nonlinear dynamics. Besides the true chaos synchronization described by the theory of identical chaos synchronization discussed in Section 10.5, one or two other synchronous phenomena are observed for each system under different operating conditions [37–39]. These other synchronous phenomena, such as chaotic driven oscillation and chaotic modulation, have different signatures compared to true identical chaos synchronization.

In this section, we address the various characteristics of chaos synchronization in unidirectionally coupled semiconductor laser systems. chaos synchronization in all three semiconductor laser systems covered in this chapter has been accomplished and their characteristics carefully studied. For each of the three different systems, we first discuss the specific requirements for true identical chaos synchronization in the system, followed by the demonstration of experimental results. The various interesting characteristics and phenomena discussed above have all been observed experimentally, but we shall not show the results for every one of them in each of the three system. Instead, a particular phenomenon is appropriately demonstrated with only one of the three systems as an example. The synchronization of the entire optical chaotic waveform, as well as the phase sensitivity of chaos synchronization, is demonstrated in the optical injection system. The existence of multiple synchronous scenarios in one system under different operating conditions is demonstrated in the optical feedback system. The anticipated and retarded synchronization as the special characteristic of a system configured with a feedback loop is clearly demonstrated in the optoelectronic feedback system. All the results shown in this section are obtained from experiments.

Before proceeding with the detailed discussions on chaos synchronization in these systems, it is necessary to define a quantitative measure of the quality of synchronization. There are a few different ways of measuring the quality of synchronization. The simplest and most commonly used measure is the correlation coefficient between the two waveforms to be synchronized. The correlation coefficient is generally defined as

$$\rho = \frac{\langle [X(t) - \langle X(t)\rangle][Y(t) - \langle Y(t)\rangle]\rangle}{\langle |X(t) - \langle X(t)\rangle|^2\rangle^{1/2} \langle |Y(t) - \langle Y(t)\rangle|^2\rangle^{1/2}}, \qquad (10.58)$$

where $X(t)$ and $Y(t)$ are the outputs of the transmitter and the receiver, respectively, and $\langle \cdot \rangle$ denotes the time average. The correlation coefficient is bounded as $-1 \leq \rho \leq 1$. A larger value of $|\rho|$ indicates a higher quality of synchronization.

10.6.1 Synchronization of Optical Injection System

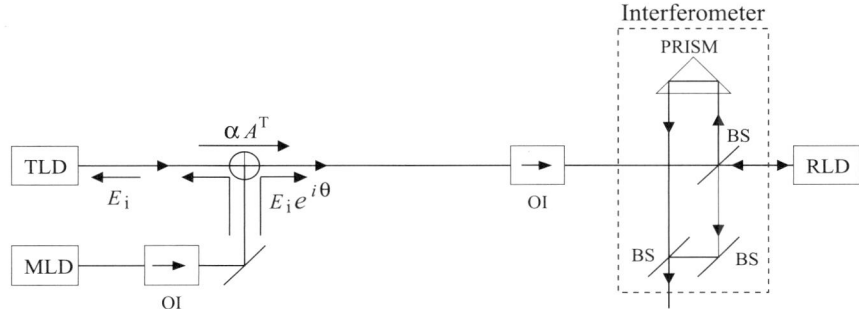

Fig. 10.15. Schematic of the experiment of complete chaos synchronization for semiconductor lasers subject to optical injection. E_i is the injection field from the master laser, MLD: master laser diode, TLD: transmitter laser diode, RLD: receiver laser diode, BS: beam splitter, OI: optical isolator.

The configuration for chaos synchronization in the optical injection system is schematically shown in Figure 10.15. The optical output of the master laser is split into two beams: the one denoted by $E_i(t)$ is injected to the transmitter laser, and the other beam denoted by $E_i(t)e^{i\theta}$ is injected into the receiver laser together with the output of the transmitter laser. The existence of the optical phase θ can be realized from the fact that the optical field of the transmitter encounters the output field of the master laser twice. The first time is when the transmitter receives the master laser output as the optical injection field to drive the transmitter into a chaotic state, and the second time is when the transmitter output is combined with the master laser output to be transmitted and injected into the receiver laser. The optical field, as the intracavity field and thus the output field, of the transmitter laser may not see the same optical phase of the master laser field at these two locations. The difference in this relative optical phase is indicated by θ as is contained in $E_i(t)e^{i\theta}$.

According to the schematic in Figure 10.15, the transmitter can be modeled by the following coupled equations in terms of the complex intracavity laser field amplitude A^T and the carrier density N^T [7,8], as has been discussed in Section 10.4:

$$\frac{dA^{\mathrm{T}}}{dt} = -\frac{\gamma_{\mathrm{c}}^{\mathrm{T}}}{2}A^{\mathrm{T}} + i(\omega_0^{\mathrm{T}} - \omega_{\mathrm{c}}^{\mathrm{T}})A^{\mathrm{T}} + \frac{\Gamma}{2}(1 - ib^{\mathrm{T}})gA^{\mathrm{T}} + F_{\mathrm{sp}}^{\mathrm{T}}$$
$$+\eta E_{\mathrm{i}}(t), \tag{10.59}$$

$$\frac{dN^{\mathrm{T}}}{dt} = \frac{J}{ed} - \gamma_{\mathrm{s}}N^{\mathrm{T}} - \frac{2\epsilon_0 n^2}{\hbar\omega_0}g|A^{\mathrm{T}}|^2. \tag{10.60}$$

where the superscript T labels the variables of the transmitter. The receiver, driven by the transmitted signal $s(t) = \alpha A^{\mathrm{T}}(t) + E_{\mathrm{i}}(t)e^{i\theta}$, with α being the coupling strength, is described by

$$\frac{dA^{\mathrm{R}}}{dt} = -\frac{\gamma_{\mathrm{c}}^{\mathrm{R}}}{2}A^{\mathrm{R}} + i(\omega_0^{\mathrm{T}} - \omega_{\mathrm{c}}^{\mathrm{R}})A^{\mathrm{R}} + \frac{\Gamma}{2}(1 - ib^{\mathrm{R}})gA^{\mathrm{R}} + F_{\mathrm{sp}}^{\mathrm{R}} + \eta s(t)$$
$$= -\left(\frac{\gamma_{\mathrm{c}}^{\mathrm{R}}}{2} - \alpha\eta\right)A^{\mathrm{R}} + i(\omega_0^{\mathrm{T}} - \omega_{\mathrm{c}}^{\mathrm{R}})A^{\mathrm{R}} + \frac{\Gamma}{2}(1 - ib^{\mathrm{R}})gA^{\mathrm{R}} + F_{\mathrm{sp}}^{\mathrm{R}}$$
$$+\eta E_{\mathrm{i}}(t)e^{i\theta} + \alpha\eta(A^{\mathrm{T}} - A^{\mathrm{R}}), \tag{10.61}$$

$$\frac{dN^{\mathrm{R}}}{dt} = \frac{J}{ed} - \gamma_{\mathrm{s}}N^{\mathrm{R}} - \frac{2\epsilon_0 n^2}{\hbar\omega_0}g|A^{\mathrm{R}}|^2, \tag{10.62}$$

where the superscript R labels the variables of the receiver. The definitions of all other variables can be found in Section 10.4.

Based on the synchronization concept proposed by Kocarev and Parlitz [33], the existence of a perfect synchronization solution, $A^{\mathrm{R}} = A^{\mathrm{T}}$, requires that $\theta = 0$, $\Delta\omega_{\mathrm{c}} \equiv \omega_{\mathrm{c}}^{\mathrm{T}} - \omega_{\mathrm{c}}^{\mathrm{R}} = 0$, and the two lasers be identical except that $\gamma_{\mathrm{c}}^{\mathrm{R}} = \gamma_{\mathrm{c}}^{\mathrm{T}} + 2\eta\alpha$ [7]. The requirement on the optical phase θ is recognized as the phase sensitivity of chaos synchronization for this system. From the steady-state condition of a free-running semiconductor laser, we further obtain the required detuning between the free-running frequencies of the transmitter and the receiver due to the difference in γ_{c} as the following,

$$\omega_0^{\mathrm{T}} - \omega_0^{\mathrm{R}} = \Delta\omega_{\mathrm{c}} + (b^{\mathrm{T}}\gamma_{\mathrm{c}}^{\mathrm{T}} - b^{\mathrm{R}}\gamma_{\mathrm{c}}^{\mathrm{R}})/2. \tag{10.63}$$

When $\Delta\omega_{\mathrm{c}} = 0$, we obtain $\omega_0^{\mathrm{T}} - \omega_0^{\mathrm{R}} = (b^{\mathrm{T}}\gamma_{\mathrm{c}}^{\mathrm{T}} - b^{\mathrm{R}}\gamma_{\mathrm{c}}^{\mathrm{R}})/2$. Because $A^{\mathrm{T}}(t)$ and $A^{\mathrm{R}}(t)$ are both complex amplitudes at the free-running frequency of the transmitter, complete chaos synchronization with $A^{\mathrm{R}} = A^{\mathrm{T}}$ requires that the fast-varying optical phase, the slowly varying phase, and the field amplitude of the receiver output be all synchronized to those of the transmitter output.

The injection strength from the master laser to the transmitter is adjusted so that the transmitter is operated in a chaotic state, the power spectrum of which is shown in Figure 10.16a. The receiver is synchronized to the transmitter when the relative optical phase difference θ is zero. When synchronization is accomplished, the detuning between the free-running frequencies of the transmitter and the receiver is around -32.6 ± 0.9 GHz, which satisfies the frequency condition of synchronization given in (10.63). The power spectrum of the synchronized receiver is shown in Figure 10.16b, which is very similar to that of the transmitter. The power spectrum of the channel signal that is the

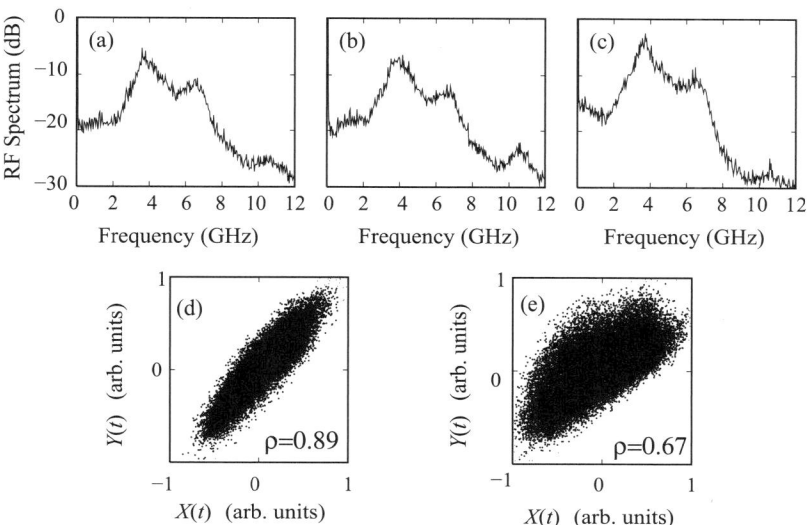

Fig. 10.16. Experimental results of chaos synchronization for the optical injection system when θ is zero: (a), (b), and (c) power spectra of the transmitter output, receiver output, and channel signal, respectively; (d) correlation plot between the receiver output $X(t)$ and the transmitter output $Y(t)$; (e) correlation plot between the receiver output $X(t)$ and the channel signal $Y(t)$.

superposition of the transmitter output and the master laser output is shown in Figure 10.16c as a reference. The quality of the experimentally achieved synchronization with $\theta = 0$ is shown in Figure 10.16d through the correlation plot between the intensities of the transmitter and receiver waveforms. The quality of synchronization is measured to be $\rho \simeq 0.89$. The correlation between the channel signal and the receiver waveform is also measured and is found to be $\rho \simeq 0.67$ as is shown in Figure 10.16e. Comparing the correlation plots in Figures 10.16d and e, we observe that the receiver is synchronized to the transmitter output, but not to the channel signal.

The synchronization on both the fast-varying optical phase and the slowly varying phase of the chaotic waveform is verified through optical interference with the results shown in Figure 10.17. The intensity of the constructive coherent interference is shown in Figure 10.17a, and that of the destructive interference is shown in Figure 10.17b. The intensity extinction ratio is larger than 5. This provides the evidence of the synchronization on both the fast-varying and slowly varying phases as well as on the field amplitude. As is expected from the theory, the entire optical field of the receiver is synchronized to that of the transmitter when chaos synchronization is achieved.

Different from the optical feedback and the optoelectronic feedback systems, this system exhibits the synchronization sensitive to the relative optical phase difference, θ, as discussed in the condition of the synchronization for this

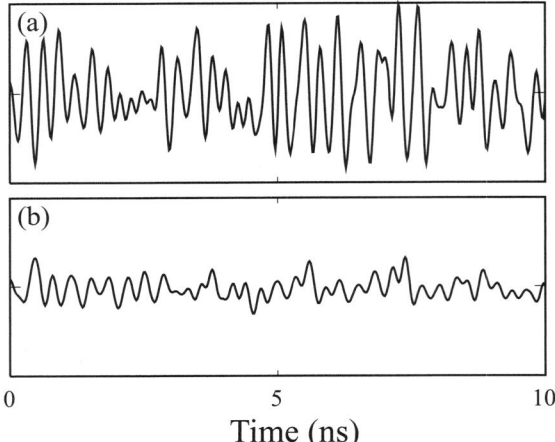

Fig. 10.17. Optical interference of synchronized fields from transmitter and receiver: (a) constructive coherent interference, (b) destructive coherent interference. (Reprinted with permission from [40], ©2005 APS.)

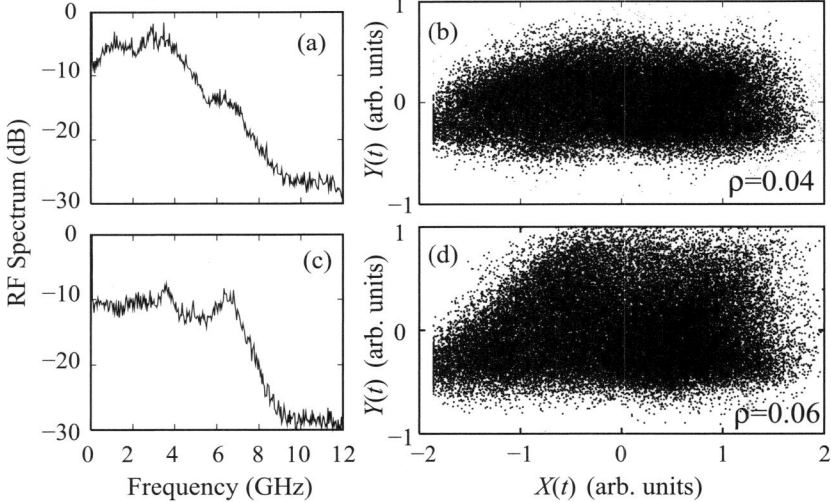

Fig. 10.18. Experimental results for the optical injection system when the receiver is desynchronized from the transmitter at $\theta = \pi$: (a) power spectrum of the receiver output, (b) correlation plot between the receiver output $X(t)$ and the transmitter output $Y(t)$, (c) power spectrum of the channel signal, (d) correlation plot between the receiver output $X(t)$ and the channel signal $Y(t)$. The power spectrum of the transmitter output is that shown in Figure 10.16a. (Reprinted with permission from [40], ©2005 APS.)

system. As expected from the theory, when we gradually tune θ away from zero without changing the other operating conditions, the receiver quickly desynchronizes from the transmitter. The power spectrum of the desynchronized receiver for $\theta = \pi$ is shown in Figure 10.18a. The correlation plot between the transmitter and the receiver is shown in Figure 10.18b. The correlation quality in this case drops to $\rho \simeq 0.04$. As a comparison, the power spectrum of the channel signal is shown in Figure 10.18c, and the quality of synchronization between the receiver and the channel signal is $\rho \simeq 0.06$.

From these observations, we see that complete synchronization of this system is achieved by the simultaneous synchronization of the fast-varying optical phase, the slowly varying phase, and the intensity of the optical chaotic waveform under the condition that the transmitter and receiver lasers are frequency-locked. Besides this synchronization characteristic, the achieved synchronization in this system is very sensitive to the relative optical phase difference between the master laser field and the transmitter laser field in the signal received by the receiver. A mismatch in this optical phase can completely desynchronize the system. This phase sensitivity is due to the fact that the receiver need two optical driving signals to achieve synchronization: One is the transmitter laser output, and the other is the master laser output.

10.6.2 Synchronization of Optical Feedback System

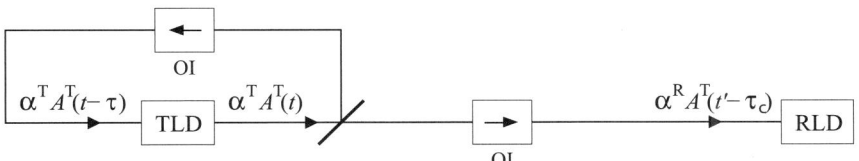

Fig. 10.19. Schematic of the experiment of chaos synchronization for semiconductor lasers with optical feedback. TLD: transmitter laser diode, RLD: receiver laser diode, OI: optical isolator.

The configuration for chaos synchronization in an optical feedback system is schematically shown in Figure 10.19. In this example, we only consider a receiver with an open-loop configuration. Assuming that the laser parameters of the transmitter and the receiver are matched as required by the theory of chaos synchronization, the transmitter configured with the optical feedback, with α^{T} being the feedback strength, can be modeled by the following coupled equations,

$$
\begin{aligned}
\frac{dA^{\mathrm{T}}}{dt} = &-\frac{\gamma_c^{\mathrm{T}}}{2}A^{\mathrm{T}} + i(\omega_0 - \omega_c)A^{\mathrm{T}} + \frac{\Gamma}{2}(1 - ib)gA^{\mathrm{T}} + F_{\mathrm{sp}}^{\mathrm{T}} \\
&+\alpha^{\mathrm{T}}\eta A^{\mathrm{T}}(t - \tau)e^{i\theta_1},
\end{aligned}
\tag{10.64}
$$

$$\frac{dN^{\mathrm{T}}}{dt} = \frac{J}{ed} - \gamma_{\mathrm{s}} N^{\mathrm{T}} - \frac{2\epsilon_0 n^2}{\hbar\omega_0} g|A^{\mathrm{T}}|^2. \tag{10.65}$$

The receiver, driven by the transmitted signal $\alpha^{\mathrm{R}} A^{\mathrm{T}}(t - T)$ with α^{R} being the coupling strength, can be described by

$$\frac{dA^{\mathrm{R}}}{dt} = -\frac{\gamma_{\mathrm{c}}^{\mathrm{R}}}{2} A^{\mathrm{R}} + i(\omega_0 - \omega_{\mathrm{c}}) A^{\mathrm{R}} + \frac{\Gamma}{2}(1 - ib)g A^{\mathrm{R}} + F_{\mathrm{sp}}^{\mathrm{R}}$$
$$+ \alpha^{\mathrm{R}} \eta A^{\mathrm{T}}(t - T) e^{i\theta_2}, \tag{10.66}$$

$$\frac{dN^{\mathrm{R}}}{dt} = \frac{J}{ed} - \gamma_{\mathrm{s}} N^{\mathrm{R}} - \frac{2\epsilon_0 n^2}{\hbar\omega_0} g|A^{\mathrm{R}}|^2, \tag{10.67}$$

where T is the propagation time of the driving signal $A^{\mathrm{T}}(t)$ from the transmitter to the receiver.

Because θ_1 is the difference between the fast-varying optical phase of the transmitter intracavity field and that of the feedback optical signal from the time delay of the feedback loop, we have $\theta_1 = \omega_0 \tau$. Similarly, we have $\theta_2 = \omega_0 T = \omega_0(\tau + \Delta t)$, where $\Delta t \equiv T - \tau$. Based on the concept of chaos synchronization, the existence of perfect synchronization requires that the rate equations of the transmitter and the receiver be identical. Therefore, the receiver should be synchronized to the transmitter with a time shift such as $A^{\mathrm{R}}(t) = A^{\mathrm{T}}(t - \Delta t)e^{i\omega_0 \Delta t}$ if they can be synchronized. From the practical point of view, the phase delay term θ_1 can vary from 0 to 2π even when the feedback delay time is fixed. This is because the optical frequency varies so fast that a small inaccuracy in the length of the feedback loop can significantly change this phase value, but such small inaccuracy in the length of the feedback loop is not significant for the dynamics of the transmitter if the phase change is not considered. Nevertheless, the receiver will self-adjust the synchronization time to response to any inaccuracy in the length of the feedback loop of the transmitter because the receiver has an open-loop configuration without feedback. Therefore, the synchronization condition is not sensitive to the optical phases, θ_1 and θ_2, though the dynamics of the transmitter is sensitive to θ_1. When $A^{\mathrm{R}}(t) = A^{\mathrm{T}}(t - \Delta t)e^{i\omega_0 \Delta t}$, the entire optical chaotic waveforms of the transmitter and the receiver are synchronized except that the synchronization can be anticipated or retarded. We use the optoelectronic feedback system as the example to demonstrate anticipated and retarded synchronization. In this setup, we present the existence of different types of synchronous chaotic phenomena that are experimentally observed [37].

In this system, there are two types of synchronous chaotic phenomena observed in experiment and theory [37, 38]. One is true chaos synchronization that defined by the concept of identical synchronization discussed in Section 10.5, and the other is defined as chaotic driven oscillation [38]. The difference between these two types of synchronous scenarios is examined with regards to the coupling strength from the transmitter to the receiver, the frequency detuning between the transmitter and the receiver, and the syn-

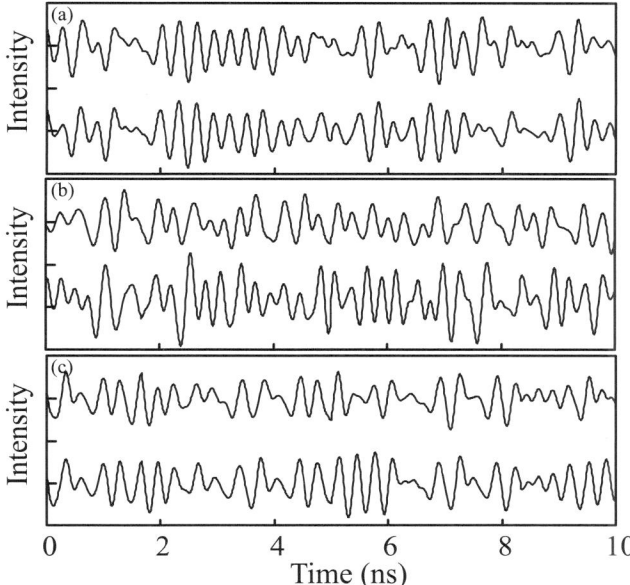

Fig. 10.20. Transmitter laser output (upper trace) and receiver laser output (lower trace) measured from the synchronization experiment for the optical feedback system. (a) $\alpha^{\mathrm{T}} = \alpha^{\mathrm{R}}$; (b) $\alpha^{\mathrm{T}} = 3.2\alpha^{\mathrm{R}}$; (c) $\alpha^{\mathrm{T}} = 10\alpha^{\mathrm{R}}$. In (a) and (c), the lower traces are shifted in time to match with the upper traces for the purpose of display. (Reprinted with permission from [37], ©2003 IEEE.)

chronization time of the receiver to the transmitter output or, equivalently, the channel signal as the optical injection field in this setup.

We first focus on the difference of these two types of synchronization states in the coupling strength. The chaos synchronization defined as the type I synchronous scenario occurs at the operating condition when $\alpha^{\mathrm{T}} = \alpha^{\mathrm{R}}$. The intensity of the transmitter (upper trace) and the synchronized intensity of the receiver (lower trace) are shown in Figure 10.20a. This synchronous phenomenon satisfies the condition of the coupling strength required by chaos synchronization. When the coupling strength increases, the system loses synchronization as is shown in Figure 10.20b, where we have $\alpha^{\mathrm{R}} \simeq 3.2\,\alpha^{\mathrm{T}}$. However, when the coupling strength is about ten times the feedback strength, synchronization between the two lasers occurs again. Their intensities are shown in Figure 10.20c, where we have the operating condition of $\alpha^{\mathrm{R}} = 10\,\alpha^{\mathrm{T}}$.

The change in the quality of synchronization along with the change of the coupling strength can be calculated by the correlation coefficient ρ. The correlation coefficient of these synchronous scenarios as a function of α^{R} is presented in Figure 10.21. As we can see in the figure, the regime, marked by I, for the chaos synchronization is much smaller than the regime, marked by II, for the chaotic driven oscillation. In a large region separating these two regimes, no synchronous phenomena are observed. At the point "A" labeled in Figure 10.21, the correlation coefficient reaches its lowest value of $\rho = 0.22$.

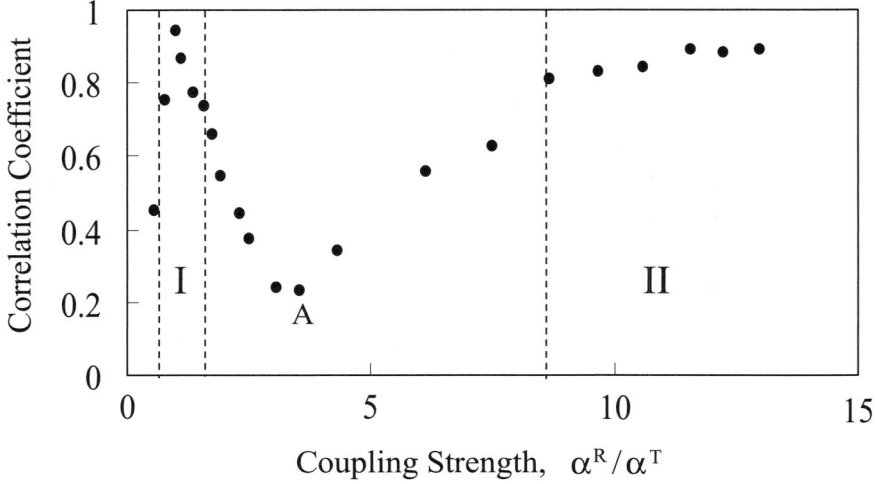

Fig. 10.21. Measured correlation function versus injection strength for chaotic injection signal in the synchronization experiment for the optical feedback system. (Reprinted with permission from [37], ©2003 IEEE.)

Because this setup is a feedback system, there should exist a time lag Δt between the chaotic waveform of the transmitter and that of the receiver when true chaos synchronization is accomplished. It is then interesting to examine if there is any difference in the characteristic of the time lag for both types of the synchronization states. To investigate the difference between these two types of synchronization states, the dependence of the time lag Δt on the delay time τ is experimentally measured, and the result is shown in Figure 10.22. As we can see from the figure, the relationship between Δt and τ in the type I synchronous scenario is that $\Delta t = T - \tau$, which is predicted by the theory of the chaos synchronization for this system. However, for the type II synchronous scenario, Δt is independent of the feedback delay time τ and is equal to T. This indicates that, for the type II synchronous scenario, the receiver duplicates the transmitter output at the moment when receiving it, as is expected in a synchronous scenario of driven oscillation.

The difference between these two types of synchronous phenomena can also be found in their tolerance to the detuning between the free-running frequencies of the transmitter and the receiver lasers. The experimental examination, shown in Figure 10.23, demonstrates that the type I synchronous scenario has very limited tolerance on this detuning frequency, whereas the type II scenario has much larger tolerance. For the type I synchronous scenario, the quality of synchronization, measured by ρ, decreases dramatically when the detuning

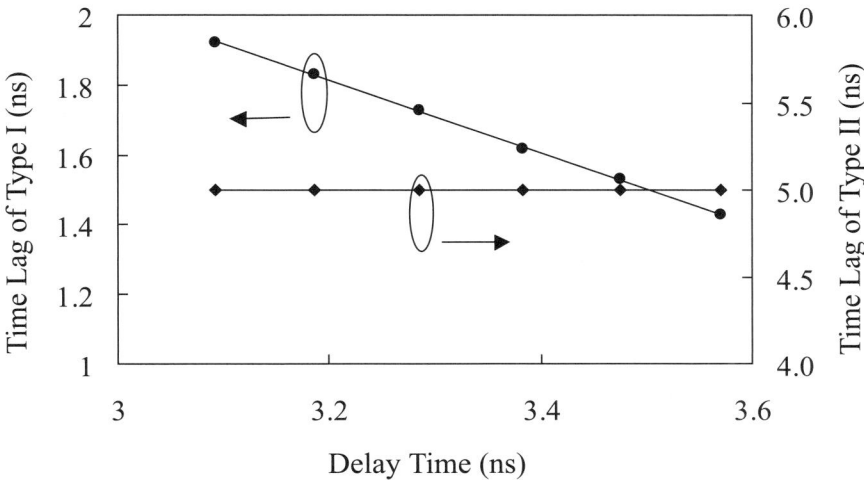

Fig. 10.22. Measured time lag (Δt) versus time delay (τ) for chaotic injection signal in the synchronization experiment for the optical feedback system. Circles: $\alpha^{\mathrm{T}} \approx \alpha^{\mathrm{R}}$, triangles: $\alpha^{\mathrm{T}} \approx 10\alpha^{\mathrm{R}}$. (Reprinted with permission from [37], ©2003 IEEE.)

frequency is tuned away from zero, as is shown in Figure 10.23a. This result satisfies the criteria of the chaos synchronization theory for this system. For the type II synchronous scenario, the range of the detuning frequency, shown in Figure 10.23b, to acquire good synchronization is not limited to the zero detuning frequency. Instead, it matches the stable locking range of the receiver subject to the optical injection from the transmitter when the transmitter is operated as a free-running laser. The range of the detuning frequency to obtain the stable locking under the injection strength of $\alpha^{\mathrm{R}} = 11.2\ \alpha^{\mathrm{T}}$ in this case is close to $-12.8\ \mathrm{GHz} < \Omega/2\pi < 0.7\ \mathrm{GHz}$. This demonstrates that the type II synchronous scenario with a high quality of synchronization is achieved over the entire stable locking regime of the receiver laser subject to the optical injection from the transmitter.

As we can see from the discussion above, besides the type I scenario of true chaos synchronization, the type II synchronous scenario of chaotic driven oscillation is also observed in the system. The type I synchronous scenario is achieved when the operating condition satisfies the requirement of chaos synchronization for this system, which means that the coupling strength has to be equal to the feedback strength, the frequency detuning between the transmitter and the receiver has to be as close to zero as possible, and the synchronization time of the receiver to the transmitter output has to change with the length of the feedback loop in the transmitter system. In contrast, the achievement of the type II synchronous scenario of chaotic driven oscillation requires a large coupling strength. It is also not sensitive to the frequency detuning as long as the receiver is injection locked in the stable locking region

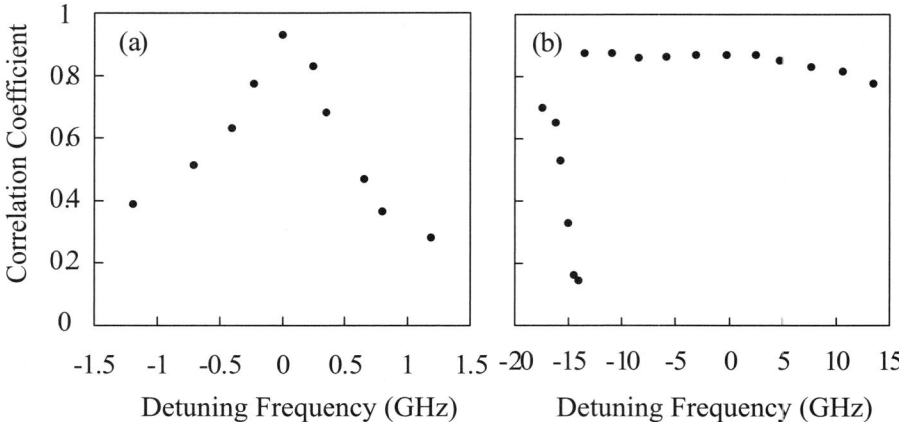

Fig. 10.23. Correlation function versus frequency detuning at the injection level (a) $\alpha^{\mathrm{T}} \approx \alpha^{\mathrm{R}}$ and (b) $\alpha^{\mathrm{T}} \approx 11.2\, \alpha^{\mathrm{R}}$ for chaotic injection signal in the synchronization experiment for the optical feedback system. (Reprinted with permission from [37], ©2003 IEEE.)

with the transmitter as the master laser to provide the optical injection field. The synchronization time of the receiver to the transmitter output in this scenario also does not change with the length of the feedback loop. Therefore, only the type I scenario matches the criteria for true chaos synchronization.

10.6.3 Synchronization of Optoelectronic Feedback System

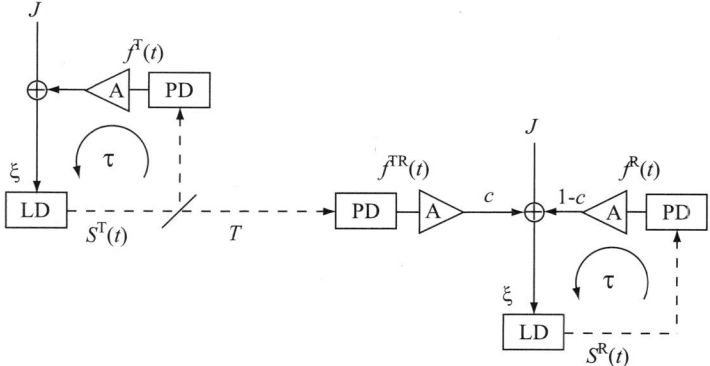

Fig. 10.24. Schematic of the experiment of anticipated and retarded synchronization for semiconductor lasers with delayed optoelectronic feedback. LD: laser diode; PD: photodetector; A: amplifier.

The schematic of two unidirectionally coupled semiconductor lasers with delayed optoelectronic feedback is shown in Figure 10.24. Their dynamics can be described by the rate equations of S and N as discussed in Section 10.4. Assuming that the laser parameters between the transmitter and the receiver are matched as required, the transmitter laser can be modeled by the following equations,

$$\frac{dS^{\mathrm{T}}}{dt} = -\gamma_{\mathrm{c}} S^{\mathrm{T}} + \Gamma g S^{\mathrm{T}} + 2\sqrt{S_0 S^{\mathrm{T}}} F_{\mathrm{s}}^{\mathrm{T}}, \tag{10.68}$$

$$\frac{dN^{\mathrm{T}}}{dt} = \frac{J}{ed}\left[1 + \xi y^{\mathrm{T}}(t-\tau)\right] - \gamma_{\mathrm{s}} N^{\mathrm{T}} - g S^{\mathrm{T}}, \tag{10.69}$$

$$y^{\mathrm{T}}(t) = \int_{-\infty}^{t} d\eta\, f^{\mathrm{T}}(t-\eta) S^{\mathrm{T}}(\eta)/S_0, \tag{10.70}$$

and the received laser can be modeled by

$$\frac{dS^{\mathrm{R}}}{dt} = -\gamma_{\mathrm{c}} S^{\mathrm{R}} + \Gamma g S^{\mathrm{R}} + 2\sqrt{S_0 S^{\mathrm{R}}} F_{\mathrm{s}}^{\mathrm{R}}, \tag{10.71}$$

$$\frac{dN^{\mathrm{R}}}{dt} = \frac{J}{ed}\left[1 + c\xi y^{\mathrm{TR}}(t-T) + (1-c)\xi y^{\mathrm{R}}(t-\tau)\right] - \gamma_{\mathrm{s}} N^{\mathrm{R}}$$
$$- g S^{\mathrm{R}}, \tag{10.72}$$

$$y^{\mathrm{TR}}(t) = \int_{-\infty}^{t} d\eta\, f^{\mathrm{TR}}(t-\eta) S^{\mathrm{T}}(\eta)/S_0, \tag{10.73}$$

$$y^{\mathrm{R}}(t) = \int_{-\infty}^{t} d\eta\, f^{\mathrm{R}}(t-\eta) S^{\mathrm{R}}(\eta)/S_0. \tag{10.74}$$

The transmitter laser has an optoelectronic feedback loop with a delay time τ. Driven by the delayed feedback signal $S^{\mathrm{T}}(t-\tau)$, the transmitter laser output $S^{\mathrm{T}}(t)$ becomes chaotic. Part of the transmitter laser output is then unidirectionally coupled to the receiver laser. In general, the receiver laser can also have its own optoelectronic feedback loop. The total driving signal to the receiver laser is $cS^{\mathrm{T}}(t-T) + (1-c)S^{\mathrm{R}}(t-\tau)$, where T is the transmission time and τ is the delay time of the feedback loop in the receiver, which is the same as that in the transmitter. The factor c can be varied from 0 to 1. When $c = 1$, the receiver has an open loop. When $c < 1$, the receiver has a closed feedback loop. Under chaos synchronization, the receiver laser is forced to follow the transmitter laser as $S^{\mathrm{R}}(t-\tau) = S^{\mathrm{T}}(t-T)$, which is equivalent to $S^{\mathrm{R}}(t) = S^{\mathrm{T}}(t-T+\tau)$. Therefore, for true chaos synchronization, there is a time shift between the outputs of the transmitter and receiver lasers. When $T > \tau$, the receiver is synchronized to the transmitter with a retardation time of $T - \tau$ in a retarded synchronization regime. When $T < \tau$, the receiver is synchronized to the transmitter with an anticipation time of $\tau - T$ in an anticipated synchronization regime [36].

Experimental results of anticipated and retarded synchronization are shown in Figure 10.25. In the experiment, the time difference is first set to be

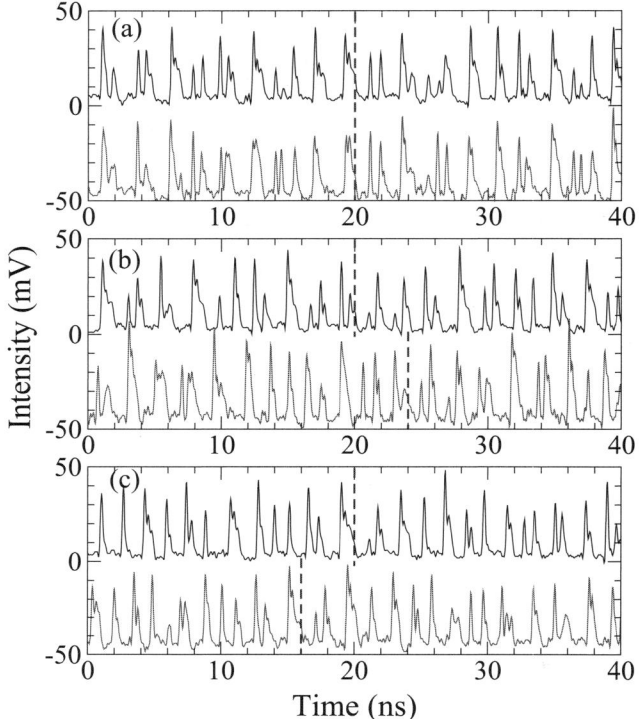

Fig. 10.25. Time series of the synchronized chaotic outputs from the transmitter laser (upper trace) and the receiver laser (lower trace) with optoelectronic feedback at $c = 0.8$. (a) Synchronization with no retardation with $T - \tau = 0.0$ ns. (b) Retarded synchronization with $T - \tau = +4.0$ ns. (c) Anticipated synchronization with $T - \tau = -4.0$ ns. (Reprinted with permission from [36], ©2003 APS.)

$T - \tau = 0.0$ ns by adjusting the transmission path and the feedback loops. Figure 10.25a shows the waveforms of the chaotic outputs of the transmitter laser (upper trace) and the receiver laser (lower trace), respectively. As can be clearly seen, the two waveforms are almost identical chaotic pulsing waveforms and the time shift between them is zero, which indicates that the receiver laser is synchronized to the transmitter laser with no retardation because $T = \tau$. In the second experiment the time difference is set to be $T - \tau = +4.0$ ns by prolonging T. Figure 10.25b shows the waveforms of the transmitter and the receiver lasers, respectively, in this situation. It is clear that the receiver laser output lags behind the transmitter laser output with a retardation time of 4.0 ns. The two lasers are now synchronized in the retarded synchronization regime. In the third experiment the time difference is set to be $T - \tau = -4.0$ ns by shortening T. Figure 10.25c shows the synchronization traces obtained in this situation. We can clearly see that the receiver laser output now leads the transmitter laser output by an anticipation time of 4.0 ns. Thus anticipated synchronization is observed between the transmitter laser and the receiver laser.

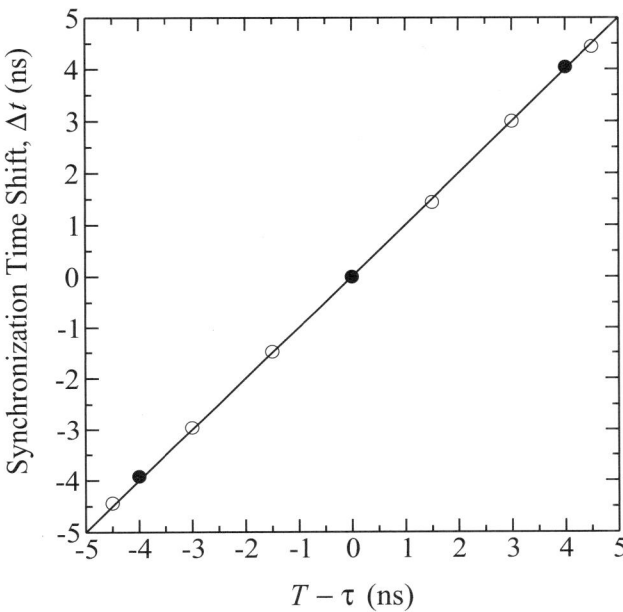

Fig. 10.26. The time shift of chaos synchronization Δt vs. the time difference of $T - \tau$ in the synchronization experiment for the optoelectronic feedback system. The open circles are from the open-loop configuration and the solid circles are from the closed-loop configuration. (Reprinted with permission from [36], ©2003 APS.)

Experiments have been conducted in both the closed-loop configuration with different c factors and the open-loop configuration with $c = 1$ [36]. The phenomena of anticipated and retarded synchronization are observed in all the cases though the quality of chaos synchronization drops dramatically as the value of the c factor drops [13]. The time shift between the outputs of the transmitter and the receiver is denoted as Δt. The experimentally measured relationship between the time shift Δt of chaos synchronization and the time difference of $T - \tau$ is summarized in Figure 10.26. The open circles are obtained from the open-loop configuration and the solid circles are obtained from the closed-loop configuration. It is clear that all the data points fall within one straight line which has a slope of 1.0. Therefore, it is further proven that the time shift of chaos synchronization is exactly $\Delta t = T - \tau$ in both regimes, where $T > \tau$ falls in the retarded synchronization regime and $T < \tau$ falls in the anticipated synchronization regime. The time shift of $T - \tau$ between the synchronized chaotic waveforms remains the same in both the open-loop and the closed-loop configurations. Therefore, the existence of two regimes of anticipated and retarded synchronization is general in the optoelectronic feedback system with unidirectional coupling.

Depending on the difference between T and τ, the two lasers fall into either the anticipated or the retarded synchronization regime. The two regimes have the same stability of chaos synchronization in the presence of small perturbations of noise and parameter mismatches. The time shift of chaos synchronization is demonstrated to be $\Delta t = T - \tau$ in both regimes, which agrees with the theoretical expectation [41–44]. The time shift of $T - \tau$ is also a proof of true chaos synchronization that is different from other phenomena such as modulation, amplification, injection locking, and driven oscillation, all of which have a different time shift related to only T [45, 46]. This point is already demonstrated by the experimentally observed difference seen in Figure 10.22 in the time delays for true chaos synchronization and chaotic driven oscillation of an optical feedback system that is discussed in the preceding subsection. Anticipated synchronization and retarded synchronization are unified phenomena under the general concept of chaos synchronization with time shift in nonlinear dynamical systems with delayed feedback.

10.7 Synchronization in the Presence of Message Encoding

In Section 10.6, the issue of chaos synchronization has been investigated in the optical injection, optical feedback, and optoelectronic feedback systems, respectively. As we have seen, in order to achieve high quality of chaos synchronization, it is very important to keep the symmetry between the transmitter and the receiver. However, in chaotic optical communication systems, the process of message encoding and decoding can further change the symmetry between the transmitter and the receiver [15]. It can either break or maintain the symmetry, depending on the configurations of the encoding and decoding schemes. Therefore, it is important to investigate chaos synchronization in the presence of message encoding. A chaotic optical communication system with optoelectronic feedback is used as an example in this investigation. Further details regarding message encoding and decoding for this system and for the other two systems are discussed in Chapter 8.

In Figure 10.27, we compare the symmetry between the transmitter and the receiver in a chaotic communication system before message encoding and after message encoding. Three major encoding and decoding schemes, namely, chaos shift keying (CSK) [47, 48], chaos masking (CMS) [49], and additive chaos modulation (ACM) [35, 50], are compared. Details of these encoding and decoding schemes in chaotic optical communication systems are discussed in Chapter 8.

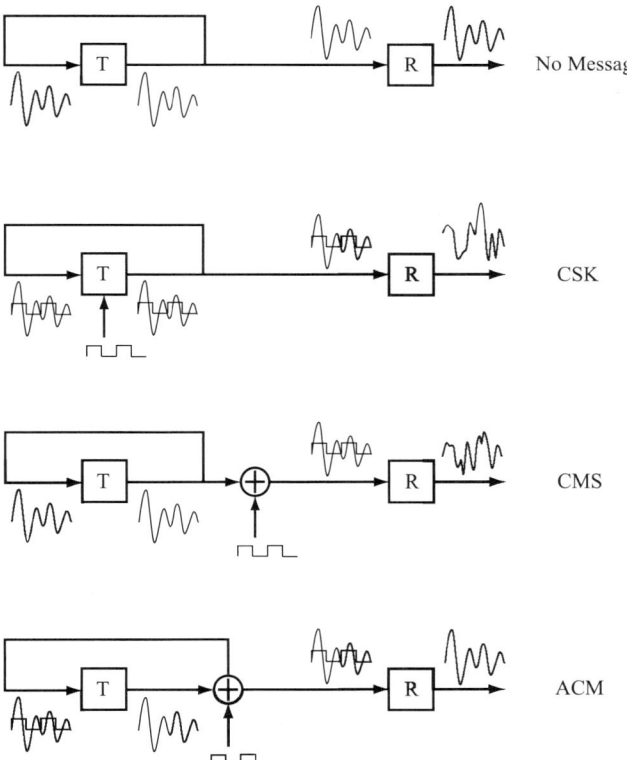

Fig. 10.27. Comparison of the symmetry between the transmitter and the receiver in a chaotic communication system before message encoding and after message encoding with the CSK, CMS, and the ACM schemes, respectively.

As is required by the theory of chaos synchronization, the transmitter and the receiver have to be identical dynamical systems. Before message encoding, the transmitter is driven by the feedback signal from the output of the transmitter itself. Meanwhile, the receiver is driven by the transmitted signal from the output of the transmitter also. Therefore, they are both driven by the same force though with a time shift, as is discussed in Section 10.6. Nevertheless, there is a symmetry of the driving force to the transmitter and that to the receiver. The system can synchronize with high quality in a chaotic state before message encoding.

When there is a message encoded through the CSK scheme, the message directly modulates a parameter of the transmitter but the corresponding parameter of the receiver is not modulated by the same message. As a result, the symmetry between the transmitter and the receiver is broken by the encoding process. This break of symmetry between the transmitter and the receiver can also be shown by the rate equations in the presence of message encoding. With the CSK schemes, the transmitter can be written as

$$\frac{dS^{\mathrm{T}}}{dt} = -\gamma_{\mathrm{c}}S^{\mathrm{T}} + \Gamma g S^{\mathrm{T}} + 2\sqrt{S_0 S^{\mathrm{T}}} F_{\mathrm{s}}^{\mathrm{T}}, \tag{10.75}$$

$$\frac{dN^{\mathrm{T}}}{dt} = \frac{J[1 + m_{\mathrm{CSK}}(t)]}{ed} \left[1 + \xi y^{\mathrm{T}}(t - \tau)\right] - \gamma_{\mathrm{s}}N^{\mathrm{T}} - g S^{\mathrm{T}}, \tag{10.76}$$

$$y^{\mathrm{T}}(t) = \int_{-\infty}^{t} d\eta f^{\mathrm{T}}(t - \eta) S^{\mathrm{T}}(\eta)/S_0, \tag{10.77}$$

where the bias current of the transmitter laser is modulated by the message $m_{\mathrm{CSK}}(t)$. The receiver, driven by the transmitted signal $S^{\mathrm{T}}(t)$, can be described by

$$\frac{dS^{\mathrm{R}}}{dt} = -\gamma_{\mathrm{c}}S^{\mathrm{R}} + \Gamma g S^{\mathrm{R}} + 2\sqrt{S_0 S^{\mathrm{R}}} F_{\mathrm{s}}^{\mathrm{R}}, \tag{10.78}$$

$$\frac{dN^{\mathrm{R}}}{dt} = \frac{J}{ed} \left[1 + \xi y^{\mathrm{TR}}(t - T)\right] - \gamma_{\mathrm{s}}N^{\mathrm{R}} - g S^{\mathrm{R}}, \tag{10.79}$$

$$y^{\mathrm{TR}}(t) = \int_{-\infty}^{t} d\eta f^{\mathrm{TR}}(t - \eta) S^{\mathrm{T}}(\eta)/S_0. \tag{10.80}$$

As can be seen, there is no current modulation on the receiver laser. The two dynamical systems are not symmetric because of the difference between (10.76) and (10.79). The quality of synchronization is expected to deteriorate in the presence of message encoding with the CSK scheme.

In the CMS scheme, the message is added on the chaotic waveform after the chaotic waveform leaves the transmitter. Thus, the message does not affect the dynamics of the transmitter. However, at the receiver, the driving force is the combined signal of the chaotic waveform and the message. Therefore, the symmetry between the transmitter and the receiver is also broken by the encoding process. With the CMS scheme, the transmitter can be written as

$$\frac{dS^{\mathrm{T}}}{dt} = -\gamma_{\mathrm{c}}S^{\mathrm{T}} + \Gamma g S^{\mathrm{T}} + 2\sqrt{S_0 S^{\mathrm{T}}} F_{\mathrm{s}}^{\mathrm{T}}, \tag{10.81}$$

$$\frac{dN^{\mathrm{T}}}{dt} = \frac{J}{ed} \left[1 + \xi y^{\mathrm{T}}(t - \tau)\right] - \gamma_{\mathrm{s}}N^{\mathrm{T}} - g S^{\mathrm{T}}, \tag{10.82}$$

$$y^{\mathrm{T}}(t) = \int_{-\infty}^{t} d\eta f^{\mathrm{T}}(t - \eta) S^{\mathrm{T}}(\eta)/S_0. \tag{10.83}$$

The encoding process is conducted after the chaotic output exits the dynamical system, and the transmitted signal to the receiver is $S^{\mathrm{T}}(t) + m_{\mathrm{CMS}}(t)$. The receiver, driven by this transmitted signal $S^{\mathrm{T}}(t) + m_{\mathrm{CMS}}(t)$, can be described by

$$\frac{dS^{\mathrm{R}}}{dt} = -\gamma_{\mathrm{c}}S^{\mathrm{R}} + \Gamma g S^{\mathrm{R}} + 2\sqrt{S_0 S^{\mathrm{R}}}F_{\mathrm{s}}^{\mathrm{R}}, \tag{10.84}$$

$$\frac{dN^{\mathrm{R}}}{dt} = \frac{J}{ed}\left[1 + \xi y^{\mathrm{TR}}(t-T)\right] - \gamma_{\mathrm{s}}N^{\mathrm{R}} - gS^{\mathrm{R}}, \tag{10.85}$$

$$y^{\mathrm{TR}}(t) = \int_{-\infty}^{t} d\eta f^{\mathrm{TR}}(t-\eta)[S^{\mathrm{T}}(\eta) + m_{\mathrm{CMS}}(\eta)]/S_0. \tag{10.86}$$

As we can see, the symmetry is also broken due to the difference between (10.83) and (10.86). The quality of synchronization is expected to deteriorate also with the CMS scheme.

In the ACM scheme, when the message is sent together with the chaotic waveform to the receiver, it is also fed back with the chaotic waveform to drive the transmitter. Therefore, both the transmitter and the receiver are driven by the same combined signal of the chaotic waveform and the message. The symmetry is maintained between the transmitter and the receiver. With the ACM scheme, the encoding process is conducted within the feedback loop of the transmitter. Therefore, the transmitter is driven by $S^{\mathrm{T}}(t) + m_{\mathrm{ACM}}(t)$, and it can be written as

$$\frac{dS^{\mathrm{T}}}{dt} = -\gamma_{\mathrm{c}}S^{\mathrm{T}} + \Gamma g S^{\mathrm{T}} + 2\sqrt{S_0 S^{\mathrm{T}}}F_{\mathrm{s}}^{\mathrm{T}}, \tag{10.87}$$

$$\frac{dN^{\mathrm{T}}}{dt} = \frac{J}{ed}\left[1 + \xi y^{\mathrm{T}}(t-\tau)\right] - \gamma_{\mathrm{s}}N^{\mathrm{T}} - gS^{\mathrm{T}}, \tag{10.88}$$

$$y^{\mathrm{T}}(t) = \int_{-\infty}^{t} d\eta f^{\mathrm{T}}(t-\eta)[S^{\mathrm{T}}(\eta) + m_{\mathrm{ACM}}(\eta)]/S_0. \tag{10.89}$$

The receiver, also driven by the transmitted signal $S^{\mathrm{T}}(t) + m_{\mathrm{ACM}}(t)$, can be described by

$$\frac{dS^{\mathrm{R}}}{dt} = -\gamma_{\mathrm{c}}S^{\mathrm{R}} + \Gamma g S^{\mathrm{R}} + 2\sqrt{S_0 S^{\mathrm{R}}}F_{\mathrm{s}}^{\mathrm{R}}, \tag{10.90}$$

$$\frac{dN^{\mathrm{R}}}{dt} = \frac{J}{ed}\left[1 + \xi y^{\mathrm{TR}}(t-T)\right] - \gamma_{\mathrm{s}}N^{\mathrm{R}} - gS^{\mathrm{R}}, \tag{10.91}$$

$$y^{\mathrm{TR}}(t) = \int_{-\infty}^{t} d\eta f^{\mathrm{TR}}(t-\eta)[S^{\mathrm{T}}(\eta) + m_{\mathrm{ACM}}(\eta)]/S_0. \tag{10.92}$$

As we can see, the system is symmetric between the transmitter and the receiver even in the presence of message encoding. Therefore, the transmitter and the receiver can be mathematically identical only when a message is encoded with the ACM scheme and when their parameters are well matched. Thus the system with the ACM encoding scheme has the potential to maintain high quality of chaos synchronization even in the presence of message encoding.

A chaotic optical communication system using semiconductor lasers with optoelectronic feedback is investigated in the experiment of chaos synchronization with message encoding. High quality of chaos synchronization is achieved

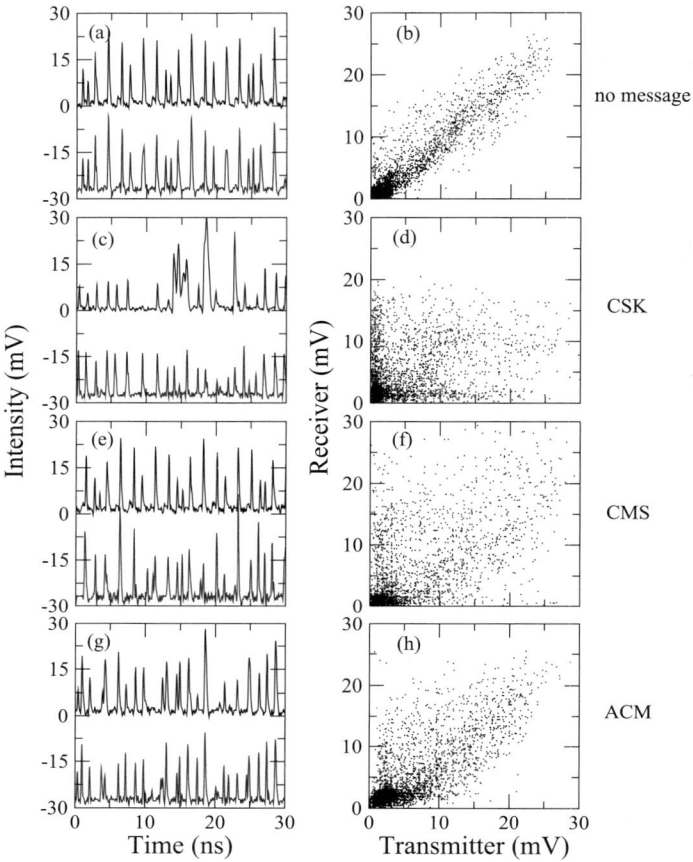

Fig. 10.28. Comparison of the time series from the transmitter (upper trace) and the receiver (lower trace) and their correlation plot, respectively, of the optoelectronic system for synchronization with and without an encoded message: (a) and (b) without the message, (c) and (d) with the message in CSK, (e) and (f) with the message in CMS, and (g) and (h) with the message in ACM.

in the system before any message is encoded. Figures 10.28a and b show the time series and the correlation plot of the outputs from the transmitter and the receiver in this situation. In Figure 10.28a, the upper trace is from the transmitter and the lower trace is from the receiver. The output from the transmitter is a chaotic pulse sequence and that from the receiver is an identical chaotic pulse sequence. The quality of synchronization is further demonstrated by the correlation plot in Figure 10.28b. When the receiver synchronizes to the transmitter, the correlation plot shows a straight line along the 45° diagonal. In Figure 10.28b, the data points are mainly distributed along the 45° diagonal, which indicates high quality of synchronization.

Figures 10.28c and d show the quality of synchronization after a message is applied with the CSK scheme. Comparing the two time series in Figure 10.28c, it is clear that the output of the receiver is very different from that of transmitter. The correlation plot further shows that the two chaotic waveforms lose synchronization, as is indicated by the widely scattered data points in Figure 10.28d.

Figures 10.28e and f show the corresponding time series and correlation plot of the transmitter and the receiver when a message is encoded through the CMS scheme. In Figure 10.28e, the two chaotic waveforms are clearly not identical chaotic waveforms. The correlation plot also spreads out over a wide area, as is shown in Figure 10.28f. Therefore, it is demonstrated that the quality of chaos synchronization dramatically deteriorates in the CMS case.

Figures 10.28g and h show the time series and the correlation plot after a message is encoded with the ACM scheme. The chaotic pulsing waveform from the transmitter in Figure 10.28g is a much more complicated chaotic pulsing waveform than that in Figure 10.28a, as the message affects the chaotic dynamics also. Nevertheless, the receiver still synchronizes with the transmitter, and the output of the receiver is an identical chaotic pulsing waveform as that of the transmitter. In the correlation plot shown in Figure 10.28h, the data points still mainly distribute along the 45° diagonal. Comparing Figures 10.28b and h, it can be seen that some deterioration of the quality of synchronization is also observed in the ACM scheme. This is caused by the increase in the complexity of the chaotic waveform generated by the transmitter when a pseudorandom message is encoded through the ACM scheme. Therefore, the synchronization error and the digitization error are both increased because of the more complicated chaotic waveform. Nevertheless, the quality of synchronization is still very high when compared with that in the CSK and the CMS schemes.

From the experimental results, it is shown that the process of message encoding and decoding can influence the quality of synchronization significantly. For different encoding and decoding schemes, the effects can be either to deteriorate or to maintain the quality of synchronization. For a chaotic communication system, it is very important to study chaos synchronization in the presence of message encoding.

10.8 Conclusions

The nonlinear dynamics and chaos synchronization of single-mode semiconductor lasers for chaotic optical communications are addressed in this chapter. Basic concepts and theoretical framework are reviewed. Experimental results are presented to demonstrate the fundamental concepts. Numerical computations are employed for mapping the dynamical states and for illustrating certain detailed characteristics of the chaotic states. The objective is to provide a complete picture of the nonlinear dynamics and chaos synchronization of the three semiconductor laser systems considered in this chapter.

Single-mode semiconductor lasers are class B lasers that have only two dynamical dimensions in solitary operating conditions. For complex nonlinear dynamics such as chaos to be possible for such as a laser, both a certain nonlinear physical mechanism to make the system nonlinear and a certain perturbational force to increase the dimension of the system to more than two are required. A semiconductor laser gain medium is inherently nonlinear. Such intrinsic nonlinearity is sufficient for a single-mode semiconductor laser to exhibit rich nonlinear dynamical characteristics if a proper perturbation is applied to sufficiently increase the dimension of the system. Among the many possible schemes for perturbing a single-mode semiconductor laser into chaos, we consider in this chapter the three different semiconductor laser systems, namely, the optical injection system, the optical feedback system, and the optoelectronic feedback system, that are of most interest for high-bit-rate chaotic optical communications. The dynamics of each of these three systems is determined by five intrinsic laser parameters, which are independent of the perturbing force, and three operational parameters, which are controllable in operation and two of which are determined by the perturbing force. These three systems have very different nonlinear dynamics, and their dynamical states have very different characteristics as well. The optical injection system is a nonautonomous system with a fixed dimension; it follows a period-doubling route to chaos. The optical feedback system is a phase-sensitive delayed-feedback autonomous system with a dimension increasing with the feedback delay time; all three known routes, namely, period-doubling, quasiperiodicity, and intermittency, to chaos are possible for this system, depending on the laser parameters and the operating condition. The optical feedback system is a phase-insensitive delayed-feedback autonomous system with a dimension also increasing with the feedback delay time; it follows a quasiperiodicity route to chaos with pulsing characteristics.

The main focus of chaos synchronization discussed in this chapter is identical synchronization in unidirectionally coupled systems, which is also the primary subject of study on synchronization for chaotic communications. The three systems considered in this chapter have different characteristics regarding chaos synchronization. For optical injection and optical feedback systems, the frequency, phase, and amplitude of the optical fields of both transmitter and receiver lasers are all locked in synchronism when synchronization is accomplished because both phase and amplitude of the laser field participate in the dynamics of these systems. Besides the complete synchronization of the entire optical field, the optical injection system exhibits synchronization sensitivity to the optical phase difference between the injection field and the transmitter laser field. The optical feedback system exhibits phase sensitivity only when the receiver has a closed-loop configuration. For the optoelectronic feedback system, chaos synchronization does not involve the locking of the optical frequency or the synchronization of the optical phase because the phase of the laser field does not participate in the nonlinear dynamics of this system. Synchronization of this system is not phase sensitive. For both optical feedback and optoelectronic feedback systems, where the transmitter is configured with a delayed feedback loop, anticipated and retarded synchronization can be observed as the difference between the feedback delay time and the propagation time from the transmitter to the receiver laser is varied.

All of these three systems also exhibit multiple synchronous scenarios regardless of the differences in their nonlinear dynamics. Besides the true chaos synchronization described by the theory of identical chaos synchronization, one or two other synchronous phenomena are observed for each system under different operating conditions. These other synchronous phenomena include chaotic driven oscillation and chaotic modulation. They have different signatures compared to true identical chaos synchronization.

The transmitted signal of any communication system in operation has to be encoded with a message, which is decoded at the received end. For a synchronized chaotic communication system, the effect of the message encoding process on the synchronization between the receiver and the transmitter is an important issue that cannot be ignored. The impact of the message encoding process on the quality of synchronization depends on many factors, including the message encoding scheme, the system configuration, and the bit rate and form of the message. For identical chaos synchronization, one fundamental requirement for maintaining a high quality of synchronization while a message is encoded is to use an encoding scheme that maintains the symmetry between the transmitter and the receiver. For this reason, an encoding scheme such as ACM that does not break the symmetry between the transmitter and the receiver is superior to one such as CSK or CMS that breaks the symmetry. It is shown in this chapter that high-quality synchronization can be maintained, with an added benefit of increasing the complexity of the chaotic waveform, when a semiconductor laser system is encoded with random message bits through the ACM scheme.

Acknowledgments

This work was supported by the U.S. Army Research Office under MURI grant DAAG55-98-1-0269.

References

1. Special issue on applications of chaos in modern communication systems, *IEEE Trans. Circuits Syst. I*, vol. 48, 2001.

2. Feature section on optical chaos and application to cryptography, *IEEE J. Quantum Electron.*, vol. 38, 2002.

3. G. D. VanWiggeren and R. Roy, Optical communication with chaotic waveforms, *Phys. Rev. Lett.*, vol. 81, pp. 3547−3550, 1998.

4. H. D. I. Abarbanel and M. B. Kennel, Synchronizing high-dimensional chaotic optical ring dynamics, *Phys. Rev. Lett.*, vol. 80, pp. 3153−3156, 1998.

5. G. D. VanWiggeren and R. Roy, Chaotic communication using time-delayed optical systems, *Int. J. Bifurcation & Chaos*, vol. 9, pp. 2129−2156, 1999.

6. L. G. Luo, P. L. Chu, and H. F. Liu, 1-GHz optical communication system using chaos in erbium-doped fiber lasers, *IEEE Photon. Technol. Lett.*, vol. 12, pp. 269−271, 2000.

7. H. F. Chen and J. M. Liu, Open-loop chaotic synchronization of injection-locked semiconductor lasers with gigahertz range modulation, *IEEE J. Quantum Electron.*, vol. 36, pp. 27−34, 2000.

8. J. M. Liu, H. F. Chen, and S. Tang, Optical communication systems based on chaos in semiconductor lasers," *IEEE Trans. Circuits Syst. I*, vol. 48, pp. 1475−1483, 2001.

9. Y. Liu, H.F. Chen, J.M. Liu, P. Davis, and T. Aida, Communication using synchronization of optical-feedback-induced chaos in semiconductor lasers, *IEEE Trans. Circuits Syst. I*, vol. 48, pp. 1484−1489, 2001.

10. Y. Liu, H. F. Chen, J. M. Liu, P. Davis, and T. Aida, Synchronization of optical-feedback-induced chaos in semiconductor lasers by optical injection, *Phys. Rev. A*, vol. 63, 031802, 2001.

11. A. Sanchez-Diaz, C. R. Mirasso, P. Colet, and P. Garcia-Fernandez, Encoded Gbit/s digital communications with synchronized chaotic semiconductor lasers, *IEEE J. Quantum Electron.*, vol. 35, pp. 292−297, 1999.

12. S. Sivaprakasam and K. A. Shore, Message encoding and decoding using chaotic external-cavity diode lasers, *IEEE J. Quantum Electron.*, vol. 36, pp. 35−39, 2000.

13. S. Tang and J. M. Liu, Synchronization of high-frequency chaotic optical pulses, *Opt. Lett.*, vol. 26, pp. 596−598, 2001.

14. S. Tang and J. M. Liu, Message encoding-decoding at 2.5 Gbits/s through synchronization of chaotic pulsing semiconductor lasers, *Opt. Lett.*, vol. 26, pp. 1843−1845, 2001.

15. J. M. Liu, H. F. Chen, and S. Tang, Synchronized chaotic optical communications at high bit rates, *IEEE J. Quantum Electron.*, vol. 38, pp. 1184−1196, 2002.

16. F. T. Arecchi, G. L. Lippi, G. P. Puccioni and J. R. Tredicce, Deterministic chaos in laser with injected signal, *Opt. Commun.*, vol. 51, pp. 308−314, 1984.

17. J. R. Tredicce, F. T. Arecchi, G. L. Lippi, and G. P. Puccioni, Instabilities in lasers with an injected signal, *J. Opt. Soc. Am. B*, vol. 2, pp. 173−183, 1985.

18. S. K. Hwang and J. M. Liu, Dynamical characteristics of an optically injected semiconductor laser, *Opt. Commun.*, vol. 183, pp. 195−205, 2000.

19. T. B. Simpson and J. M. Liu, Spontaneous emission, nonlinear optical coupling, and noise in laser diodes, *Opt. Commun.*, vol. 112, pp. 43−47, 1994.

20. J. M. Liu and T. B. Simpson, Four-wave mixing and optical modulation in a semiconductor laser, *IEEE J. Quantum Electron.*, vol. 30, pp. 957−965, 1994.

21. J. M. Liu, C. Chang, and T. B. Simpson, Amplitude noise enhancement caused by nonlinear interaction of spontaneous emission field in laser diodes, *Opt. Commun.*, vol. 120, pp. 282−286, 1995.

22. T. B. Simpson, J. M. Liu, A. Gavrielides, V. Kovanis, and P. M. Alsing, Period-doubling route to chaos in semiconductor lasers subject to optical injection, *Appl. Phys. Lett.*, vol. 64, pp. 3539−3541, 1994.

23. R. Lang and K. Kobayashi, External optical feedback effects on semiconductor injection laser properties, *IEEE J. Quantum Electron.*, vol. 16, pp. 347−355, 1980.

24. J. Mork, B. Tromborg, and P. L. Christiansen, Bistability and low-frequency fluctuations in semiconductor lasers with optical feedback: a theoretical analysis, *IEEE J. Quantum Electron.*, vol. 24, pp. 123−133, 1998.

25. I. Fischer, G. H. M. Van Tartwijk, A. M. Levine, W. Elsasser, E. Gobel, and D. Lenstra, Fast pulsing and chaotic itinerancy with a drift in the coherence collapse of semiconductor lasers, *Phys. Rev. Lett.*, vol. 76, pp. 220−223, 1996.

26. Y. H. Kao, N. M. Wang, and H. M. Chen, Mode description of routes to chaos in external-cavity coupled semiconductor lasers, *IEEE J. Quantum Electron.*, vol. 30, pp. 1732−1739, 1994.

27. M. W. Pan, B. P. Shi, and G. R. Gray, Semiconductor laser dynamics subject to strong optical feedback, *Opt. Lett.*, vol. 22, pp. 166−168, 1997.

28. T. Sano, Antimode dynamics and chaotic itinerancy in the coherent collapse of semiconductor-lasers with optical feedback, *Phys. Rev. A*, vol. 50, pp. 2719−2726, 1994.

29. Y. Liu, Y. Takiguchi, P. Davis, T. Aida, S. Saito, and J. M. Liu, Experimental observation of complete chaos synchronization in semiconductor Lasers, *Appl. Phys. Lett.*, vol. 80, pp. 4306−4308, 2002.

30. S. Tang and J. M. Liu, Chaotic pulsing and quasiperiodic route to chaos in a semiconductor laser with delayed optoelectronic feedback, *IEEE J. Quantum Electron.*, vol. 37, no. 3, pp. 329−336, 2001.

31. F. Y. Lin and J. M. Liu, Nonlinear dynamics of a semiconductor laser with delayed negative optoelectronic feedback, *IEEE J. Quantum Electron.*, vol. 39, pp. 562−568, 2003.

32. L. M. Pecora and T. L. Carroll, Synchronization in chaotic systems, *Phys. Rev. Lett.*, vol. 64, no. 8, pp. 821−824, 1990.

33. L. Kocarev and U. Parlitz, General Approach for Chaotic Synchronization with Applications to Communication, *Phys. Rev. Lett.*, vol. 74, pp. 5028−5031, 1995.

34. L. M. Pecora, T. L. Carrol, G. A. Johnson, D. J. Mar, and J. F. Heagy, Fundamentals of synchronization in chaotic systems, concepts, and applications, *Chaos*, vol. 7, pp. 520−543, 1997.

35. S. Tang, H. F. Chen, S. K. Hwang, and J. M. Liu, Message encoding and decoding through chaos modulation in chaotic optical communications, *IEEE Trans. Circuits Syst. I*, vol. 49, pp. 163−169, 2002.

36. S. Tang and J. M. Liu, Experimental verification of anticipated and retarded synchronization in chaotic semiconductor lasers, *Phys. Rev. Lett.*, vol. 90, pp. 194101, 2003.

37. Y. Liu, P. Davis, Y. Takiguchi, T. Aida, S. Saito, and J.M. Liu, Injection locking and synchronization of periodic and chaotic signals in semiconductor lasers, *IEEE J. Quantum Electron.*, vol. 39, pp. 269−278, 2003.

38. J. Ohtsubo, Chaos synchronization and chaotic signal masking in semiconductor lasers with optical feedback, *IEEE J. Quantum Electron.*, vol. 38, pp. 1141−1154, 2002.

39. S. Tang and J. M. Liu, Chaos synchronization in semiconductor lasers with optoelectronic feedback, *IEEE J. Quantum Electron.*, vol. 39, pp. 708−715, 2003.

40. H. F. Chen and J. M. Liu, Complete phase and amplitude synchronization of broadband chaotic optical fields generated by semiconductor lasers subject to optical injection, *Phys. Rev. E*, vol. 71, pp. 046216-1−7, 2005.

41. V. Ahlers, U. Parlitz, and W. Lauterborn, Hyperchaotic dynamics and synchronization of external-cavity semiconductor lasers, *Phys. Rev. E*, vol. 58, pp. 7208−7213, 1998.

42. H. U. Voss, Anticipating chaotic synchronization, *Phys. Rev. E*, vol. 61, pp. 5115−5119, 2000.

43. C. Masoller, Anticipation in the synchronization of chaotic semiconductor lasers with optical feedback, *Phys. Rev. Lett.*, vol. 86, pp. 2782−2785, 2001.

44. H. U. Voss, Dynamic long-term anticipation of chaotic states, *Phys. Rev. Lett.*, vol. 87, pp. 014102, 2001.

45. A. Murakami and J. Ohtsubo, Synchronization of feedback-induced chaos in semiconductor lasers by optical injection, *Phys. Rev. A*, vol. 65, p. 033826, 2002.

46. I. Fischer, Y. Liu, and P. Davis, Synchronization of chaotic semiconductor laser dynamics on subnanosecond time scales and its potential for chaos communication, *Phys. Rev. A*, vol. 62, 011801(R), 2000.

47. U. Parlitz, L. O. Chua, L. Kocarev, K. S. Halle, and A. Shang, Transmission of digital signals by chaotic synchronization, *Int. J. Bifurcation & Chaos*, vol. 2, pp. 973−977, 1992.

48. J.-B. Cuenot, L. Larger, J.-P. Goedgebuer, and W. T. Rhodes, Chaos shift keying with an optoelectronic encryption system using chaos in wavelength, *IEEE J. Quantum Electron.*, vol. 37, pp. 849−855, 2001.

49. L. Kocarev, K. S. Halle, K. Eckert, L. O. Chua, and U. Parlitz, Experimental demonstration of secure communications via chaotic synchronization, *Int. J. Bifurcation & Chaos*, vol. 2, pp. 709−713, 1992.

50. C. W. Wu and L. O. Chua, A simple way to synchronize chaotic systems with applications to secure communication systems, *Int. J. Bifurcation & Chaos*, vol. 3, pp. 1619−1627, 1993.

11

Performance of Synchronized Chaotic Optical Communication Systems

Shuo Tang, How-Foo Chen, and Jia-Ming Liu

Summary. Chaotic optical communication is a novel communication scheme that utilizes optical chaotic waveform to transmit messages at a high bit rate. Its potential applications include secure communications and spread-spectrum communications. In a chaotic optical communication system, a nonlinear dynamical system is used to generate the optical chaotic waveform for message transmission. Messages are encoded through chaos encryption where the messages are mixed with the chaotic waveform. Message recovery is achieved by comparing the received signal with a reproduced chaotic waveform which synchronizes with the chaotic waveform from the transmitter. Details are discussed in this chapter regarding each of the above basic issues. Furthermore, we also review the experiment of chaotic optical communication at 2.5 Gb/s, which has the highest bit rate in any chaotic communication systems ever reported in the literature. This system uses semiconductor lasers with delayed optoelectronic feedback to generate chaotic pulses. Three major encoding and decoding schemes, namely, chaos masking, chaos shift keying, and chaos modulation, are implemented and compared in this 2.5 Gb/s chaotic optical communication system. The chaos modulation scheme is found to have the best performance. To investigate the potential applications of chaotic optical communications at an even higher bit rate, numerical simulations are carried out on chaotic optical communication systems operating at 10 Gb/s. In this numerical study, three different systems using semiconductor lasers with optical injection, optical feedback, or optoelectronic feedback, respectively are investigated. It is shown that chaotic optical communication at 10 Gb/s is feasible with high-speed semiconductor lasers.

11.1 Introduction

Chaotic communication is a novel communication scheme that has attracted much interest in the last decade [1, 2] because of its potential applications in secure communications and in spread-spectrum communications. This scheme is based on the transmission of messages encoded on chaotic waveforms generated by nonlinear dynamical systems. Chaotic optical communication uses fast chaotic optical waveforms to transmit messages at high bit rates.

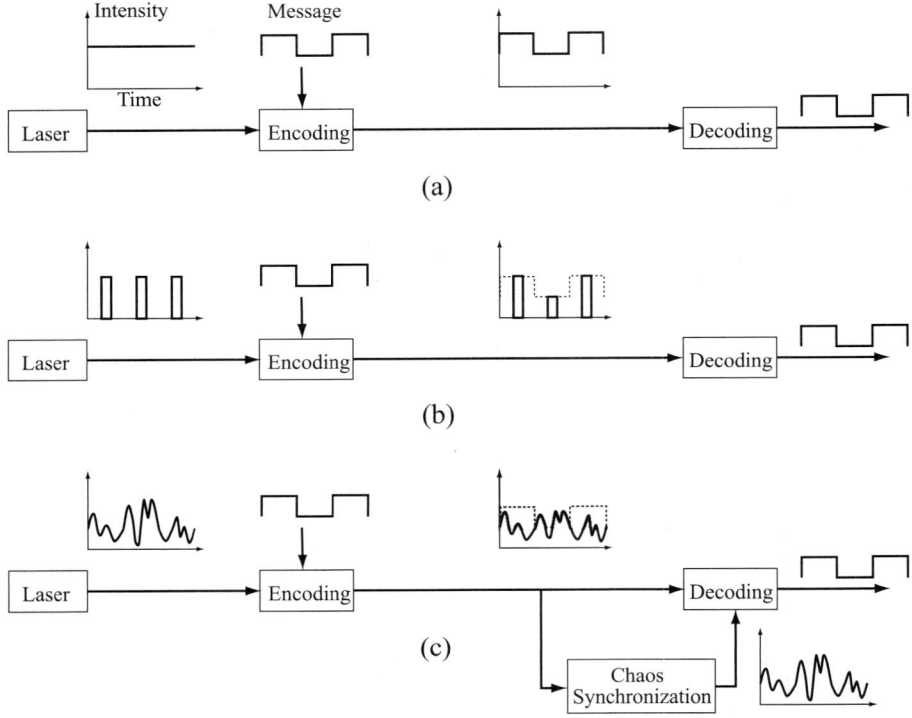

Fig. 11.1. Schematics of (a) and (b) two different conventional optical communication systems and (c) chaotic optical communication system.

In order to explain how a chaotic optical communication system works, we compare it with conventional optical communication systems. In conventional optical communication systems, the optical carrier either has a constant optical intensity or a series of optical pulses with a constant peak value as indicated in Figures 11.1a and b, respectively. When a message is intensity modulated on the optical carrier, the optical intensity takes on two distinct levels that respectively represent a "1" bit or a "0" bit being transmitted. Message recovery is achieved by detecting the optical intensity and comparing it with an appropriate decision threshold. In contrast, in a chaotic optical communication system, the intensity of the optical carrier does not maintain a constant level but fluctuates chaotically, as shown in Figure 11.1c. Messages can be encoded by modulating a certain parameter of the chaotic waveform such as the intensity. For example, the intensity of the chaotic waveform is not modified when a "1" bit is transmitted, but the intensity is lowered by a certain ratio when a "0" bit is transmitted. However, because the intensity of the chaotic waveform fluctuates originally, the message cannot be directly detected by simply measuring the intensity of the received chaotic waveform.

In order to decode the message in a chaotic communication system, a replica of the original chaotic waveform is needed at the receiver. For the example shown in Figure 11.1c, the message can be recovered by dividing the received chaotic waveform with the reproduced chaotic waveform at the receiver. The replication of the chaotic waveform can be achieved by chaos synchronization, which is discussed in Chapter 10.

Chaotic waveforms have some unique characteristics such as noiselike time series and broadband spectrum [3]. Figure 11.2 shows the time series and the power spectrum of a chaotic pulsing waveform generated by a semiconductor laser with optoelectronic feedback [4]. As we can see, the pulse intensity of the chaotic waveform is very noiselike and fluctuates chaotically. Based on this property, chaotic communication can be applied to secure communications [5], where messages can be hidden within the chaotic fluctuations of the waveforms. An eavesdropper who can only record the modulated chaotic waveform cannot recover the messages without the knowledge of the original chaotic waveform. Another important characteristic of chaotic waveform is its broadband spectrum, as is seen in the power spectrum shown in Figure 11.2. The power of a chaotic waveform is distributed over a broad spectral range. Therefore, chaotic communications potentially can have less interference among different communication channels within the same transmission media. The inherent properties of chaotic waveforms also hint at the potential application in multiple-access networks. Other potential benefits of chaotic communications include efficient use of the bandwidth of a communication channel [6], utilization of the intrinsic nonlinearities in communication devices, large-signal modulation for efficient use of the carrier power, reduced number of components in a system, and security of communication.

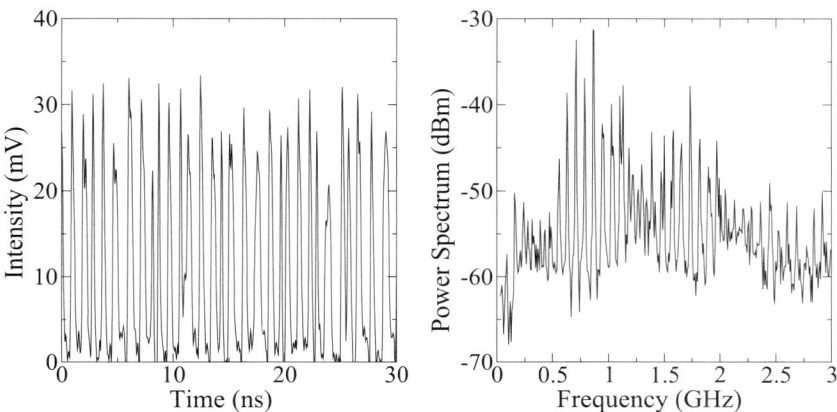

Fig. 11.2. Time series and power spectrum of a chaotic pulsing waveform.

A brief comparison between conventional and chaotic communications is given in Table 11.1. As can be seen, compared to the matured technology of conventional communications, chaotic communication is an emerging field. However, motivated by the potential benefits from the broadband and noise-like chaotic carrier, chaotic communications have attracted much interest recently, and their performances are being improved rapidly. We expect to see chaotic communications becoming an important supplement to conventional communications, with unique features that cannot be accomplished by conventional communications along. The implementation of nonlinear systems in chaotic optical communications opens up many possibilities that cannot be achieved with linear systems along.

Table 11.1. Conventional Versus Chaotic Communications

	Conventional	Chaotic
History	Over 100 years old; matured technology	Less than 10 years old; emerging technology
Industrial base	Heart of worldwide information technology	None existing
Transmission bandwidth	Transmitting at information bandwidth (BW) or at wide BW (e.g., spread spectrum)	Transmitting at wide BW
Carrier waveform	CW or pulse	Chaotic waveform or pulse
Bit rate	10 Gb/s, 40 Gb/s [7]	2.5 Gb/s [8]
Bit error rate	10^{-9} [7]	10^{-3} at 2.5 Gb/s [8], 10^{-5} at 126 Mb/s [9]

Chaotic communications have been investigated using electric circuits [10], which are generally bandwidth-limited to the kilohertz region. To achieve fast chaotic communication, laser systems are of great interest because of their ability to carry high-bit-rate messages in the gigahertz region. Different laser systems have been investigated for chaotic optical communications. Table 11.2 summarizes the current status of chaotic optical communications. Both semiconductor lasers [11–16] and fiber ring lasers [17–19] have been used successfully in high-bit-rate chaotic optical communications. Semiconductor lasers with optical injection [11], optical feedback [12, 13], or optoelectronic feedback [14–16] have been investigated to generate fast chaotic waveforms. The chaotic fluctuations can be in optical intensity, wavelength, or polarization. From the literature, the highest bit rate ever achieved in chaotic optical communications is 2.5 Gb/s using semiconductor lasers with optoelectronic feedback. Experimental results on this transmission of 2.5 Gb/s chaotic optical communication are reviewed in detail in Section 11.3.

Table 11.2. Status of Chaotic Optical Communications

System	Chaos	Chaos Bandwidth	Bit rate	Ref.
Optical injection	Intensity	3−10 GHz		[11]
Optical feedback	Intensity	3 GHz	1.5 GHz	[12, 13]
Optoelectronic feedback	Intensity	3 GHz	2.5 Gb/s	[8, 14, 15]
	Wavelength	kHz	4 kHz	[16]
Fiber ring laser	Intensity	1.5 GHz	250 Mb/s, 1 GHz	[17, 18]
	Polarization	1.5 GHz	80 Mb/s	[19]

From the above discussions, it is seen that the field of chaotic optical communications is growing fast because of the advantages brought by chaotic waveforms and nonlinear dynamical systems. In this chapter, we give an introduction to chaotic optical communications and review our research results on high-bit-rate chaotic optical communications. The arrangement of this chapter is as follows. The configurations and the general issues of a chaotic optical communication system are discussed in Section 11.2. The experiment on chaotic optical communication at 2.5 Gb/s [8] is described in Section 11.3. The effect of message encoding and decoding on the performances of chaotic optical communication systems is discussed in Section 11.4. Simulations on chaotic optical communications at 10 Gb/s have been investigated [22], and the results are summarized in Section 11.5. Finally, conclusions and discussions on future research are included in Section 11.6.

11.2 General Issues on Chaotic Optical Communications

The schematic block diagram of a synchronized chaotic optical communication system has been shown in Figure 11.1c. Basically, on the transmitter side, a chaotic communication system includes a chaos generator that generates a chaotic waveform, and a chaos encoding scheme that encodes messages on the chaotic waveform. On the receiver side, an identical chaos generator is required for chaotic communications based on chaos synchronization. The decoding scheme is the inverse of the encoding process. Each of these important parts of a chaotic optical communication system are discussed in the following subsections.

11.2.1 Generation of Chaotic Waveforms

As shown in Figure 11.1c, in a chaotic optical communication system, a chaotic waveform is used to encode messages for transmission. Therefore, to configure a chaotic optical communication system, we first need a chaos generator to generate fast chaotic optical waveforms. The chaos generator is basically a nonlinear dynamical system. To generate a chaotic optical waveform, we can use a semiconductor laser system or any other optical dynamical system. Because semiconductor lasers are the choice of light sources for optical communications, we focus on using semiconductor lasers in this chapter for chaotic optical communications. Semiconductor lasers with external perturbations such as optical injection, optical feedback, or optoelectronic feedback have been found to be able to generate fast chaotic optical waveforms [11–16]. Details of the nonlinear dynamics of semiconductor lasers and the characteristics of the chaotic states generated by such semiconductor lasers are covered in Chapter 10. Semiconductor lasers can generate either continuous or pulsed chaotic optical waveforms that vary on a subnanosecond time scale [23]. Such chaotic optical waveforms are especially suitable for high-bit-rate chaotic optical communications.

11.2.2 Synchronization of Chaotic Optical Communications

In a chaotic communication system, the received chaotic waveform carries the information about the encoded messages. However, the messages are encrypted with the fluctuations of the chaotic waveform. In order to recover the messages, a replica of the original chaotic waveform before message encoding is required on the receiver side. This replicating process needs an identical chaos generator on the receiver side that is synchronized to the chaos generator on the transmitter side [24, 25]. Message recovery is then achieved by comparing the received signal with the reproduced chaotic waveform. The quality of chaos synchronization is very important for the performance of a synchronized chaotic communication system. Although chaotic communication systems that do not require chaos synchronization have also been proposed, most of the chaotic communication systems studied so far need chaos synchronization. Therefore, we focus on chaotic communications based on chaos synchronization.

Figure 11.3 shows how chaos synchronization and message recovery can be achieved in a chaotic communication system. A nonlinear dynamical system, whose system function is denoted by G^T, is used to generate a chaotic output indicated by $S^T(t)$. Message encoding is achieved through additive chaos modulation where the message is added on the chaotic waveform as $S^T(t) + m(t)$. The added signal, $S^T(t) + m(t)$, is then split into two paths, where one is fed back to drive the G^T dynamical system, and the other is transmitted to the receiver. Therefore, the output from the transmitter can be written as $S^T(t) = G^T[S^T(t) + m(t)]$. On the receiver side, an identical nonlinear system, denoted by G^R, is needed for chaos synchronization. The driving signal to G^R is the received signal, which is $S^T(t) + m(t)$. The output of G^R can be written as $S^R(t) = G^R[S^T(t) + m(t)]$. Comparing the expressions of $S^T(t)$ and $S^R(t)$, we can see that they can be synchronized if G^T and G^R are identical nonlinear dynamical functions with identical parameters. Finally, message recovery is

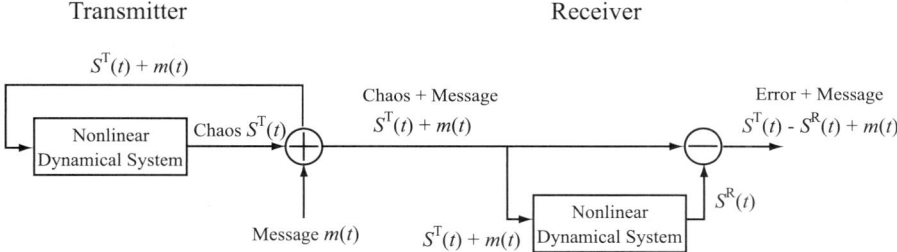

Fig. 11.3. Block diagram of a synchronized chaotic communication system illustrating the process of chaos synchronization and message recovery.

achieved by subtracting $S^{\mathrm{R}}(t)$ from the received signal $S^{\mathrm{T}}(t) + m(t)$. The recovered message is $S^{\mathrm{T}}(t) - S^{\mathrm{R}}(t) + m(t)$, which contains the original message plus some residual synchronization errors in $S^{\mathrm{T}}(t) - S^{\mathrm{R}}(t)$. For simplicity, the time delay of signal propagation due to its finite speed is not considered in this analysis. As we show later, even with the consideration of the time delay in a real system, the synchronization process is still the same except with a time shift in $S^{\mathrm{T}}(t)$ and $S^{\mathrm{R}}(t)$.

In order to achieve chaos synchronization, the receiver dynamical system needs to be identical to the transmitter dynamical system. The receiver is further coupled to the transmitter through the transmission path. Both the transmitter and the receiver dynamical systems need to be driven by the same force. When the parameters of the receiver are respectively identical to those of the transmitter and the coupling strength is strong enough, the receiver can synchronize to the transmitter in all dynamical states including the chaotic state. Therefore, the output of the receiver can reproduce the original chaotic waveform from the transmitter. Details of chaos synchronization in semiconductor lasers are discussed in Chapter 10.

However, as the receiver nonlinear dynamical system may not have exactly the same parameters as those of the transmitter dynamical system, synchronization of the two chaotic waveforms generated by the transmitter and the receiver is not perfect but has errors. These synchronization errors cause errors in the recovery of the message because the message is recovered by comparing the two chaotic waveforms. In a chaotic communication system based on synchronization, the system performance largely depends on the quality of chaos synchronization.

11.2.3 Encoding and Decoding Schemes

In Figure 11.1c, after the chaos generator, an encoding scheme is required to encode messages on the chaotic waveform. Many chaotic encoding and decoding schemes have been investigated. Due to the special requirement on chaos synchronization in chaotic optical communication systems, the encoding and decoding schemes are categorized differently from those used in conventional optical communication systems. They can be classified into three major categories: chaos masking [26], chaos shift keying [27], and chaos modulation [10, 28, 29], shown in Figures 11.4a−c, respectively.

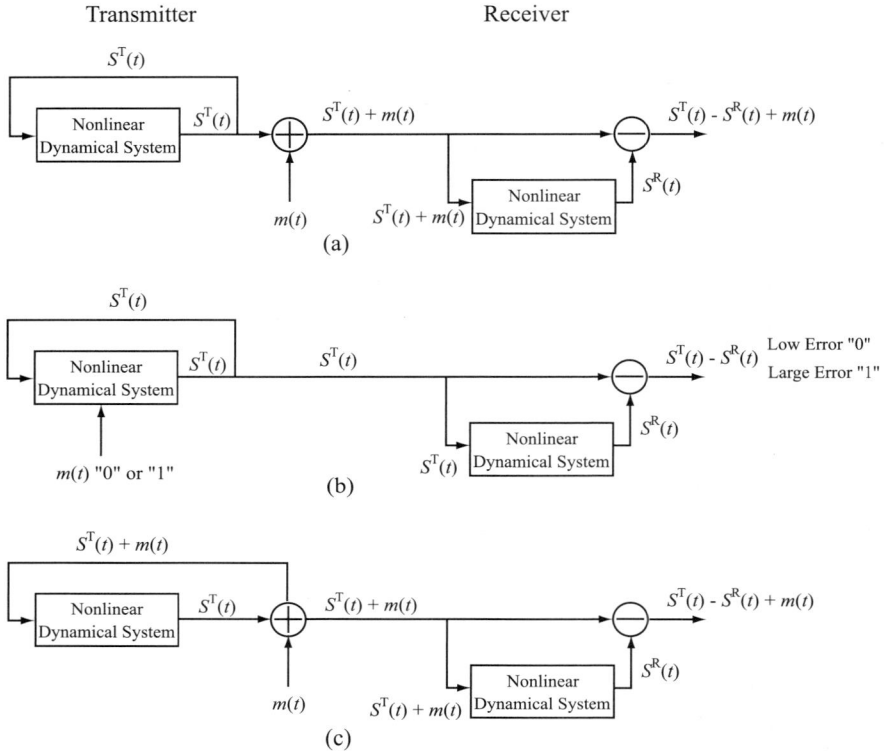

Fig. 11.4. Schematics of the (a) chaos masking, (b) chaos shift keying, and (c) chaos modulation schemes in a chaotic communication system.

In the chaos masking scheme, a message is encoded onto the chaotic waveform after the chaotic waveform exits the dynamical system of the transmitter. Therefore, the chaotic transmitter is independent of the message and is only driven by the dynamics of the transmitter itself, whereas the receiver is driven not only by the chaotic waveform from the transmitter but also by the encoded message. Because of this difference in the driving forces, the receiver cannot be perfectly synchronized to the transmitter in this system. Therefore, the message signal is required to be small compared to the chaotic waveform in order to keep the synchronization errors small [24, 30].

In chaos shift keying, a certain parameter of the transmitter is modulated by the binary bits of the message. As a result, the binary bits "1" and "0" are mapped into two different chaotic states. The receiver can detect the message by tracking one of the chaotic states as either synchronized or unsynchronized, with low or high levels of synchronization errors, respectively. When the transmitter is switched between the two chaotic states, the receiver cannot synchronize with the transmitter all the time. The bit rate of the modulating message is limited for a system using chaos shift keying because it is not steadily synchronized.

Chaos modulation is the scheme that has been used to explain the importance of chaos synchronization in Figure 11.3. Chaos modulation can overcome the limitations on the amplitude and the bit rate of the message in the other two encoding schemes discussed above. In chaos modulation, when a message is encoded onto the chaotic waveform of the transmitter and sent to the receiver, it is also injected into the dynamics of the transmitter. Different from the chaos masking scheme, now both the transmitter and the receiver are driven by the same force that includes the message. Consequently, when their parameters are all matched, the transmitter and the receiver potentially can perfectly synchronize with each other all the time in the presence of the variations of the message [10, 24, 30]. From the viewpoint of synchronization, the message can have a large amplitude and a very high bit rate while still allowing the system to maintain perfect synchronization.

In both the chaos masking and the chaos modulation schemes, the message can also be encoded on the chaotic carrier through other operations, such as multiplication. The decoding is achieved by comparing the received signal with the reproduced chaotic waveform. This decoding can be either subtraction or division, depending on whether the encoding process is addition or multiplication. Depending on the applications of chaotic optical communications, different encoding and decoding schemes can be selected. For good quality of chaotic communication, minimization of synchronization errors is very important.

11.2.4 Channel Noise and Bit-Error Rate

In a chaotic communication system, once chaos synchronization is achieved between the transmitter and the receiver, message encoding and decoding can be implemented. Because the additive white Gaussian noise (AWGN) channel constitutes the most basic component of a communication link, the investigation of the system performance under AWGN is of great importance. Different from that of a conventional communication system, the performance of a chaotic communication system strongly depends on the quality of synchronization. Channel noise does not only contaminate the signal in the process of transmission but also seriously influences the quality of synchronization of the system by generating synchronization errors. Such synchronization errors are generated because the identity of the driving forces to both the transmitter and the receiver is corrupted by the presence of channel noise. Therefore, the system performance strongly depends on the robustness of synchronization under the influence of channel noise, which can be very different for different systems as we will show later in our simulation of three chaotic optical communication systems.

Bit-error rate (BER) is a standard performance measurement of a communication system. It is often measured in a form as $\log$ BER versus signal-to-noise ratio (SNR). In a chaotic communication system, while additional energy has to be used to transmit the chaotic carrier waveform, the information is only included in the energy of the message. Therefore, it is common to treat the chaotic waveform as carrier and include only the message energy in the calculation of SNR. When the message is disturbed by noise, the signal levels designated to bits "1" and "0" spread out and overlap. A bit error is detected when a "1" bit is mistaken as a "0" bit, or vice versa. As is discussed above, the bit errors caused by the channel noise come from both the contamination of the message by the channel noise during transmission and the synchronization errors generated by the injection of the channel noise into the receiver.

For chaotic communications, these issues related to chaos generation, encoding and decoding schemes, and chaos synchronization are very important. They are addressed both experimentally and numerically in the following sections.

11.3 Experiment of Chaotic Optical Communication at 2.5 Gb/s

In this section, we review the optical communication experiment using a fast chaotic pulsing semiconductor laser system with delayed optoelectronic feedback [8]. The encoding and decoding scheme used in the experiments is basically chaos modulation discussed above. The chaos modulation scheme can be either additive chaos modulation or multiplicative chaos modulation, depending on whether the message is encoded through adding or multiplying the message onto the chaotic waveform. In this experiment, the message is encoded through incoherent addition with the chaotic waveform.

The schematic experimental setup of the chaotic communication system using chaotically pulsing semiconductor lasers is shown in Figure 11.5. In this setup, the transmitter laser has an optoelectronic feedback loop which drives the laser into chaotic pulsing when the feedback delay time τ is carefully adjusted [4]. The receiver laser operates in an open-loop configuration and is driven by the signal from the transmitter. Message $m(t)$ is encoded by means of incoherent addition onto the output of the transmitter laser as $S^{\mathrm{T}}(t) + m(t)$. When $m(t)$ is encoded onto $S^{\mathrm{T}}(t)$ and the combined signal is sent to the receiver laser, it is also fed back to the transmitter laser. Therefore, the transmitter laser is driven by $S^{\mathrm{T}}(t - \tau) + m(t - \tau)$ because of the delay time in the feedback loop, and the receiver laser is driven by $S^{\mathrm{T}}(t - T) + m(t - T)$ because of the time delay in transmission, where T is the transmission time.

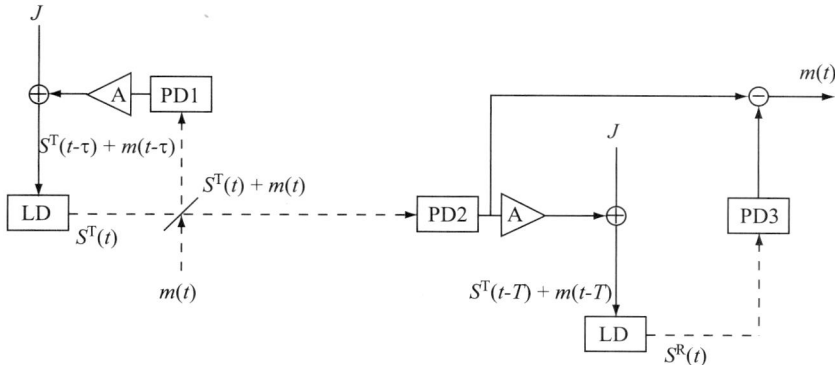

Fig. 11.5. Schematic setup of the chaotic communication system with chaotically pulsing semiconductor lasers for encoding and decoding messages through additive chaos modulation. LD: Laser Diode; PD: Photodetector; A: Amplifier.

With parameters matched between the transmitter and the receiver lasers, the receiver is synchronized to the transmitter with a time shift $T - \tau$. Therefore, the receiver laser output is $S^{R}(t) = S^{T}(t-T+\tau)$, and the message is recovered as $S^{T}(t - T + \tau) + m(t - T + \tau) - S^{R}(t) = m(t - T + \tau)$.

In the experiment, the transmitter and the receiver lasers are InGaAsP/InP single-mode DFB lasers emitting at the same wavelength of 1.299 μm. These two lasers are carefully chosen from the same batch with the closest characteristics. In the experiment, both lasers are temperature stabilized at 21.00 °C. The photodetectors are InGaAs photodetectors with a 6-GHz bandwidth. The amplifiers are Avantek SSF86 amplifiers with a 3-dB pass band of 0.4 to 3 GHz. The chaotic waveforms are measured with a Tektronix TDS 694C digitizing real-time oscilloscope with a 3-GHz bandwidth and a 1×10^{10} Samples/s sampling rate. The 2.5 Gb/s pseudorandom digital bits are generated by an HP 70843A pseudorandom pattern generator.

With the transmitter and the receiver steadily synchronized [14], recovery of message utilizing the chaotic pulse system is investigated. Figure 11.6 shows the recovery of a 500 MHz pulse stream. This pulse stream is generated by an HP 33004A comb generator. The top trace is the received signal, $S^{T}(t - T + \tau) + m(t - T + \tau)$, measured at PD2 in Figure 11.5 and shifted by a time τ. The second trace, measured at PD3, is the local receiver laser output, which synchronizes with and duplicates the chaotic pulse output of the transmitter laser as $S^{R}(t) = S^{T}(t-T+\tau)$. The time shift between the transmitter and the receiver has been matched in the oscilloscope. Shown in the third trace, signal recovery is achieved by subtracting the receiver laser output (the second trace) from the received signal (the top trace). A decision threshold can be set at the position of the dashed line. The recovered signal shows very good quality of decoding compared with the original signal, which is also shown in Figure 11.6 in the bottom trace as a reference. Limited by the size of the record length of

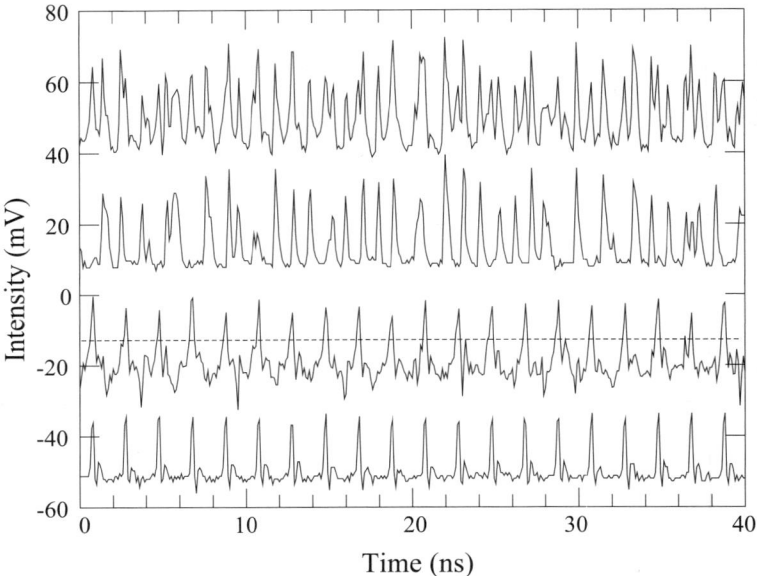

Fig. 11.6. Transmission of a pulse stream at 500 MHz repetition rate in a chaotic communication system using semiconductor lasers with optoelectronic feedback. Time series of received signal (top), receiver laser output (second), recovered signal (third), and encoded signal (bottom). (Reprinted with permission from [8], ©2001 OSA.)

the digitizing sampling oscilloscope, which is 120,000 samples for TDS 694C, we can only record about 6000 pulses in one shot. Within the 6000 pulses, no error bit is been detected, indicating a BER less than 1.7×10^{-4}.

To show the potential communication capacity of this chaotic pulse system, a pseudorandom sequence at a bit rate of 2.5 Gb/s is transmitted [8]. Figure 11.7 shows the results of message recovery, where each time series has the same meaning as that of the corresponding trace in Figure 11.6. The 2.5 Gb/s NRZ message is generated by an HP 70843A pseudorandom pattern generator. Even though there are some fluctuations in the recovered time series, by setting a decision threshold at the position of the dashed line, it is clear that the message can be successfully recovered at this high bit rate. Thresholding is an advantage of systems transmitting messages in digital bits, compared with analog messages. The performance can also be largely improved by modifying the shape and the format of the bit sequence to better fit the characteristics of the chaotic pulse carrier. Adding a low-pass filter at the receiver end can smooth out the fast fluctuating noise and further improve the system performance. To our knowledge, this system of transmitting a message at 2.5 Gb/s has the highest bit rate in any chaotic communication systems reported in

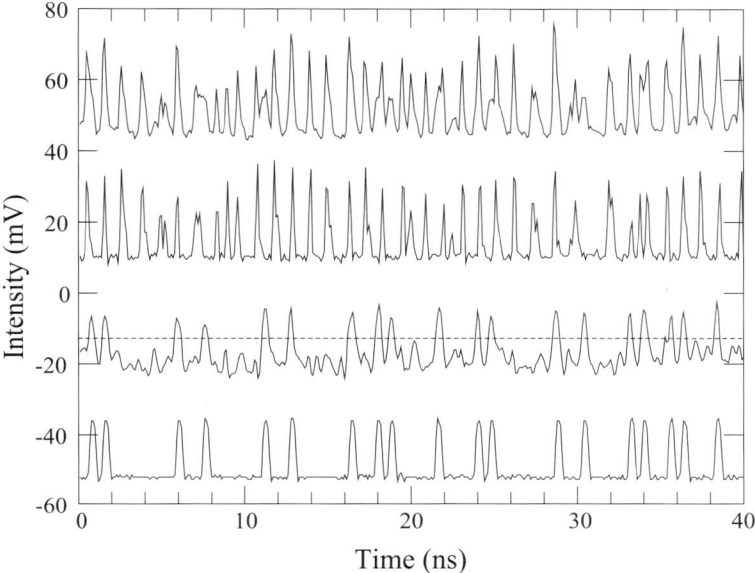

Fig. 11.7. Transmission of a pseudorandom NRZ bit sequence at a bit rate of 2.5 Gb/s in a chaotic communication system using semiconductor lasers with optoelectronic feedback. Each time series has the same meaning as that of the corresponding trace in Figure 11.6. (Reprinted with permission from [8], ©2001 OSA.)

the literature. This bit rate fits the standard of commercial digital OC-48 system. Therefore, the experiment has the significance in proving that chaotic communications can be implemented with commercial standards.

The eye diagram of the recovered message transmitted through this chaotic optical communication system is shown in Figure 11.8a. A clear eye opening is obtained, which indicates the good quality of message recovery. As a reference, Figure 11.8b also shows the eye diagram obtained with the same setup but with the semiconductor laser operated in a nonchaotic stable state. As can be seen, a large amount of errors already exists in the nonchaotic system because of the limited modulation bandwidth of the semiconductor laser and the large noise from the electronic components, which are not intentionally optimized for this purpose. These errors can be reduced by using semiconductor lasers that have larger modulation bandwidths and electronic components that have much less noise. The deterioration of the eye opening between Figures 11.8a and b is the real deterioration that is caused by the introduction of a chaotic carrier. This deterioration can be improved by reducing the synchronization errors between the transmitter and the receiver.

High-bit-rate digital messages at up to a bit rate of 2.5 Gb/s have been successfully transmitted [8]. The chaos modulation scheme is demonstrated to

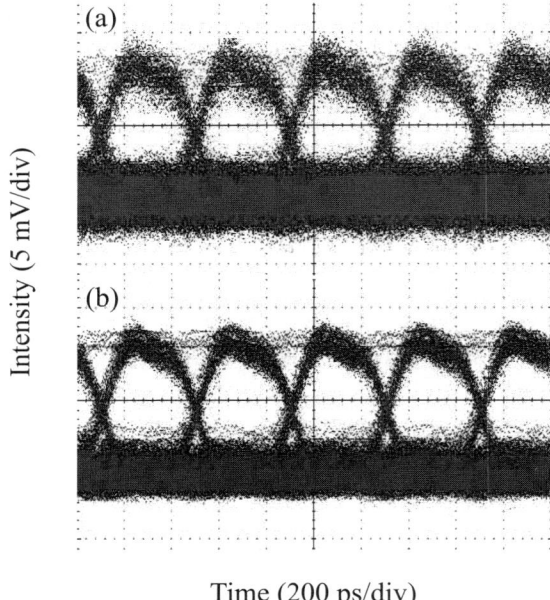

Time (200 ps/div)

Fig. 11.8. (a) Experimentally measured eye diagram of decoded 2.5 Gb/s NRZ messages in a chaotic optical communication system using semiconductor lasers with optoelectronic feedback. (b) Reference eye diagram using the same setup but with the semiconductor laser operated in a nonchaotic stable state.

have the advantage in maintaining high-quality synchronization through all the time of message transmission. The communication quality and capacity are both shown to be very high with this chaotically pulsing semiconductor laser system, which indicates that chaotic communication with a bit rate of several Gb/s is feasible with semiconductor lasers.

11.4 Comparison of Different Encoding and Decoding Schemes

Many systems based on either semiconductor lasers [11–16] or fiber ring lasers [17–19] have been proposed and studied for chaotic optical communications. Several encoding and decoding schemes have been considered and demonstrated for chaotic communications. The most important ones include chaos shift keying (CSK), chaos masking (CMS), and chaos modulation. For chaos modulation, possible encoding and decoding schemes include additive chaos modulation (ACM) and multiplicative chaos modulation (MCM). Because of the differences in these message encoding/decoding schemes, they have very different effects on the system performance in regard to the dynamics, synchronization, and communication performances. In this section, we review experimental demonstrations of the effects of message encoding and decoding on the chaotic dynamics, chaos synchronization, and chaotic communication performances of a chaotic optical communication system.

11.4.1 Experimental Setup

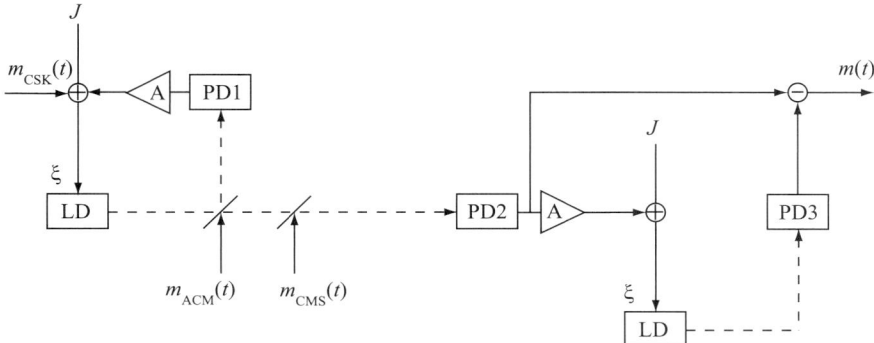

Fig. 11.9. Schematic experimental setup of the chaotic optical communication system using semiconductor lasers with optoelectronic feedback. Also shown are the message encoding and decoding schemes: CSK, chaos shift keying; CMS, chaos masking; ACM, additive chaos modulation. The solid lines indicate the electronic paths. The dashed lines indicate the optical paths. LD: laser diode; PD: photodetector; A: amplifier; $m(t)$: message.

Figure 11.9 shows the schematic experimental setup of our chaotic optical communication system, which is similar to the setup in Figure 11.5. Three major encoding and decoding schemes, namely, CSK, CMS, and ACM are investigated and compared because they are all additive in nature on the transmitter side and are subtractive in nature on the receiver side. The message that is encoded and decoded in the system is a sequence of pseudorandom digital bits at a bit rate of 2.5 Gb/s. This bit rate matches the OC-48 standard of optical communications.

In the CSK scheme, the digital message $m_{CSK}(t)$ directly modulates the current injected to the transmitter laser. Thus the current injected to the transmitter laser switches between two distinct levels depending on whether a "0" bit or a "1" bit is transmitted. Meanwhile, the receiver laser is biased at a fixed current level that is equivalent to the current level of the transmitter laser at which a "0" bit is transmitted. Thus the two semiconductor lasers are biased at the same current level when a "0" bit is transmitted, but they are biased at different levels when a "1" bit is transmitted. By measuring the synchronization errors between the transmitter and the receiver, the message can be recovered. A low level of synchronization error corresponds to a "0" bit, and a high level of synchronization error corresponds to a "1" bit. The synchronization errors are measured by the difference between the outputs from PD2 and PD3 in Figure 11.9.

In the CMS scheme, the message $m_{CMS}(t)$ is added on the chaotic waveform after it leaves the optoelectronic feedback loop of the transmitter on its way to the receiver. Thus the message does not affect the feedback loop of the transmitter, and the chaotic waveform is solely generated by the optoelectronic feedback loop of the transmitter. At the receiver, the output from PD2 is the combined signal of the chaotic waveform and the message, and the output from PD3 is the reproduced chaotic waveform that is regenerated by the receiver laser. By subtraction, the receiver can recover the message.

In the ACM scheme, the message $m_{ACM}(t)$ is added on the chaotic waveform within the optoelectronic feedback loop of the transmitter. When the message is sent together with the chaotic waveform to the receiver, it is also sent to drive the transmitter laser through the optoelectronic feedback loop. Therefore, the dynamics of the transmitter laser is also affected by the encoded message. At the receiver, message recovery is achieved through the same process as that in the CMS scheme. However, because the message is sent to drive both the transmitter and the receiver lasers in the ACM scheme but is not sent to drive the transmitter laser in the CMS scheme, the effects of message encoding and decoding are very different in these two schemes.

11.4.2 Effects on Chaotic Dynamics

When there is no message encoded, the chaotic waveform is solely generated by the transmitter laser with the optoelectronic feedback. Details of the nonlinear dynamics and the characteristics of the chaotic states are discussed in Chapter 10. Figures 11.10a−c show the time series, power spectrum, and phase portrait, respectively, of a typical chaotic pulsing waveform that is generated by the transmitter laser without any encoded message.

In order to have a fair comparison among the different encoding and decoding schemes, we defined a normalized message strength, m, as follows,

$$m = \frac{\text{average intensity of message}}{\text{average intensity of chaotic waveform}}. \tag{11.1}$$

In this definition, the average intensity of the message is measured in the transmission channel after PD2 when no chaotic waveform is transmitted. Similarly, the average intensity of the chaotic waveform is measured at the same position when no message is transmitted. In Figures 11.10d−l, m is set to be 0.7 for all the three schemes when a message is encoded.

Figures 11.10d−f show the corresponding characteristics of the chaotic waveform after a pseudorandom message is encoded through the CSK scheme in the chaotic optical communication system. In Figure 11.10d, the time series shows a train of pulses with some additional irregular fluctuates in both the

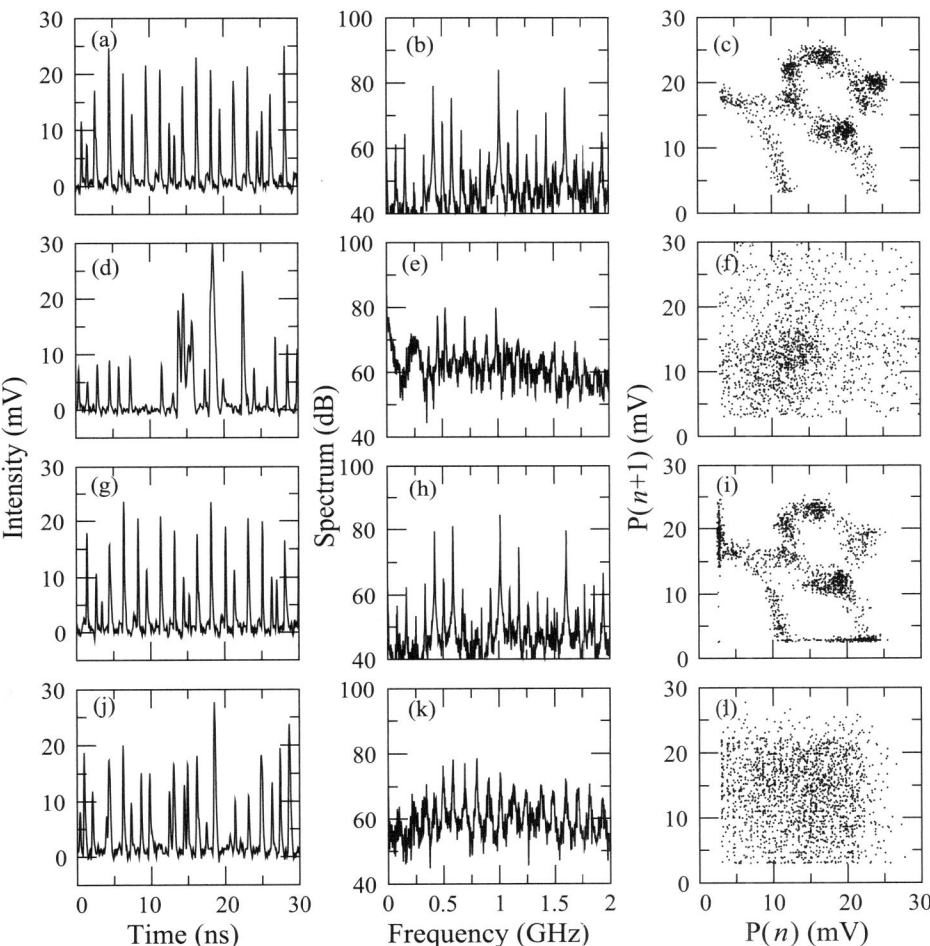

Fig. 11.10. Time series (first column), power spectra (second column), and phase portraits (third column), respectively, of the chaotic waveforms in a chaotic optical communication system with optoelectronic feedback obtained with and without an encoded message: (a)−(c) without the message, (d)−(f) with the message in CSK, (g)−(i) with the message in CMS, and (j)−(l) with the message in ACM. The normalized message strength is $m = 0.7$. (Reprinted with permission from [20], ©2003 IEEE.)

pulse intensity and the pulse interval. In Figure 11.10e, the background of the power spectrum increases to a much higher level than that in Figure 11.10b. At the same time, the spiky characteristic of the spectrum in Figure 11.10b is much suppressed in Figure 11.10e. Both the increase of the background and the suppression of the spikes in the spectrum are evidences of increased complexity of the chaotic waveform. Furthermore, from the phase portrait shown in Figure 11.10f, a very scattered distribution of the data points is observed. Therefore, from the time series, the power spectrum, and the phase portrait, the complexity of the chaotic state is shown to be increased after a pseudorandom message is encoded through the CSK scheme.

In the CMS scheme, the message is added on the chaotic waveform after the waveform leaves the optoelectronic feedback loop, as can be seen from the setup in Figure 11.9. Therefore, the message is not injected into the dynamics that generates the chaotic waveform of the transmitter. No increase in the complexity of the chaotic waveform is expected in the case of CMS. Figures 11.10g–i show the time series, the power spectrum, and the phase portrait, respectively, after a pseudorandom message is encoded through the CMS scheme. As we can see, Figures 11.10g–i are very similar to the corresponding ones in Figures 11.10a–c, which indicates that indeed the characteristics of the chaotic state is not changed by the encoding process in the CMS scheme.

Different from the CMS scheme, in the ACM scheme, when the message is added on the chaotic waveform and is sent to the receiver, the message is also sent to the optoelectronic feedback loop and further drives the dynamics of the transmitter laser. Figures 11.10j–l show the characteristics of the chaotic waveform after a message is encoded in this scheme. As we can see, the time series in Figure 11.10j is more irregular than that in Figure 11.10a. The power spectrum in Figure 11.10k has a much higher background and much less spiky nature than that in Figure 11.10b. Figure 11.10l also has a much more scattered phase portrait than that in Figure 11.10c. Therefore, the message encoding process in the ACM scheme increases the complexity of the dynamics of the transmitter laser in the chaotic optical communication system.

11.4.3 Effect on Chaos Synchronization

For the chaotic optical communication system shown in Figure 11.9, the receiver laser is identical to the transmitter laser, and the driving force to the transmitter and that to the receiver are the same when there is no encoded message. Therefore, under this condition, the receiver can synchronize with the transmitter.

For chaotic communications, it is important to investigate the effect of message encoding and decoding on chaos synchronization. The issue of maintaining high quality of synchronization in the presence of message encoding is already addressed in Chapter 10. The quality of synchronization can be maintained or its deterioration minimized if the symmetry between the transmitter and the receiver is not broken by the message encoding process. In both the CSK and CMS schemes, the symmetry between the transmitter and the receiver is broken by the process of message encoding. In the ACM scheme, however, the symmetry between the transmitter and the receiver is maintained because the message is injected into the dynamics of the transmitter when it is sent to the receiver.

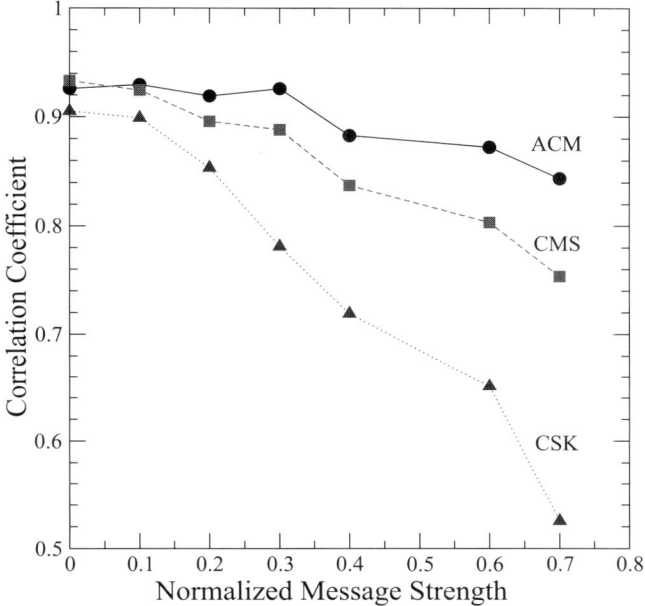

Fig. 11.11. Experimentally measured correlation coefficient between the transmitter and the receiver in a chaotic optical communication system using semiconductor lasers with optoelectronic feedback for three encoding schemes, under various strengths of the message. (Reprinted with permission from [20], ©2003 IEEE.)

The gradual change of the quality of synchronization in the presence of message encoding is investigated with the increase in the message strength for the three encoding and decoding schemes. The quality of synchronization is quantified by the calculation of the correlation coefficient ρ [21] between the outputs of the transmitter and the receiver. The correlation coefficient, which is defined in (9.106) of Chapter 9 and in (10.58) of Chapter 10, is bounded as $-1 \leq \rho \leq 1$. A larger value of $|\rho|$ means a higher quality of synchronization. In Figure 11.11, the correlation coefficient is calculated for different message strengths for each of the three encoding and decoding schemes. For $m = 0.0$, there is in fact no encoded message. Due to the slight difference in the system configurations and in the operating conditions for CSK, CMS, and ACM schemes, the correlation coefficients are slightly different for the three schemes when $m = 0.0$. Nevertheless, this difference is very small and the correlation coefficients are above 0.9 for all the three schemes when $m = 0.0$. For the CSK scheme, as the message strength is gradually increased from $m =$

0.0 to $m = 0.7$, the correlation coefficient is observed to decrease dramatically from 0.9 to 0.5. For the CMS scheme, the correlation coefficient also drops from above 0.9 to below 0.8 when the message strength is increased from $m = 0.0$ to $m = 0.7$. The highest quality of synchronization is achieved with the ACM scheme when there is an encoded message. Even at $m = 0.7$, the correlation coefficient is still as high as 0.84 for the ACM scheme. The slight decrease of the quality of synchronization in the ACM scheme with the increase in m is mainly caused by the significant increase in the complexity of the chaotic state as is discussed in the preceding subsection. Therefore, the effects of message encoding on chaos synchronization are very different among the three schemes.

11.4.4 Performances of Chaotic Communications

In the preceding subsections, the different effects of message encoding and decoding on chaotic dynamics and chaos synchronization are discussed for the three schemes of CSK, CMS, and ACM. Because of such differences, the chaotic communication performances are also different among the three schemes.

In the CSK scheme, the bias current of the transmitter laser is modulated by the digital message but that of the receiver laser is not modulated. Therefore, the two lasers can synchronize when a "0" bit is transmitted, and they are desynchronized when a "1" bit is transmitted. By measuring the synchronization errors, the message can be recovered. Figure 11.12a shows the measured synchronization errors representing the recorded message when the message as shown in Figure 11.12d is transmitted through the CSK scheme. The dashed line shows the estimated threshold between the "0" bits and the "1" bits. Comparing Figures 11.12a and d, we see that a few bits can be recovered, but there are still many bits that cannot be recovered.

In the CMS scheme, the message is added on the chaotic waveform and is sent to the receiver. The receiver synchronizes to the transmitter and reproduces the chaotic waveform. By subtraction, the message can be recovered. However, the message is contaminated by the synchronization errors caused by the asymmetry between the transmitter and the receiver in the presence of the encoded message. Because these synchronization errors are significant, the communication performance of the CMS scheme is also poor. Figure 11.12b shows the recovered message through the CMS scheme. Comparing Figure 11.12b with the reference message in Figure 11.12d, we see that the residual synchronization errors are still large, and some bits can be recovered while others cannot.

In the ACM scheme, the quality of synchronization is basically maintained even in the presence of an encoded message. As the communication quality depends sensitively on the quality of synchronization, the communication performance is expected to be high with this high quality of synchronization. Figure 11.12c shows the recovered message using the ACM scheme. It can be

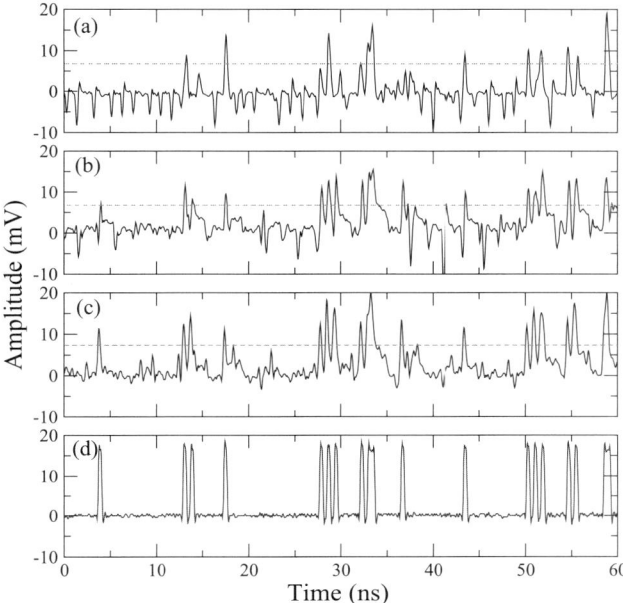

Fig. 11.12. Recovered message in (a) the CSK scheme, (b) the CMS scheme, (c) the ACM scheme, and (d) the reference input message in a chaotic optical communication system using semiconductor lasers with optoelectronic feedback. The dashed lines indicate the estimated threshold between "0" and "1". The normalized strength of the input message is $m = 0.7$. (Reprinted with permission from [20] ©2003 IEEE.)

seen that all bits are reliably recovered. Because synchronization in the system is maintained all the time, the highest bit rate that can be transmitted with ACM is not limited by the resynchronization time of the system.

To further compare the communication performances of the three different schemes, the BER is measured for each of the three schemes. In this measurement, 12,000 bits are recorded and counted for each case. Figure 11.13 shows the BER measured at the message strength of $m = 0.7$ for each of the three encoding schemes. The BER is worse than 1×10^{-2} for both the CSK and the CMS schemes. The ACM scheme has the best BER, which is between 1×10^{-2} and 1×10^{-3}. For comparison, the BER of a traditional optical communication system with the same semiconductor laser under direct current modulation and direct detection using the same experimental setup is also shown in Figure 11.13 as the open circle. It is seen that even when there is no chaos in the system, the BER is already worse than 1×10^{-4}. This large BER is mainly caused by the limited modulation bandwidth of the semiconductor laser and the large noise of the electronic system used in this experimental

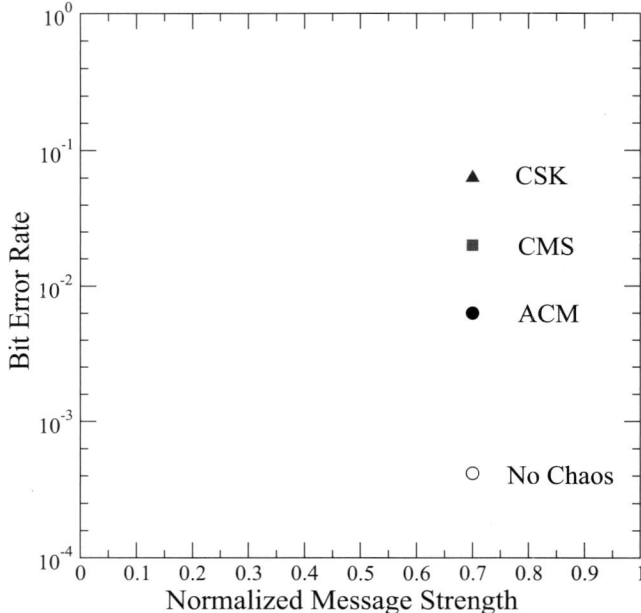

Fig. 11.13. Measured BER of a chaotic optical communication system using semiconductor lasers with optoelectronic feedback for the three chaotic message encoding schemes. The BER of the system without chaos is also indicated by the open circle. The normalized strength of the input message is $m = 0.7$. (Reprinted with permission from [20], ©2003 IEEE.)

setup, which are not intentionally optimized for high-speed, low-BER communication applications. Furthermore, with the semiconductor lasers operating in the broadband chaotic state, the chaotic optical communication system has a higher demand on the bandwidth of the system, including the electronic components and the detection equipment. Therefore, with high-speed electronics, the performance of the chaotic optical communication system can be much improved. The system performance can be further improved with better matched and less noisy semiconductor lasers and other components in the system.

11.4.5 Summary

Chaotic optical communication at 2.5 Gb/s is experimentally investigated and demonstrated using the three major chaotic encoding and decoding schemes, namely, CSK, CMS, and ACM. The effects of message encoding and decoding on chaotic dynamics, chaos synchronization, and chaotic communication performances are investigated and compared among the three schemes.

The chaotic dynamics of the system is influenced when the message is injected into the dynamics through the process of message encoding. This is the case in the CSK and ACM schemes. It is found that a small amount of message injected into the chaotic dynamics can increase the complexity of the chaotic state significantly because of the random nature of the message. This feature increases the practicality of secure communications using chaotic systems because such systems of increased complexity are difficult for eavesdroppers to attack. However, not every chaotic encoding scheme has this feature. In the case of CMS, the chaotic dynamics is found not to be influenced by the encoded message, which is clear because the message is not injected into the chaotic dynamics.

The quality of chaos synchronization deteriorates when the process of message encoding and decoding breaks the symmetry between the transmitter and the receiver. In the cases of CSK and CMS, the quality of synchronization is found to drop dramatically with the increase in the message strength because of this reason. For chaotic communications that rely on high-quality chaos synchronization, an encoding and decoding scheme that can maintain the symmetry between the transmitter and the receiver is most desirable. The ACM scheme has this characteristic, and it is found to have the highest quality of synchronization among the three schemes in the presence of an encoded message.

The performances of chaotic communications are directly related to the quality of synchronization. Through comparison, it is found that the ACM scheme has the best communication performance among the three schemes, which is in agreement with the fact that it also has the highest quality of synchronization among the three schemes. The system performance is currently limited by multiple conditions, such as the speed of the electronic devices and the equipment, parameter mismatch between the transmitter and the receiver, and noise in the system, etc. Nevertheless, the performances of the three most popular chaotic message encoding and decoding schemes are clearly demonstrated and compared.

11.5 Chaotic Optical Communications at 10 Gb/s

Motivated by the consideration of developing chaotic communication systems at bit rates higher than 2.5 Gb/s [8], we choose the OC-192 standard bit rate of 10 Gb/s for the system performance studies through simulation [22]. We study several basic, yet important, issues regarding high-bit-rate chaotic optical communications based on the synchronization of semiconductor laser chaos. The emphasis is on the system performance measured by the BER in the presence of channel noise and laser noise. Three chaotic optical communication systems using semiconductor lasers with optical injection, optical feedback, or optoelectronic feedback, respectively, are investigated in the simulation. For each system, the three encoding and decoding schemes, namely,

CMS, CSK, and ACM are implemented and compared. The details of the modeling and the characteristics of the nonlinear dynamics of each system are addressed in Chapters 9 and 10. In this chapter, we focus on the communication performances and the BER calculations.

In a chaotic communication system, the synchronization errors of a system are contributed by the following two forms of errors: synchronization deviation, which is associated with the accuracy of synchronization, and desynchronization burst, which is associated with the robustness of synchronization. The synchronization deviation is simply the synchronization error when the system is synchronized, but not perfectly and precisely. Desynchronization bursts are characterized by sudden desynchronization between the transmitter and the receiver. A desynchronization burst can cause a large, abrupt difference between the waveforms of the transmitter and the receiver. Because a system takes some finite time to resynchronize after a desynchronization burst, the bits that follow a desynchronization burst within the resynchronization time are destroyed. In the following subsections, we will see how synchronization deviation and desynchronization bursts have different effects on the system performance.

11.5.1 Performance of Optical Injection System

According to the configuration in Figure 11.14a, the transmitter can be modeled by the following coupled equations in terms of the complex intracavity laser field amplitude A^T and the carrier density N^T [11, 23],

$$\frac{dA^T}{dt} = -\left(\frac{\gamma_c^T}{2} + \eta\alpha\right) A^T + i(\omega_0 - \omega_c)A^T + \frac{\Gamma}{2}(1 - ib^T)gA^T + F_{sp}^T$$
$$+ \eta\{\alpha A^T(t) + A_i e^{-i\Omega t}[1 + m_{ACM}(t)]\}, \tag{11.2}$$

$$\frac{dN^T}{dt} = \frac{J[1 + m_{CSK}(t)]}{ed} - \gamma_s N^T - \frac{2\epsilon_0 n^2}{\hbar\omega_0}g|A^T|^2, \tag{11.3}$$

whereas the receiver, driven by the transmitted signal $s(t)$, can be described by

$$\frac{dA^R}{dt} = -\frac{\gamma_c^R}{2}A^R + i(\omega_0 - \omega_c)A^R + \frac{\Gamma}{2}(1 - ib^R)gA^R + F_{sp}^R + \eta s(t), \tag{11.4}$$

$$\frac{dN^R}{dt} = \frac{J}{ed} - \gamma_s N^R - \frac{2\epsilon_0 n^2}{\hbar\omega_0}g|A^R|^2. \tag{11.5}$$

In this model, $A_i e^{-i\Omega t}$ is the optical injection field with a detuning frequency of $\Omega = 2\pi f$, η is the injection rate of the optical field into the laser, and α defines the coupling strength between the transmitter and the receiver. The transmitted signal has the form $s(t) = \alpha A^T(t) + A_i e^{-i\Omega t}[1 + m(t)]$ for both ACM and CMS schemes, but the form $s(t) = \alpha A^T(t) + A_i e^{-i\Omega t}$ for the CSK scheme. The subscript of the encoding message $m(t)$ indicates the encoding

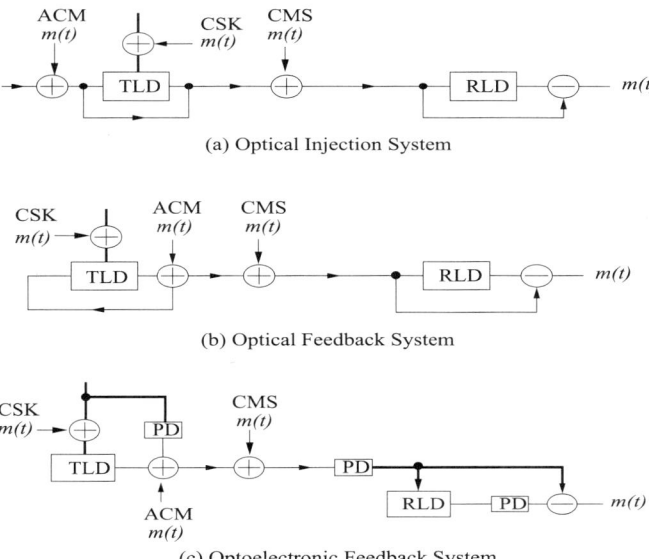

Fig. 11.14. Schematics of the chaotic optical communication system using semiconductor lasers with (a) optical injection, (b) optical feedback, and (c) optoelectronic feedback. CMS: chaos masking; CSK: chaos shift keying; ACM: additive chaos modulation. TLD: transmitter laser diode; RLD: receiver laser diode; PD: photodetector. (Reprinted with permission from [22], ©2002 IEEE.)

scheme used. When a particular scheme is used, $m(t)$ with a subscript of the other schemes should be set to zero. As can be seen from the mathematical model, when CMS is applied, the encoding message, $m(t)$, is injected into the receiver but not into the transmitter. When CSK is applied, only the transmitter laser, but not the receiver laser, is current modulated with the encoding message. By comparison of the rate equations above for different encoding schemes, it can be clearly seen that the transmitter and the receiver are never mathematically identical when a message is encoded through the CMS or CSK scheme. They can be identical only when a message is encoded through the ACM scheme and when their parameters are properly matched.

As a general characteristic of chaotic communications based on synchronization, the recovered message is contaminated by the channel noise and the synchronization errors. Therefore, the performance of the message recovery is critically determined by the quality of synchronization, which depends on the noise in the system and the encryption method used. The performance of this system is first investigated by examining the recovered message in the time domain for each encryption method, shown in Figure 11.15, to provide an understanding of the generation of the error bits. In order to reveal the effect

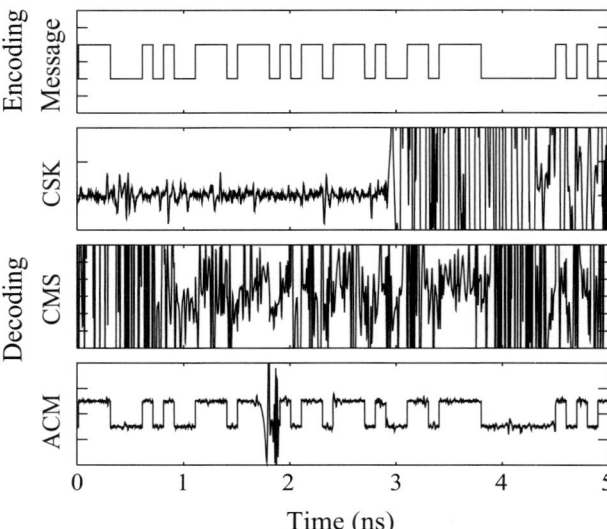

Fig. 11.15. Time series of the decoded messages of the three different encryption schemes in the optical injection system. CSK, chaos shift keying; CMS, chaos masking; ACM, additive chaos modulation. (Reprinted with permission from [22], ©2002 IEEE.)

of the channel noise on the message recovery, the data shown in this figure are obtained when the laser noise is not considered. It is observed that message recovery is almost impossible for the CSK scheme because message encoding with CSK causes frequent desynchronization bursts and the success of message decoding for this encryption scheme is determined by the resynchronization time. At a bit rate of 10 Gb/s, resynchronization is difficult to achieve within the short time of the bit duration of $T_b = 0.1$ ns. The performance can be improved at a low bit rate when the bit duration gets longer than the resynchronization time. As for CMS, the synchronization errors mainly arises from the breaking of the mathematical identity between the transmitter and the receiver by the encoded message. Since the encoding message used here is small in comparison to the transmitter output, the encoded message acts only as a perturbation on the synchronization. Therefore, the recovered message shows some resemblance to the pattern of the encoding message. Better message recovery can be expected if a low-pass filter is used. The performance of ACM is the best among the three encryption methods because message encoding by ACM does not break the mathematical identity between the transmitter and the receiver. The error bits are contributed by synchronization errors caused by the channel noise, as well as by the laser noise when it is considered. Whether synchronization deviation or desynchronization bursts dominate in the generation of error bits depends on the amount of noise present. We can

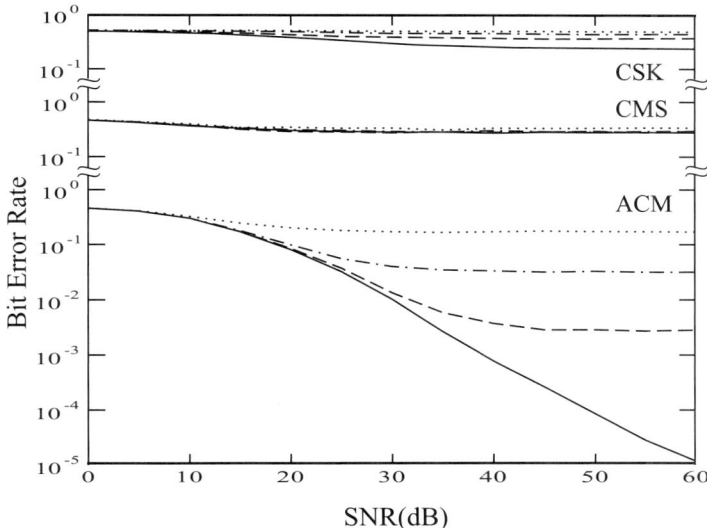

Fig. 11.16. BER versus SNR for the three different encryption schemes in the optical injection system. Solid curves are obtained when the laser noise is absent. Dashed curves are obtained at a laser noise level equivalent to $\Delta\nu = 100$ kHz for both the transmitter and receiver lasers. Dot-dashed curves are obtained when $\Delta\nu = 1$ MHz. Dotted curves are obtained when $\Delta\nu = 10$ MHz. CSK, chaos shift keying; CMS, chaos masking; ACM, additive chaos modulation. (Reprinted with permission from [22], ©2002 IEEE.)

see that the single error bit seen in the ACM decoded message in Figure 11.15 is generated by the occurrence of a desynchronization burst.

The system performance measured by the BER as a function of channel SNR for the optical injection system is shown in Figure 11.16 for each of the three encryption schemes. We observe that CSK and CMS have similar performance when the laser noise is not considered. The performance of CMS is barely affected by the laser noise because the breaking of the mathematical identity between the transmitter and the receiver caused by the encoded message has a much more significant effect on creating synchronization errors than the perturbation of the laser noise to the system. The performance of CSK is, however, deteriorated by the laser noise because the noise further increases the desynchronization probability and the resynchronization time. As for the performance of the ACM scheme, a BER lower than 10^{-5} can be obtained for a SNR larger than 60 dB under the condition that the laser noise is absent. However, in the presence of the laser noise indicated by laser linewidth, the system performance deteriorates quickly as the intrinsic laser noise increases.

11.5.2 Performance of Optical Feedback System

According to the configuration in Figure 11.14b, the transmitter can be modeled by the following coupled equations [12],

$$
\frac{dA^{\mathrm{T}}}{dt} = -\frac{\gamma_{\mathrm{c}}^{\mathrm{T}}}{2}A^{\mathrm{T}} + i(\omega_0 - \omega_{\mathrm{c}})A^{\mathrm{T}} + \frac{\Gamma}{2}(1 - ib^{\mathrm{T}})gA^{\mathrm{T}} + F_{\mathrm{sp}}^{\mathrm{T}}
$$
$$
+ \eta[\alpha A^{\mathrm{T}}(t - \tau) + m_{\mathrm{ACM}}(t - \tau)], \tag{11.6}
$$
$$
\frac{dN^{\mathrm{T}}}{dt} = \frac{J[1 + m_{\mathrm{CSK}}(t)]}{ed} - \gamma_{\mathrm{s}}N^{\mathrm{T}} - \frac{2\epsilon_0 n^2}{\hbar\omega_0}g|A^{\mathrm{T}}|^2, \tag{11.7}
$$

whereas the receiver, driven by the transmitted signal $s(t)$, can be described by

$$
\frac{dA^{\mathrm{R}}}{dt} = -\frac{\gamma_{\mathrm{c}}^{\mathrm{R}}}{2}A^{\mathrm{R}} + i(\omega_0 - \omega_{\mathrm{c}})A^{\mathrm{R}} + \frac{\Gamma}{2}(1 - ib^{\mathrm{R}})gA^{\mathrm{R}} + F_{\mathrm{sp}}^{\mathrm{R}} + \eta s(t - T) \tag{11.8}
$$
$$
\frac{dN^{\mathrm{R}}}{dt} = \frac{J}{ed} - \gamma_{\mathrm{s}}N^{\mathrm{R}} - \frac{2\epsilon_0 n^2}{\hbar\omega_0}g|A^{\mathrm{R}}|^2. \tag{11.9}
$$

Here τ is the feedback delay time, and η is the injection rate. Because the feedback strength has to be equal to the coupling strength between the transmitter and the receiver for the existence of perfect chaos synchronization, the parameter α is used for both quantities. The transmitted signal has the form $s(t) = \alpha A^{\mathrm{T}}(t) + m(t)$ for both ACM and CMS schemes and the form $s(t) = \alpha A^{\mathrm{T}}(t)$ for the CSK scheme. When CMS is applied, the encoding message $m(t)$ is sent to the receiver but is not fed back to the transmitter. When CSK is applied, only the transmitter laser, but not the receiver laser, is current modulated with the encoding message. By comparison of the rate equations above for different encoding schemes, we see that the transmitter and the receiver can be mathematically identical in the presence of a message only when ACM is applied and when their parameters are well matched.

The performance of this system for 10 Gb/s communications is also first investigated by examining the recovered message in the time domain, shown in Figure 11.17. It is observed that message recovery is not possible for the CSK scheme because the resynchronization time after a desynchronization burst is much longer than the bit duration as is also the situation for the use of CSK in the optical injection system. The performance can be improved at a low bit rate when the bit duration gets longer than the resynchronization time. The performance of CMS in this system is similar to that of CMS in the optical injection system because any disturbance due to CMS encoding that causes synchronization errors only affects the receiver in the same manner for both systems. The bit error mainly arises from the mismatch between the transmitter and the receiver due to message encoding. Because the encoded message is considered as a perturbation to the synchronization, the recovered message shows the image of the encoding message. Therefore, better message recovery is expected if a low-pass filter is used. The performance of ACM is

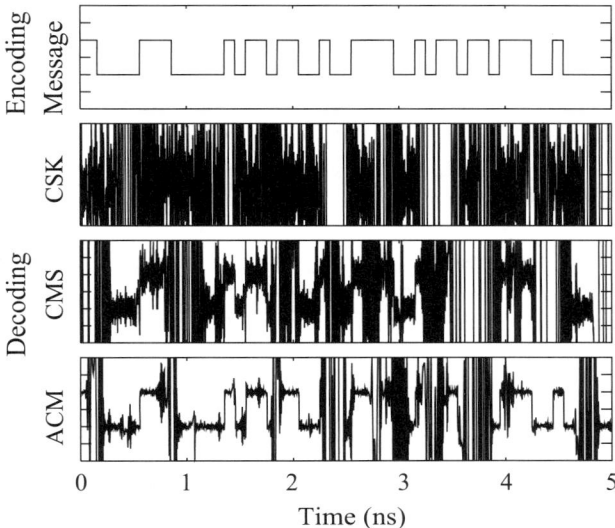

Fig. 11.17. Time series of the decoded messages of the three different encryption schemes in the optical feedback system. CSK, chaos shift keying; CMS, chaos masking; ACM, additive chaos modulation. (Reprinted with permission from [22], ©2002 IEEE.)

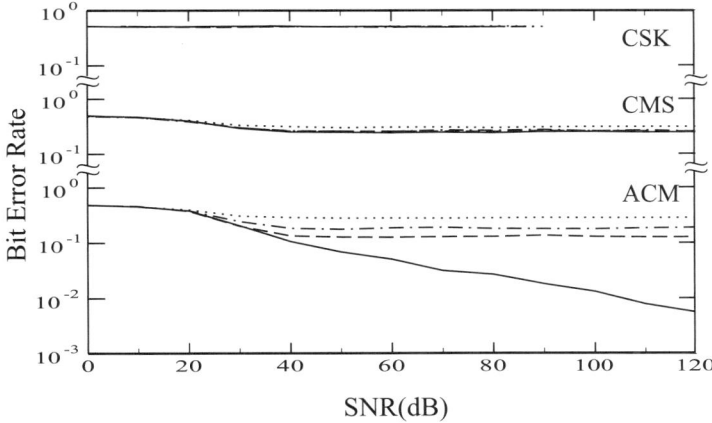

Fig. 11.18. BER versus SNR for the three different encryption schemes in the optical feedback system. Solid curves are obtained when the laser noise is absent. Dashed curves are obtained at a laser noise level equivalent to $\Delta\nu = 100$ kHz for both the transmitter and receiver lasers. Dot-dashed curves are obtained when $\Delta\nu = 1$ MHz. Dotted curves are obtained when $\Delta\nu = 10$ MHz. CSK, chaos shift keying; CMS, chaos masking; ACM, additive chaos modulation. (Reprinted with permission from [22], ©2002 IEEE.)

the best among the three encryption schemes because message encoding by ACM does not break the mathematical identity between the transmitter and the receiver. The error bits of ACM are caused mainly by desynchronization bursts triggered by both the channel noise and the laser noise. In Figure 11.17, we see that the error bits are almost all generated by desynchronization bursts. More error bits than those in the optical injection system are observed.

The system performance measured by the BER as a function of channel SNR for the optical feedback system is shown in Figure 11.18. From Figure 11.18, we find that message recovery for the CSK scheme is not possible at the high bit rate studied here because the resynchronization time after a desynchronization burst has to be shorter than the bit duration for a following bit to be recoverable. The performance of CMS in this system is similar to that of CMS in the optical injection system, and it is barely affected by the laser noise for the same reason as that mentioned above for the optical injection system. As for the performance of the ACM scheme, a BER lower than 10^{-3} cannot be obtained even when the channel SNR is as large as 120 dB. This is caused by the frequent occurrence of desynchronization bursts in this system even at an extremely low level of channel noise. In the presence of laser noise, the BER is always higher than 10^{-1} at a laser noise level characterized by a linewidth of $\Delta\nu \geq 100$ kHz, and the BER saturates at a higher value for a higher level of laser noise.

11.5.3 Performance of Optoelectronic Feedback System

To investigate the system performance with different encoding and decoding schemes, the configuration in Figure 11.14c is used for the simulation in this subsection. To specify the different encoding and decoding schemes, the transmitter can be modeled by the following coupled equations in terms of the photon density S^{T} and the carrier density N^{T} [23],

$$\frac{dS^{\mathrm{T}}}{dt} = -\gamma_{\mathrm{c}}S^{\mathrm{T}} + \Gamma g S^{\mathrm{T}} + 2\sqrt{S_0 S^{\mathrm{T}}}F_{\mathrm{s}}^{\mathrm{T}}, \tag{11.10}$$

$$\frac{dN^{\mathrm{T}}}{dt} = \frac{J[1 + m_{\mathrm{CSK}}(t)]}{ed}\left[1 + \xi y^{\mathrm{T}}(t - \tau)\right] - \gamma_{\mathrm{s}}N^{\mathrm{T}} - gS^{\mathrm{T}}, \tag{11.11}$$

$$y^{\mathrm{T}}(t) = \int_{-\infty}^{t} d\eta f^{\mathrm{T}}(t - \eta)[S^{\mathrm{T}}(\eta) + m_{\mathrm{ACM}}(\eta)]/S_0, \tag{11.12}$$

whereas the receiver, driven by the transmitted signal $s(t)$, can be described by

$$\frac{dS^{\mathrm{R}}}{dt} = -\gamma_{\mathrm{c}}S^{\mathrm{R}} + \Gamma g S^{\mathrm{R}} + 2\sqrt{S_0 S^{\mathrm{R}}}F_{\mathrm{s}}^{\mathrm{R}}, \tag{11.13}$$

$$\frac{dN^{\mathrm{R}}}{dt} = \frac{J}{ed}\left[1 + \xi y^{\mathrm{TR}}(t - T)\right] - \gamma_{\mathrm{s}}N^{\mathrm{R}} - gS^{\mathrm{R}}, \tag{11.14}$$

$$y^{\mathrm{TR}}(t) = \int_{-\infty}^{t} d\eta f^{\mathrm{TR}}(t - \eta)s(\eta)/S_0. \tag{11.15}$$

The transmitted signal has the form $s(t) = S^{\mathrm{T}}(t) + m(t)$ for both ACM and CMS schemes and the form $s(t) = S^{\mathrm{T}}(t)$ for the CSK scheme. The subscript of the encoding message $m(t)$ in the equations indicates the encoding scheme used. The transmitter and the receiver can be mathematically identical only when a message is encoded through the ACM scheme and when their parameters are well matched. It is assumed in the simulation that $f^{\mathrm{T}}(t) = f^{\mathrm{R}}(t) = \delta(t)$.

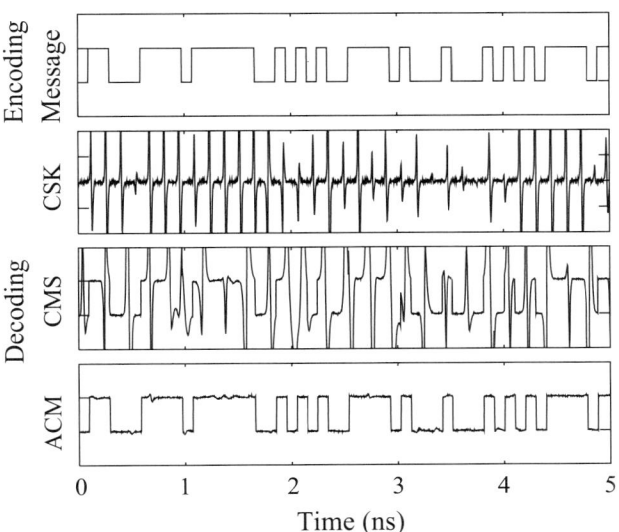

Fig. 11.19. Time series of the decoded messages of the three different encryption schemes in the optoelectronic feedback system. CSK, chaos shift keying; CMS, chaos masking; ACM, additive chaos modulation. (Reprinted with permission from [22], ©2002 IEEE.)

The performance of this system for 10 Gb/s communications is first investigated by examining the recovered message in the time domain, shown in Figure 11.19. Because the resynchronization time after a desynchronization burst has to be shorter than the bit duration for a following bit to be recoverable, the recovery of the high-bit-rate message in CSK is not possible. As is seen in Figure 11.19, almost all the bits in the recovered message for CSK are destroyed by desynchronization bursts. The performance can be improved at a low bit rate when the bit duration gets longer than the resynchronization time. For CMS in this system, the breaking of the identity between the transmitter and the receiver lasers causes synchronization deviation in the form of timing errors in the synchronization of the chaotic pulses. For the performance of CMS, the errors in message recovery are primarily generated by these timing errors. The recovered message is thus contaminated by frequent spikes.

Whether the use of a low-pass filter can improve the CMS performance for this system is a question for further investigation. The performance of ACM is the best among the three encryption schemes. The error bits in the recovered message for ACM are caused by synchronization errors in the form of synchronization deviation due to the channel noise and the laser noise. No desynchronization bursts are observed in this system when the ACM encryption scheme is applied.

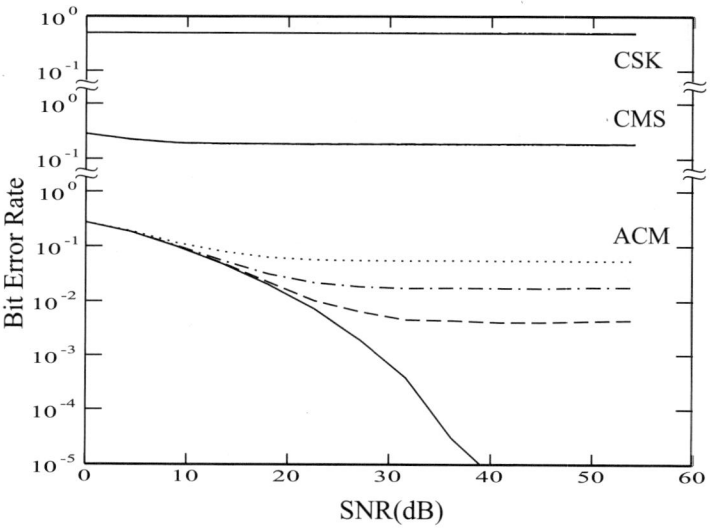

Fig. 11.20. BER versus SNR for the three different encryption schemes in the optoelectronic feedback system. Solid curves are obtained when the laser noise is absent. Dashed curves are obtained at a laser noise level equivalent to $\Delta\nu = 100$ kHz for both the transmitter and receiver lasers. Dot-dashed curves are obtained when $\Delta\nu = 1$ MHz. Dotted curves are obtained when $\Delta\nu = 10$ MHz. CSK, chaos shift keying; CMS, chaos masking; ACM, additive chaos modulation. (Reprinted with permission from [22], ©2002 IEEE.)

The system performance measured by the BER as a function of channel SNR for the optoelectronic feedback system is shown in Figure 11.20. From Figure 11.20, we find that message recovery at this high bit rate is not possible for the CSK and the CMS schemes. As for the performance of the ACM scheme, a BER lower than 10^{-5} can be obtained when the SNR is larger than 38 dB, which is much better than the performances of the CSK and the CMS schemes. A difference between the optoelectronic feedback system and the optical injection or optical feedback system is that the optoelectronic feedback system is not sensitive to optical phase. Consequently, the performance of the optoelectronic feedback system is observed to be better than those of the optical injection and optical feedback systems.

11.5.4 Error Reduction with Filter

The bit errors from different sources behave very differently when a filter is used to filter out the noise and the synchronization errors. How much BER can be reduced by the filter depends on which type of error source dominates the system BER performance. The system BER performances, shown in Figure 11.21, under different encryption schemes are obtained by choosing SNR = 30 dB and a bit rate equal to 10 Gb/s. The BER performances of the three different systems are marked by different symbols: circles for the optical injection system, squares for the optical feedback system, and triangles for the optoelectronic feedback system. The solid symbols mark the BER after the filter, and the open symbols mark the BER before the filter.

For the CMS and CSK encryption schemes, most of the bit errors are generated by the desynchronization bursts because the encoding of a message through such schemes breaks the mathematical symmetry of the system so that true synchronization cannot be accomplished. Consequently, the filter has very limited, if any, improvement on the BER performance of the CMS and CSK schemes, as is seen in Figure 11.21. For the ACM encryption scheme, true synchronization is possible because encoding of a message by ACM does not

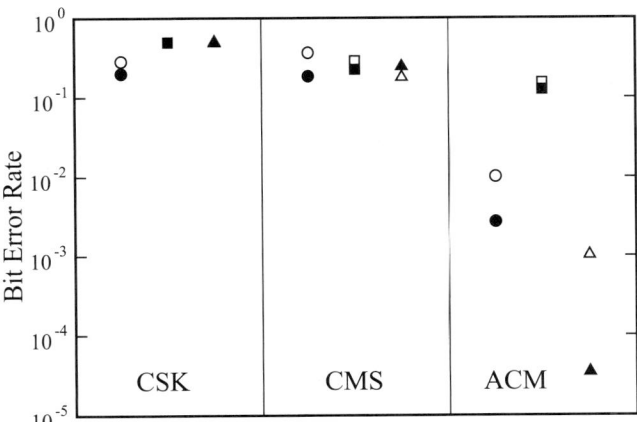

Fig. 11.21. BER for the optical injection system, marked as circles, the optical feedback system, marked as squares, and the optoelectronic feedback system, marked as triangles, of the three different encryption schemes. The solid symbols mark the BER after the filter, and the open symbols mark the BER before the filter. (Reprinted with permissions from [22], ©2002 IEEE.)

break the mathematical identity between the transmitter and the receiver. In this situation, the effectiveness of filtering on the reduction of BER depends on how frequently desynchronization bursts occur and how large synchronization deviation is on average in a particular operating condition. For the ACM performance of the three systems at 10 Gb/s, we see in Figure 11.21 that the filter reduces the BER for both the optical injection and the optoelectronic feedback systems, but it hardly reduces the BER for the optical feedback system.

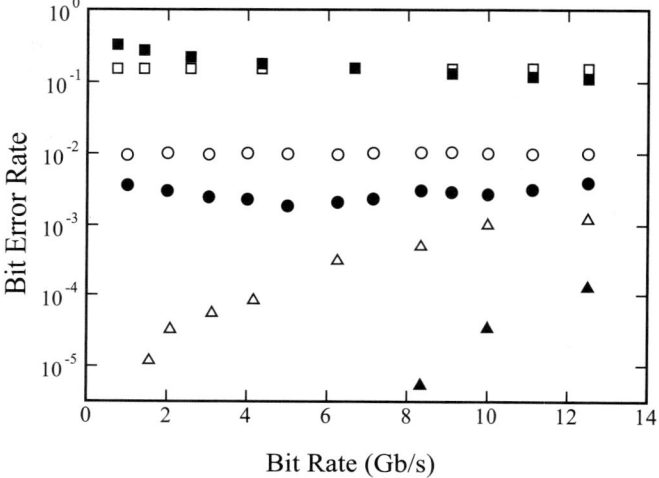

Fig. 11.22. BER versus bit rate for the three systems under ACM. The meanings of the symbols are the same as those in Figure 11.21. (Reprinted with permission from [22], ©2002 IEEE.)

The effect of the filter on ACM for the three systems over a range of bit rates is presented in Figure 11.22. For the optical feedback system, there is no significant improvement in the BER. For the optical injection and the optoelectronic feedback systems, the BER is improved over the entire range of bit rates after filtering. For the optoelectronic feedback system , the BER after the filter continues to decrease as the bit rate is lowered and the bandwidth of the filter is reduced accordingly.

11.5.5 Summary

The inherent advantage of any optical communication system is its ability to handle high-bit-rate communications. Chaotic optical communications at the OC-192 standard bit rate of 10 Gb/s are possible when high-speed semiconductor lasers are used. Three semiconductor laser systems, namely, the optical injection system, the optical feedback system, and the optoelectronic feedback system, that are capable of generating broadband, high-frequency chaos for high-bit-rate communications are considered. The performance of each system at 10 Gb/s is numerically studied for the three encryption schemes of CSK, CMS, and ACM. Channel noise and laser noise at the realistic levels of common semiconductor lasers from both the transmitter and the receiver are considered in the study. It is found that these noise sources have very significant effects on the system performance at high bit rates primarily because they cause synchronization errors in the forms of synchronization deviation and desynchronization bursts in these synchronized chaotic communication systems. Among the three encryption schemes, it is found that at high bit rates only the performance of ACM with low-noise lasers is acceptable because ACM allows true synchronization in the process of message encoding by maintaining the mathematical identity between the transmitter and the receiver. Both CSK and CMS cause significant desynchronization bursts or synchronization deviation in the systems because they break the identity between the transmitter and the receiver in the process of message encoding.

Our study results clearly point out that for synchronized high-bit-rate chaotic optical communications, it is very important to reduce as much noise as possible from all sources so that noise-induced desynchronization bursts and synchronization deviation can be minimized. It is equally important to select a system that is robust against noise and an encryption scheme such as ACM that allows the system to remain stably synchronized when messages are encoded. Proper filtering can improve the performance by reducing the errors caused by synchronization deviation.

11.6 Conclusions

Chaotic optical communication is an emerging field of growing interest. This new communication scheme utilizes, instead of avoiding, nonlinear effects in dynamical systems so that it has many new features such as noiselike time series and broadband spectrum. Based on these features, chaotic optical communications have potential applications in secure communications and spread-spectrum communications.

In this chapter, the basic concepts of chaotic optical communications are presented. In a chaotic optical communication system, a nonlinear dynamical system is used to generate the optical chaotic waveform for message transmission. Messages are encoded through chaos encryption where the messages are mixed with the chaos. Message recovery is achieved by comparing the received signal with a reproduced chaotic waveform that synchronizes with the chaotic waveform from the transmitter. To achieve chaos synchronization, an identical nonlinear dynamical system is required on the receiver side, which is coupled to the nonlinear dynamical system on the transmitter side. Details of chaos generation, chaos encoding and decoding, and chaos synchronization are discussed in this chapter.

Furthermore, we also review our achievement on the experiment of chaotic optical communication at 2.5 Gb/s [8], which has the highest bit rate in any chaotic communication systems ever reported in the literature. Three major encoding and decoding schemes, namely, chaos masking, chaos shift keying, and chaos modulation are implemented and compared in this 2.5 Gb/s chaotic optical communication system. The system uses semiconductor lasers with delayed optoelectronic feedback. The chaos modulation scheme is found to have the best performance.

To investigate the potential applications of chaotic optical communications at an even higher bit rate, numerical simulations are carried out on chaotic optical communication systems operated at 10 Gb/s. In this numerical study, three systems using semiconductor lasers with optical injection, optical feedback, or optoelectronic feedback are investigated using the three encoding and decoding schemes. It is shown that chaotic optical communication at 10 Gb/s is feasible using fast semiconductor lasers. The implementation of proper filtering further improves the system performance.

It is believed that nonlinear dynamical systems will bring new features to optical communications. However, many practical issues related to chaotic optical communications still need to be investigated. Such issues include the effect of fiber dispersion and nonlinearity in a fiber transmission system, parameter mismatch in semiconductor lasers and in other components, the effect of spectrum spreading with chaotic encryption, and so on. Multiuser applications also need to be addressed for chaotic communications. For example, different chaotic waveforms can be used as the natural spreading codes for different users. The issues regarding how to decode messages among multiple users are subjects to be investigated.

Acknowledgments

This work was supported by the U.S. Army Research Office under MURI grant DAAG55-98-1-0269.

References

1. Special issue on applications of chaos in modern communication systems, *IEEE Trans. Circuits Syst. I*, vol. 48, 2001.

2. Feature section on optical chaos and application to cryptography, *IEEE J. Quantum Electron.*, vol. 38, 2002.
3. E. Ott, *Chaos in Dynamical Systems* (Cambridge University Press, 1993).
4. S. Tang and J. M. Liu, Chaotic pulsing and quasiperiodic route to chaos in a semiconductor laser with delayed optoelectronic feedback, *IEEE J. Quantum Electron.*, vol. 37, pp. 329–336, 2001.
5. F. Dachselt and W. Schwartz, Chaos and cryptography, *IEEE Trans. Circuits Syst. I*, vol. 48, pp. 1498–1509, 2001.
6. C. C. Chen, K. Yao, K. Umeno, and E. Biglieri, Design of spread-spectrum sequences using chaotic dynamical systems and ergodic theory, *IEEE Trans. Circuits Syst. I*, vol. 48, pp. 1110–1114, 2001.
7. G. P. Agrawal, *Fiber-Optic Communication Systems* (Wiley-Interscience, New York, 2002).
8. S. Tang and J. M. Liu, Message encoding-decoding at 2.5 Gbits/s through synchronization of chaotic pulsing semiconductor lasers, *Opt. Lett.*, vol. 26, pp. 1843–1845, 2001.
9. G. D. VanWiggeren and R. Roy, Optical communication with chaotic waveforms, *Phys. Rev. Lett.*, vol. 81, pp. 3547–3550, 1998.
10. C. W. Wu and L. O. Chua, A simple way to synchronize chaotic systems with applications to secure communication systems, *Int. J. Bifurcation & Chaos*, vol. 3, pp. 1619–1627, 1993.
11. H. F. Chen and J. M. Liu, Open-loop chaotic synchronization of injection-locked semiconductor lasers with gigahertz range modulation, *IEEE J. Quantum Electron.*, vol. 36, pp. 27–34, 2000.
12. Y. Liu, H. F. Chen, J. M. Liu, P. Davis, and T. Aida, Communication using synchronization of optical-feedback-induced chaos in semiconductor lasers, *IEEE Trans. Circuits Syst. I*, vol. 48, pp. 1484–1489, 2001.
13. S. Sivaprakasam and K. A. Shore, Message encoding and decoding using chaotic external-cavity diode lasers, *IEEE J. Quantum Electron.*, vol. 36, pp. 35–39, 2000.
14. S. Tang and J. M. Liu, Synchronization of high-frequency chaotic optical pulses, *Opt. Lett.*, vol. 26, pp. 596–598, 2001.
15. S. Tang, H. F. Chen, S. K. Hwang, and J. M. Liu, Message encoding and decoding through chaos modulation in chaotic optical communications, *IEEE Trans. Circuits Syst. I*, vol. 49, pp. 163–169, 2002.
16. J. P. Goedgebuer, L. Larger, and H. Porte, Optical cryptosystem based on synchronization of hyperchaos generated by a delayed feedback tunable laser diode, *Phys. Rev. Lett.*, vol. 80, pp. 2249–2252, 1998.
17. G. D. VanWiggeren and R. Roy, Chaotic communication using time-delayed optical systems, *Int. J. Bifurcation & Chaos*, vol. 9, pp. 2129–2156, 1999.
18. L. G. Luo, P. L. Chu, and H .F. Liu, 1-GHz optical communication system using chaos in erbium-doped fiber lasers, *IEEE Photon. Technol. Lett.*, vol. 12, pp. 269–271, 2000.
19. G. D. VanWiggeren and R. Roy, Communication with dynamically fluctuating states of light polarization, *Phys. Rev. Lett.*, vol. 88, 097903, 2002.
20. S. Tang and J. M. Liu, Effects of message encoding and decoding on synchronized chaotic optical communications, *IEEE J. Quantum Electron.*, vol. 39, pp. 1468–1475, 2003.
21. S. Haykin, *Communication Systems* (John Wiley & Sons, New York, 1994).

22. J. M. Liu, H. F. Chen, and S. Tang, Synchronized chaotic optical communications at high bit rates, *IEEE J. Quantum Electron.*, vol. 38, pp. 1184−1196, 2002.

23. J. M. Liu, H. F. Chen, and S. Tang, Optical communication systems based on chaos in semiconductor lasers, *IEEE Trans. Circuits Syst. I*, vol. 48, pp. 1475−1483, 2001.

24. L. Kocarev and U. Parlitz, General approach for chaotic synchronization with applications to communication, *Phys. Rev. Lett.*, vol. 74, pp. 5028−5031, 1995.

25. L. M. Pecora and T. L. Carroll, Synchronization in chaotic systems, *Phys. Rev. Lett.*, vol. 64, pp. 821−824, 1990.

26. L. Kocarev, K. S. Halle, K. Eckert, L. O. Chua, and U. Parlitz, Experimental demonstration of secure communications via chaotic synchronization, *Int. J. of Bifurcation & Chaos*, vol. 2, pp. 709−713, 1992.

27. U. Parlitz, L. O. Chua, L. Kocarev, K. S. Halle, and A. Shang, Transmission of digital signals by chaotic synchronization, *Int. J. of Bifurcation & Chaos*, vol. 2, pp. 973−977, 1992.

28. K. S. Halle, C. W. Wu, M. Itoh, and L. O. Chua, Spread spectrum communication through modulation of chaos, *Int. J. of Bifurcation & Chaos*, vol. 3, pp. 469−477, 1993.

29. M. Itoh, H. Murakami, and L. O. Chua, Communication systems via chaotic modulations, *IEICE Trans. Fundamentals*, vol. E77-A, pp. 1000−1006, 1994.

30. U. Parlitz, L. Kocarev, T. Stojanovski, and H. Preckel, Encoding messages using chaotic synchronization, *Phys. Rev. E*, vol. 53, pp. 4351−4361, 1996.

Index

A-CDMA, *see* CDMA
ACM, *see* Additive chaos modulation
Active/passive decomposition, 195
Adaptive parametric control, 65
Additive chaos modulation, 330, 331, 333–335, 337, 354–375
Additive Chaos Modulation (ACM), 15
Additive mixing, 194
Atmospheric turbulence, 32, 46
Attractor, 212
Autonomous system, 285, 304, 308, 336

BER, 50, *see* Bit-error rate
Bernoulli shift, 98
Bifurcation, 163, 183, 188
    diagram, 256, 265, 266, 268, 305
    flip, 184
    Neimark–Sacker, 184
    tangent, 184
Binary phase shift keying, 66, 280
Bit-error rate, 81, 243, 244, 275, 278–280, 283, 349, 350, 361, 363, 367, 370, 372, 373
BPSK, 34, 135, *see* Binary phase shift keying
Brownian motion, 247

CDMA, 135
    asynchronous (A-CDMA), 141
    chaotic, 138
    chip-synchronous (CS-CDMA), 143
    system models, 140
Channel capacity, 93

Channel noise, 243, 244, 246, 275–281, 283, 349, 350, 363, 365, 366, 370, 372, 375
Chaos, 163, 165, 184, 188, 197
    transient, 163, 165, 173, 183, 188
Chaos Masking, 60
Chaos masking, 330–335, 337, 354–373, 375
Chaos Masking (CMS), 15
Chaos shift keying, 197, 330–332, 334, 335, 337, 354–373, 375
Chaos synchronization, 30, *see* Synchronization
Chaotic attractor, 243, 245, 259, 263, 264, 266, 271, 281, 282
Chaotic CDMA system performances, 148
Chaotic communication, 60, 192, 193, 243, 244, 246, 275, 281, 285, 341, 343–346, 349, 350, 353, 358, 360, 363
Chaotic driven oscillation, 316, 322, 323, 325, 330, 337
Chaotic modulation, 316, 337
Chaotic optical communication, 278, 285–288, 298, 315, 330, 333, 335, 341, 344–347, 350, 353, 355, 356, 358, 362, 363, 375, 376
Chaotic optimal CS-CDMA sequences, 146
Chaotic Pulse Position Modulation (CPPM), 12, 31, 60

Chaotic pulsing, 285, 310, 328, 335, 343, 350, 356
Chebyshev polynomial maps, 144, 146
Class A laser, 290
Class B laser, 291–293, 315, 336
Class C laser, 291
CMS, *see* Chaos masking
Code Division Multiplexing Access, *see* CDMA
Coding function, 237, 238
Control
    of chaos, 197
    of transient chaos, 163, 185, 188
Controlling chaos, 62
Correlation coefficient, 243, 271, 274, 281–283, 316, 317, 323, 324, 359, 360
Correlation Delay Shift Keying (CDSK), 5
Correlation dimension, 243, 245, 258, 263, 264, 266, 268, 281, 282
Correlation integral, 264, 266
Coupling strength, 269, 273, 274, 314, 318, 322, 323, 325, 347, 364, 368
Cryptanalysis, 192, 200, 203, 216
CS-CDMA, *see* CDMA, 146
CSK, *see* Chaos shift keying

Decoding scheme, 347, 354
Delayed-feedback system, 285, 308, 336
Desynchronization burst, 275, 364, 366–368, 370–375
Differential Chaos Shift Keying (DCSK), 4, 60
Distributed Dynamics Encryption, 212
Duffing equation, 198

Embedding, 203
    dimension, 203, 209, 263, 266
    space, 263, 264
Encoding scheme, 347, 354
Encryption, 191, 192, 203, 211
    chaotic, 193
    public key, 212
Ergodic dynamical systems, 143
Eye diagram, 353

FCC indoor emission mask, 94
FDMA, 135

Feedback delay time, 285, 303–305, 308, 316, 322, 324, 336, 337, 350, 368
Feedback strength, 303, 304, 308, 321, 323, 325, 368
Fiber laser, 287, 344
Finite precision issue of chaotic CDMA sequences, 149
Fixed point, 163, 175, 183, 184
    indecisive, 165, 181, 183
    unequivocal, 165, 181, 183, 185
Fokker−Planck equation, 247
Free distance, 104
Free-space laser communication, 46
Frequency Division Multiplexing Access, *see* FDMA
Frequency hopping, 67

Gaussian noise, 83, 246, 247, 255, 275, 349
Generating partition, 236

Hidden Markov model, 218

Impulse radio, 92
Injection locked, 300, 325
Interference, 73
Intermittency, 264, 285, 305, 336
Interpolated frequency hopping, 68
Invariant measure, 232
Ito integral, 243, 246, 250–254, 282

Jacobian matrix, 184, 261, 269

Known ciphertext attack, 216–218, 220

Langevin equation, 247, 248, 252
Laser
    classification, 290
    dynamics, 275, 287, 290, 291, 293, 295, 314, 315
    noise, 255, 257, 258, 266, 280, 302, 311, 363, 366, 367, 370, 372, 375
Lebesgue spectrum of a dynamical system, 145
LFF, *see* Low-frequency fluctuation
Lorentz system, 263
Low-density parity-check code, 163
Low-frequency fluctuation, 305, 306
Lyapunov exponent, 184, 195–197, 215, 243, 245, 259–261, 264–266, 268, 270, 274, 281, 282

conditional, 196, 203
global, 261–263, 265
largest, 263, 265, 282
local, 268, 313
transverse, 243, 245, 263, 268–270,
     281, 282

Markov chain, 232
Markov process, 247
Markovian, 247
MCM, *see* Multiplicative chaos
     modulation
Message decoding, 277, 286, 287, 330,
     335, 345, 347, 349, 354–356, 358,
     360, 362, 363, 366
Message encoding, 275, 285, 287,
     330–333, 335, 337, 345–347, 349,
     354–356, 358–360, 362, 363, 366,
     368, 370, 375
Modulation
   binary frequency shift keying
     modulation, 75
   chaotic frequency modulation, 68
   amplitude, 42, 49
   chaotic frequency modulation, 61
   Differential BPSK Modulation, 74
   FM demodulator, 61
   multilevel, 204
   parameter modulation, 60
Multiplicative chaos modulation, 354
Multiplicative Chaos Modulation
     (MCM), 15
Multiuser, 53
Multiuser CFM Communication
     System, 79

Nonautonomous system, 285, 300, 336

OC-192, 363, 375
OC-48, 353, 355
Optical Communications Based on
     Nonlinear Dynamics, 15
Optical feedback, 285, 287, 292, 298,
     302, 304–307, 310, 315, 316, 319,
     321, 330, 336, 337, 341, 344–346,
     363, 368, 370, 372–376
Optical injection, 243, 245, 254, 264,
     271, 272, 278, 282, 285, 287, 292,
     297–303, 315, 317, 325, 330, 336,

341, 344–346, 363, 364, 367, 368,
     370, 372–376
Optoelectronic feedback, 285, 287, 292,
     298, 307–310, 314–316, 319, 322,
     326, 327, 330, 333, 336, 337, 341,
     343–346, 350, 356–359, 363, 370,
     372–376

Parameter
   match, 243, 245, 268, 278, 282
   mismatch, 66, 243, 245, 270, 275, 282,
     330, 363, 376
PCTH
   bit-error-rate using MLSE, 113
   definition, 96
   frame-by-frame bit-error-rate, 110
   multiaccess, 96
   receiver, 100
   transfer function, 105
Period
   doubling, 257, 264, 285, 302, 306, 336
   four, 265, 305
   one, 265, 300, 302, 305
   two, 265, 300, 302, 305
Phase space, 199
PLL-Based Synchronization Scheme, 69
Probability of detection, 51
Pseudo-orbits, 232
Pseudo-random sequences, 52

QR decomposition, 262
Quantization, 51, 52, 199, 218–220, 222
Quasiperiodic pulsing, 310
Quasiperiodicity, 264, 285, 306, 336

Random Finite Approximation, 232
Rate equation, 247, 255, 278, 293, 294,
     322, 327, 332, 365, 368
Rayleigh fading channel, 149
Rician fading channel, 149
Riemann–Stieltjes integral, 244, 246,
     249, 250, 282
Route to chaos, 257, 264, 268, 285, 302,
     305, 306, 336
Runge–Kutta method, 254
Rssler oscillator, 78

Scintillations, 47, 48
Semiconductor laser, 46, 243, 245, 246,
     254, 256, 264, 271, 272, 275, 278,

282, 285, 287, 291–309, 313–316,
318, 327, 333, 335–337, 341,
343, 344, 346, 347, 350, 353–355,
361–363, 375, 376

Shannon limit, 163

Signal-to-noise ratio, 29, 47, 138, 243,
275, 278–280, 283, 350, 367, 370,
372

SNR, 138, *see* Signal-to-noise ratio

Spatial capacity, 93

Spread spectrum, 59, 135

Stochastic differential equation, 243,
244, 246, 247, 254, 276, 282

Stochastic integral, 243, 246, 249–252,
254

Stochastic nonlinear system, 243, 244,
281

Stochastic process, 247, 249–251

Strange attractor, 197

Stratonovich integral, 243, 246, 250–254,
282

Symbolic dynamics, 199

Symmetric Chaos Shift Keying (SCSK),
6

Synchronization
    anticipated, 285, 316, 322, 327, 329,
    330, 337
    condition, 41
    deviation, 275, 364, 366, 371, 372,
    374, 375

error, 34, 199, 243, 271, 275, 281, 282,
335, 347, 349, 350, 353, 355, 360,
364–368, 372, 373, 375

generalized, 271, 312

identical, 30, 285, 312, 322, 336

of chaos, 30, 60, 194, 195, 243, 244,
246, 269, 270, 274, 275, 281,
285–287, 298, 311–318, 321–327,
329–331, 333, 335–337, 343,
345–347, 349, 350, 355, 358, 360,
362, 376

quality of, 243–245, 271, 272, 274,
275, 277, 279, 281, 282, 285, 286,
315–317, 319, 321, 323–325, 329,
330, 332–335, 337, 346, 347, 349,
358–360, 363, 365

retarded, 285, 316, 322, 327–330, 337

robustness of, 243–245, 268, 271, 274,
275, 281, 282, 349, 364

TDMA, 135

Time Division Multiplexing Access, *see*
TDMA

Transient chaos, *see* Chaos

Turbo code, 163, 167, 183

Turbo decoding algorithm, 163, 164

Type I synchronous scenario, 323–326

Type II synchronous scenario, 323–325

Ultra-wideband, 60, 92

Wiener process, 247–251